M

For
Barbara Noelle Slattery
and
Grace (Bergin) Slattery

*

MILITARY AVIATION IN IRELAND

1921–45

MICHAEL C. O'MALLEY

UNIVERSITY COLLEGE DUBLIN PRESS

PREAS CHOLÁISTE OLLSCOILE BHAILE ÁTHA CLIATH

First published 2010
by University College Dublin Press
Newman House
86 St Stephen's Green
Dublin 2
Ireland
www.ucdpress.ie

© Michael C. O'Malley 2010

ISBN 978-1-906359-48-5 hb
ISBN 978-1-906359-49-2 pb

CIP data available from the British Library

The right of Michael C. O'Malley to be identified as the
author of this work has been asserted by him

Typeset in Scotland in Adobe Caslon and
Bodoni Oldstyle by Ryan Shiels
Printed in England on acid-free paper by
CPI Antony Rowe, Chippenham, Wilts.

Contents

—

Illustrations

—

Preface

—

While the thesis on which this book is based required my full attention for about five years, the production of the work has in fact taken in excess of 25. However, the cultural and other influences that gave rise to the undertaking go back even further. In February 1961 I was fortunate to be awarded an Air Corps short service cadetship. I had applied to join this particular corps having been informed and influenced by three years of FCA experience with infantry and artillery units.

Very early in my cadet training at the Curragh Camp my 11 classmates and I soon realised that the Air Corps was different. While in theory it was just another army corps, it existed very much on the periphery of a largely infantry army whose influence dominated the ethos and culture of the Defence Forces. This difference was brought home more forcefully when the class made its transition, after a mercifully short 12 weeks of infantry training, to the Air Corps station at Gormanston, County Meath. In terms of military culture the contrast between the Curragh Camp and Gormanston was profound. At this stage, only 16 years after the Emergency the air station still retained the some-what relaxed and independent attitude it had inherited from the pre-1956 Fighter Squadron, an attitude that had evolved out of that squadron's travails in Baldonnell (1939–43) and Rineanna (1943–5).

In August 1962 I became a qualified pilot officer and a very junior member of the officers' mess at Baldonnell. I soon became aware of the folklore and apocryphal stories that pointed to the Corps' place in the history of aviation in Ireland. Part of the duties of a junior officer was to be the recipient of many stories of aviation derring-do. Many stories were based on actual flying incidents and often suggested some degree of departure from best practice. The stories lost nothing in being told and retold while, in any event, the hero (or villain) was usually retired or not inclined to correct any exaggerated departure from the truth. Taken in their entirety the emphasis of these stories largely reflected the distinct culture of the Corps and helped to define and illustrate what made it different and separate from the more traditional Army Corps. To a certain extent the Corps culture and ethos reflected its RAF roots and caused varying degrees of prejudice from those who were not in receipt of flying pay. On reflection it can be said that the folklore did not adequately reflect the reality of the genesis and history of a minority corps.

Stories relating to events and characters of bygone days abounded, but there remained little or no folk memory of the Civil War and the more difficult aspects of the Emergency years were inclined to be ignored or forgotten. Although apocryphal tales, some humorous, some cautionary, survive from the years of the Emergency, the overriding impression is that of a time when the Air Corps, particularly the pilot group, was more than a little frustrated and demoralised although a good military sense of humour saw them through. In the case of pilot officers the sources of that demoralisation would be explained by the proceedings of the investigation of 1941–2. The fundamental cause emanated from the command and direction of the Air Corps by army officers unsuited to the task. A controversy relating to qualification to wear pilot's wings (and receipt of flying pay) was central to the disharmony of the Emergency period. My interest in a balanced study of, and detailed research into, the history of the Air Corps has been greatly influenced by the necessity to elucidate the nuances of the situation that caused antipathy between the pilot group and their commanding officers and gained for the pilot body an undeserved reputation for indiscipline.

A major aspect that influenced my decision to study the history of the country's military aviation was that it has produced practically no academic work while the perceived history is largely based on anecdotal evidence and apocryphal stories. One is struck by the paucity of memoirs or records from retired officers. Similarly, few collections of private papers survive to give insight into early events and flying careers. I am aware of one instance where a retired commanding officer made a bonfire of his personal papers. As a result of this general situation, while a certain sense of inadequacy and injustice pervades the folklore of the period under review, the recorded history, such as it is, reflects little of the policies, events and issues that in effect conspired to make the Air Corps what it had become by Emergency time. In this respect the organs that might be expected to best reflect the history of the Air Corps fail to do so. *The Army to-day* is typical of the Defence Forces publications that provide little insight into the Air Corps. Published in 1945 this handbook did little to identify the technical and professional nuances that make the Air Corps distinctive. In particular, the roles and functions of the Air Corps during the Emergency were completely ignored. In a similar manner the series of Defence Forces handbooks from 1968 to 1988 do not adequately reflect the influences that shaped the early and evolving Air Corps. To a large extent these publications concentrate on Baldonnell and its association with pioneering flights and historic civil aviation events rather than its record as the main base for the nation's military aviation. The 1988 edition was particularly unbalanced in its treatment of the Air Corps with important aspects omitted, and the material published was somewhat inaccurate and incomplete. While the Defence Forces magazine, *An Cosantóir*, has been even-handed and generous in its coverage of Air Corps matters, the *Irish Sword* (1949–), promulgating the academic study of Irish military history, has not been supplied with many articles reflecting the history of the State's military aviation. Although it appears that there was, and still is, an innate reticence among pilots to record the events of their careers, this does not fully explain the absence of academic works. Major and minor academic works on the Defence Forces, and related matters such as defence policy and practice, reflect a similar lack of balance in relation to military aviation. My view is that these works reflect inadequate research on what is, admittedly, a relatively minor military discipline in the Defence Forces. As a result of this, works on the Defence Forces and defence policy reflect observations, comment and conclusions with regard to the efficacy or otherwise of the Air Corps, during the Emergency in particular, that are to a surprising extent based on ill-informed perception.

From about 1985, as considerations of family and career allowed scope, it was possible to begin to find the time and opportunities to satisfy my growing inquisitiveness about various aspects of history. Initially, during some years of spasmodic and somewhat aimless research, I found myself concentrating on the history of my local area and came to recognise the need for a more formal approach to my historical researches. In 1995–6 I completed the Diploma in Local History course at NUI Maynooth, concentrating on the sources for the early history of the civil parish of Oughterard (Kildare). From 1998 to 2002 I undertook the modular degree course in Local and Community Studies at the same college. The modular format afforded me the opportunity to research and study selected aspects of the history of the Air Corps and Baldonnell. Formal training in the rudiments of research and writing helped develop an appreciation of the history of military aviation as a national subject and sparked my interest in researching and recording that history in order to better understand the real facts behind the folklore of military aviation. My doctoral thesis has, I hope, helped to fill the void caused by the paucity of academic research and study of the subject. I trust that this particular work will provide the general reader with an accessible and balanced view of the history of the military aviation of the period. The fact that I had a fortunate and rewarding career in a unique corps has brought a sense of duty to the project – a duty that I have been eager and willing to undertake.

Acknowledgements

—

As this work has taken much longer that the nine years of my formal education in matters historical it is appropriate that I acknowledge relevant assistance received over a considerable period. My introduction to military archives was in the Red House in the early 1980s when the late Commandant Peter Young showed me the growing amount of material that would later become the Military Archives. Though Peter and I were separated by corps ethos and loyalties he assisted me greatly in my early efforts to appreciate the value of the material relating to military aviation. In particular he introduced me to those more historical documents that sparked my interest in the Military Air Service and the Air Corps. I am particularly grateful for the access he granted me to the personal files of a number of officers who served during the early years. In the context of the early history an insight into the careers of individual officers has proved to be absolutely essential to an understanding of the establishment of the State's military aviation and of its fortunate survival when the civil and military authorities had little interest in a new military discipline. Such unique access to personal files has, of course, long since ceased.

My thanks go to the late Madeline O'Rourke and to her husband Colm. Their interest in, and willingness to collect and research and share with others the history of the Air Corps was, and still is, unique. I appreciate also the personal comments of the late Agnes Russell, in June 2004, regarding her late father Colonel Charlie Russell. Regrettably she had to tell me that her father's private papers, which would have been of considerable interest, had been lost in a domestic fire many years previously. Similarly I owe a debt of gratitude to Ms Aine Broy for an understanding of aspects of the brief Air Service career of Colonel Ned Broy. Thanks to the kindness of Captain Eoin Hassett I had extended access to the small, but very important, collection of private papers of his late father, Lieutenant Colonel P. J. (Laddie) Hassett who served in the Air Corps from 1926 to 1935.

The late Colonel W. J. Keane was, to the best of my knowledge, probably the first to undertake research into the early Air Corps. In 1951, coinciding with the 25th anniversary of the first cadet class, Billy Keane interviewed some of the early officers and NCOs and conducted some very valuable research into the early years. The notes and copied documents contained in his collection of papers, now in the Military Archives, have proved to be uniquely relevant in documenting the period 1921 to 1924. As Cadet Billy Keane he displayed great historical foresight in the summer of 1928. Being at a loose end at the end of his wings course he wandered out to the dump near the western boundary of the aerodrome at Baldonnell and viewed the mangled wreckage of the crashed aircraft that had been written off charge since 1922. Amongst the wreckage he found the related airframe and engine log books. Realising their future worth, he took possession of them and kept them safe for many years by storing them under his bed. These log books and others of the period, though not well kept, I found essential to piecing together the details of the early years.

I gratefully acknowledge the kind co-operation and assistance of many archives. I refer particularly to the National Archives and Military Archives. In addition I was fortunate to be granted access, by Commandant Rory O'Connor, School Commandant of my aeronautical *alma mater*, to the confidential course files from the 1930s.

Over the years I have been in receipt of assistance of various kinds from my former colleagues in the Air Corps. I am grateful to Brigadier General Barney McMahon for Tosti Gogan's photographs from 1922–3. I am particularly grateful to Captain Tony Kearney for Andy Woods's excellent photographs. I wish to express my sincere gratitude to Dr Ian Speller of the Department of History, NUI Maynooth, my PhD supervisor, for his expert guidance, criticism, direction and advice. I also acknowledge the foundation in history provided by the various academics at Maynooth, particularly Professor R. V. Comerford, during my diploma, degree and doctorate courses. My grateful thanks go to Dr Sean Swords, formerly of the Air Support Company, Signal Corps at Baldonnell and of Trinity College, for his assistance and guidance in the matter of aeronautical communications and direction finding. My sincere thanks also go to Marie O'Malley and Ann Donoghue for patient assistance and guidance in the matters of computers and word-processing. I am particularly grateful to Adrian Slattery for the fine drawings based on originals sourced during my research.

The publication of this work could not have proceeded without the generous material assistance of my esteemed friends and classmates, Lieutenant Colonel John P. Kelly and Captain Thomas P. Murphy.

I must acknowledge the professional expertise and judgement of UCD Press. I am extremely grateful for the excellent assistance, advice and guidance of Barbara Mennell and Noelle Moran that has kept me on the straight and narrow, and greatly reduced the stress of the final stages of publication.

MICHAEL C. O'MALLEY
Straffan, April 2010

Abbreviations

—

A/	Acting
AA	Anti-aircraft (artillery)
AAC	Army Air Corps
AAS	Army Air Service
AC	Air Corps
ACF/	Air Corps flying (file)
AC HQ	Air Corps Headquarters
ACS	Air Corps School(s)*
ACS	Assistant Chief of Staff*
ADC	Aircraft Disposal Company
Adjt	Adjutant
AFO	Army Finance Officer
AG	Adjutant General
AM	Air Ministry
AOC	Air Officer Commanding (UK)
Arty	Artillery
AS	Air Service
ATC	Air Traffic Control
AVRO	A. V. Roe & Co. Ltd
BF	Bristol Fighter
BTNI	British Troops Northern Ireland
CAB	Cabinet Office
Capt.	Captain
CAS	Chief of Air Staff (UK)
Cav	Cavalry
CFI	Chief Flying Instructor
CIC	Commander-in-Chief
CID	Committee of Imperial Defence (UK)
Coy	Company
CO	Commanding Officer
COD	Council of Defence
COE	Corps of Engineers

* The sense should be clear from the context

COGS	Chief of General Staff
COI	Court of Inquiry
Col.	Colonel
Comd	Command – one of four designated military regions
Comdt	Commandant
CORO	Commissioned Officers Record Office (DFHQ, 2005)
COS	Chief of Staff
CP Sqn	Coastal Patrol Squadron
C/S	Chief of Staff
CSO	Chief Staff Officer
CV	Curriculum Vitae
DCA	Director of Civil Aviation
DCAS	Deputy Chief [of] Air Staff (UK)
DEA	Department of External Affairs
DF	Department of Finance*
DF	Defence Forces*
DF, D/F	Direction Finding*
DFHQ	Defence Forces Headquarters
DFR	Defence Force Regulation
DMA	Director of Military Aviation
DO	Dominions Office
DOD	Department of Defence (Departmental Secretariat and General Staff)
DS	Director of Signals
DT	Department of An Taoiseach (Prime Minister)
EC	Executive Council
E. Comd.	Eastern Command
EDP	Emergency Defence Plan
FCA	Forsa Cosanta Áitiúil (Local Defence Force)
FIN	prefix for early Finance files
FS	Fighter Squadron
G2	Intelligence Branch, GHQ
GHQ	General Headquarters
GOC	General Officer Commanding
GOCF	General Officer Commanding the Forces
GRO	General Routine Order
HC	High commissioner
HMSO	His Majesty's Stationery Office
HQ	Headquarters

* The sense should be clear from the context

DI & C	Department of Industry and Commerce
IFS	Irish Free State
IRA	Irish Republican Army
JU88	Junkers 88
L/LOPs	Look-out post
LS	Local strength
Lt, Lieut.	Lieutenant
MA	Military Archives
Maj.	Major
Maj. Gen.	Major General
MAS	Military Air Service
MFD	Minister for Defence
MOF	Minister for Finance
MP	Mulcahy Papers
MS	Military secretary
NAI	National Archives of Ireland
NCO	Non-Commissioned Officer
NI	Northern Ireland
NLI	National Library of Ireland
NUIM	National University of Ireland, Maynooth
OC, O/C	Officer Commanding
OIC	Officer In Charge
OPW	Office of Public Works
OTC	Officer Training Corps
PE	Peace Establishment
PG	Provisional Government
PMG	Postmaster General
PO	Post Office
PREM	Prime Minister's Office [TNA]
P+T	Posts and Telegraph
QMG	Quartermaster General
RAF	Royal Air Force
RDF	Radio Direction Finding
R & MB Sqn	Reconnaissance and Medium Bomber Squadron
RFC	Royal Flying Corps
RSP	Reinforced steel planking
R/T	Radio Telephony
SAR	Search and Rescue
SBAC	Society of British Aircraft Constructors

* The sense should be clear from the context

SC	School Commandant*
SC	Signal Corps*
S. Comd.	Southern Command
SDR	Staff Duties Record
Sec.	Secretary
Sec. DOF	Secretary Department of Finance
Sgt	Sargeant
SIS	Secret Intelligence Service (UK)
Sigs	Signal Corps
Sqn	Squadron
TDS	Training Depot Station
TI	Training Instruction
TNA	The National Archives (Kew)
QM	Quartermaster
QMG	Quartermaster General
UCDA	University College Dublin Archives
VCAS	Vice Chief of Air Staff (UK)
Vol	Volunteer
W. Comd.	Western Command
WE	War Establishment
WOMs	Wireless Operator Mechanics
W/T	Wireless telegraphy

* The sense should be clear from the context

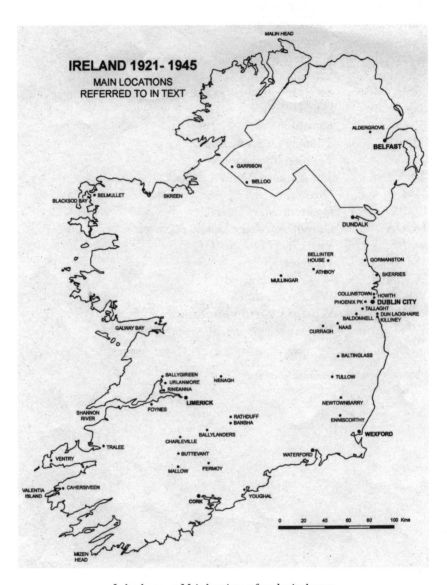

Ireland 1921–45: Main locations referred to in the text

ONE

EARLY AVIATION IN IRELAND

—

In keeping with the ideology of the time and the state of the weaponry and military equipment of the early twentieth century, the latter stage of the military occupation of Ireland was dominated by British Army infantry, backed up by artillery and cavalry units. Notwithstanding the basic nature of the military aviation there was by 1922, side by side with a great proliferation of military (army) barracks, posts and naval bases, a significant number of aerodromes, airfields and landing grounds and a relatively small element of the Royal Air Force (RAF) that operated in support of British ground troops in Ireland. In the years 1919 to 1921 various and varying elements of several RAF squadrons were dispersed to, and moved between, about a dozen different aerodromes and airfields on the island. Despite the presence of elements of Nos 2, 100, 105, 106, 117 and 141 Squadrons, distributed mainly in single detached flights to various locations during the IRA campaign, armed military aircraft were not to play a significant role during the War of Independence. When the RAF eventually withdrew in 1922 they abandoned or handed over a large number of aerodromes and airfields that probably represented a level of infrastructure well in excess of that warranted by the air operations carried out during and after the War of Independence. As the Free State's provisional administration assumed governance of the country, the nascent Free State Army monitored the British military withdrawal and took over many military installations. However, the vast bulk of the aviation infrastructure inherited from the British would hold little interest for the incoming administration.

ARMY MANOEUVRES 1913

Royal Flying Corps aircraft had originally been deployed in Ireland in August–September 1913 in the context of British Army manoeuvres. This deployment involved personnel and aircraft of No 2 Squadron of the Royal Flying Corps that had its base at Montrose in Scotland. On 27 August 1913 a detachment of six pilots and six aircraft commenced the journey to an army camp near Limerick. Two aircraft initially failed to get past Stranraer. The detachment, consisting of five aircraft, 12 officers, NCOs and 92 men, eventually arrived at a 56-acre field at Rathbane House, three miles from Limerick city, on 1 September 1913. The sixth machine arrived on 5 September. The detachment was tasked with supplying reconnaissance support to the opposing sides in the Army manoeuvres of September 1913.[1]

Sundry flights, not directly connected with the exercises, took place over the next week and the appearance of aircraft at various locations in Limerick, Tipperary and Offaly naturally gave rise to great local interest. Due to an accident, putting one machine totally out of action and causing technical problems with another, only four aircraft were available for the manoeuvres which started on 9 September. Two aircraft each were allotted to support the brown and white armies that totalled some 14,000 men. The manoeuvres were held in the Limerick, Tipperary, Offaly and the lower Shannon area. Inclement weather greatly curtailed the participation of aircraft. However, the limited support given to the respective armies the air exercise was deemed to have been a success. On 23 September, a few days after the end of the exercises, five aircraft left Rathbane for their home base. Bad weather delayed the crossing to Scotland so it was 26 September before the machines reached their base. From November 1915, up to the end of 1917, Royal Flying Corps activity in Ireland was confined to a little known trade school that functioned in the Curragh, County Kildare.[2]

THE GREAT WAR

One of the major military aviation activities in Ireland during the Great War was that of the US Naval Air Service, which operated from sites that had been selected and partially developed by the British admiralty following the USA's late entry into the War. Four seaplane patrol stations and a kite balloon station were established in Ireland by the United States Navy. The first and the principal patrol and training base taken over was that at Aghada, Queenstown (Cobh). While the first Naval Air Service personnel arrived on 6 February 1918, the completion of construction, using materials shipped in from the US, did not commence until March or April 1918. The first eight seaplanes were taken on charge in July 1918. Although the first test flight took place on 3 August 1918, the first operational flight only took place on 30 September. A total of 125 hours of operational flight were completed by 11 November 1918.[3]

The Wexford Naval Air Station at Ferrybank, started by the Admiralty in December 1917, was completed using contracted civilian labour. Seaplanes for anti-submarine duties were supplied from the main base at Queenstown and arrived on 18 September 1918. The station's function was to patrol from

Queenstown to the southern part of the Irish Sea. By the Armistice, over 300 hours of anti-submarine patrols had been flown. The Lough Foyle station was developed near Muff, County Donegal. A civilian firm, on contract to the Admiralty, started work on 5 January 1918. Despite initial difficulties with the transportation of materials, progress was reported as being rapid from May onwards. After a period of training and testing the first operational patrol took place on 3 September 1918. At the end of operations at Lough Foyle about 100 hours of patrol flying had taken place. The development of the Naval Air Station at Whiddy Island mirrored that at the other locations. Patrol flying, that started about 18 September 1918, totalled about 65 hours at the end of operations.

After a total of 762 hours of training and operational flying completed in the last three months of the War, the four naval air stations were finally decommissioned with the Armistice of 11 November 1918. The flying boats were dismantled and shipped home. Similarly the timber buildings were dismantled and those that could not be disposed of locally were shipped back to the US and the sites returned to the original owners.[4]

Starting at an earlier stage in the War, anti-submarine operations had been carried out by airships. The main bases in Ireland included Ballyliffin (Donegal) and Larne (Antrim) where Airship Patrol Stations were developed. The principal buildings were the airship hangars. Johnstown Castle (Wexford) and Malahide Castle (Dublin) were airship mooring out-stations with no facilities other than a mooring base. Johnstown Castle was an out-station used by airships patrolling the Irish Sea from their main base at Pembroke Patrol Station. In a similar manner Malahide Castle was used by airships operating out of Anglesey.[5]

TRAINING DEPOT STATIONS

The first flying training element of the RFC in Ireland was set up at the British Army camp at the Curragh, in December 1917, in advance of a major expansion in the number of training squadrons dispersed throughout the British Isles. These squadrons were required to train new pilots to support the RFC aspect of the British war effort in France. The scale of the expansion created a need for the development of many new aerodromes in Britain and Ireland. Four particular sites that had previously been identified by the lands officer of Irish Command were subsequently approved by Major Sholto Douglas as the locations for training aerodromes. Starting in about November 1917, training depot stations were developed at Gormanston (No 22 TDS) in County Meath and Baldonnell (No 23 TDS), Cookstown/ Tallaght (No 25 TDS), and Collinstown (No 24 TDS), all in county Dublin.[6]

The developments at Gormanston and Baldonnell would have been typical of the scores of aerodromes developed by the British War Office and the Air Council. Each aerodrome represented a major building undertaking even though the individual buildings, that were common to all training depot stations, were of a standard design and basic specifications. In taking over the lands at Gormanston the President of the Air Council displaced four tenant farmers of the Pepper Estate of east Meath. The farmers' interests in the lands were formally acquired in 1919 at a total cost to the Air Council of £11,410, while a yearly rent was still payable to the landlord.[7] The aerodrome buildings were erected in the north-western corner of a 218-acre site, comprising parts of the townlands of Irishtown and Richardstown and straddling the public road from Gormanston village to Mosney and Laytown. While a minor part of work was carried out by Royal Engineers, under the supervision of Lieutenant R.W. Dunn, the main construction contract was completed by the Dublin branch of the building firm McLaughlin and Harvey Ltd. Their construction operation at Gormanston started about 12 November 1917 when the wage bill for the first week was less than 20 pounds. When the building effort was at its maximum, in May 1918, the wage bill topped £3,000 per week.[8]

The first major aspect of the development was the removal of hedges and ditches in order to consolidate some 26 small fields into a large grass aerodrome. Another early development was the laying down of a temporary narrow gauge railway. Approximately 3,000 yards of line was laid from a new siding at Gormanston station, in a generally north-west direction, across the fields on the south side of the aerodrome, across the aerodrome itself, and through the building site to the sand and gravel pits at Rynking Wood to the north of the site. This line was used to transport building materials from main line trains to the site and to bring sand and gravel in from Rynking Wood. Huge quantities of timber, bricks, breeze blocks and other materials, mainly supplied from Dublin, were delivered where required. One of the main uses for the timber was the construction, on site, of scores of Belfast trusses that would be the main support elements of the roofs of the aeroplane sheds or hangars. The first order for nails, of various lengths, was for 25 hundredweight. Subsequent orders were for as much as two or three hundredweight at a time.[9] No 22 TDS Gormanston, with Major R. F. S. Morton in command, was formally established on 1 August 1918, though the building phase was far from complete. An oblique aerial photograph taken by the RAF on 26 October 1918 shows that the main buildings were still not complete. The three pairs of the aeroplane sheds and the repair shed appear largely complete but still lacked doors. Bessoneau temporary hangars of canvas on timber frames were in use to house the Avro 504K training aircraft. The photograph verifies that Royal

Air Force training operation, with No 26 and No 69 Training Squadrons, was then in progress with no less than 13 aircraft to be seen on the aerodrome.[10]

The development at Baldonnell followed a similar process, pattern and plan. Most of the 230 acres of land taken over by the President of the Air Council in late 1917 had formally been part of the 403-acre estate owned by the Grierson family, located mainly in the townlands of Baldonnell Upper and Lower. These lands had been sub-divided by the Land Commission in about 1909. On 1 November 1919 the Air Council bought the leasehold interest of Patrick King Joyce in 42 acres, mainly in the townland of Kilbride. At various dates between May 1919 and February 1920 title to the Land Commission holdings of Charles Kenny, P. J. Nugent, John Connor and J. Dowd (about 188 acres in total) was transferred to the Air Council.

The main contractor at Baldonnell was J. & R. Thompson (Dublin) Limited whose clerk of works was Mr Andrew Lynn (formally of the parent Belfast firm). Thompsons undertook the construction of some 90 individual buildings ranging in size from 700,000 cubic feet of the pairs of aeroplane sheds to the six-cubicle toilet blocks. Similar to Gormanston, the installation of a temporary narrow gauge (two foot) railway was one of the first developments. This line was laid from the nearest main-line station, Lucan South, across the fields to Miltown and followed the road leading to the west side of the aerodrome, entering the site at a point very close to the present officers' mess. Two 0.6.0WT locomotives, numbers 1311 and 1313, were supplied to the War Department by Hudswell Clarke & Co. of Leeds. At Baldonnell the locomotives operated at the front and rear of the wagon train. Materials and workers from Dublin were transferred from the main line and brought to the centre of the site at the point near where the garrison church now stands. A spur line to the south of the aeroplane sheds facilitated the delivery of timber, for the Belfast trusses, closer to the buildings under construction.[11] On 1 April 1918, the day the RFC and the Royal Naval Air Service were amalgamated to become the Royal Air Force, Baldonnell was reported to be 25 per cent finished and predicted to be complete by 1 June. This proved over optimistic as on 1 August 1918, the day on which No 23 TDS Baldonnell was formally established, progress of the various elements of the contract was reported to average about 72 per cent. An RAF vertical photograph taken on 5 September 1918 shows that not only were the hangars incomplete (a number of Bessoneau canvas aircraft hangars were in use), but also that a large number of bell tents were in use to accommodate personnel.[12] While the building programme continued to progress the contract had still not been completed by 11 November 1918, but accounts reveal that is was finished soon after.

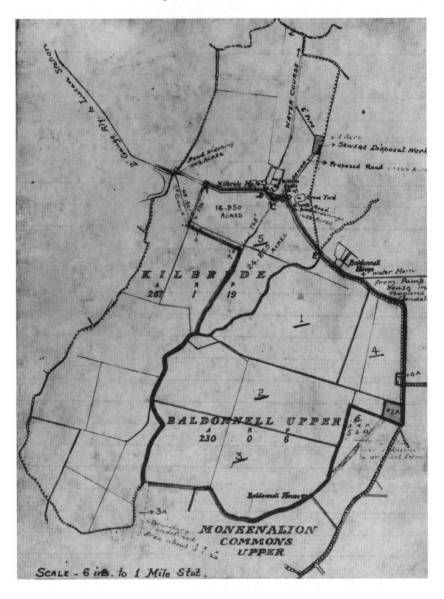

The lands and rights-of-way acquired for No 23 Training Depot Station, Baldonnell in November 1917 (NAI).

It is not clear to what extent the training regime at Baldonnell had developed before the Armistice. The plan was for No 31 and No 51 Training Squadrons from Wyton to become training flights at Baldonnell. However, with the Armistice the nascent training regimes at all four training depot stations were wound down before they could be fully established. By October 1919 the training depot status of the four aerodromes had ceased; most of the flying units were subsequently dispersed back to Britain, and some 420 officers and 4,005 other ranks would be demobilised. Tallaght, Collinstown and Gormanston were placed on a care and maintenance basis while Baldonnell became the headquarters for the remnants of the RAF units still remaining in Ireland.[13]

AIR OPERATIONS

As early as May 1918, in response to popular opposition to conscription in the absence of Home Rule, RAF elements had been sent to forward landing grounds at Omagh (No 105 Squadron) and Fermoy (No 106 Squadron). These squadrons carried out reconnaissance and communications duties in support of ground units. In 1919, in an attempt to exercise military control in Ireland, the viceroy had recommended to Lloyd George that aircraft should be deployed to strongly defended aerodromes in the hope that patrolling by aircraft armed with bombs and machine guns would counteract the military activities of Sinn Féin. As a result, seven flights of No 2 and No 100 Squadrons, RAF, mainly equipped with Bristol Fighters, were posted to Baldonnell and dispersed to airfields at Fermoy, Castlebar and Oranmore during the War of Independence. However, initially the government did not permit the aircraft to use bombs or machine guns mainly because of the difficulty of identifying – from the air – civilian-clad Irregular soldiers operating amongst the general population. The greatest opposition to the use of armed aircraft came from Major General Sir Hugh Trenchard who opposed their use unless a state of war had been declared.[14] Even though permission was granted to arm aircraft in March 1921 caution dictated that 'the only real use which the Army found for the RAF was in transporting senior officers and in running an air mail service once the roads and ordinary mails had become dangerous'.[15] Townshend asserts that the utility of RAF aircraft was limited due to the lack of communications with air bases or ground forces. He also suggests that 'nothing was done to create air camps or increase the number of landing grounds' and that this had contributed to a lack of effectiveness.[16] Whatever about the lack of communications, the records listing the many RAF facilities would suggest that aerodrome infrastructure was not wanting. In addition to the four former training depot stations there were four other 'Class A' aerodromes – Fermoy (Cork), Omagh (Tyrone), Oranmore (Galway) and Aldergrove (Antrim). They also had 43 'Class B' and 'Class C' airfields. The RAF also had the use of over 60 sites, many on the estates of the Anglo-Irish gentry, generally located near a military or Royal Irish Constabulary barracks. The latter were marked with a large cross and were designated and listed as landing grounds.[17] Townshend cites the future Field Marshal Montgomery (when Brigade Major, 17 Brigade, Cork) as saying that the RAF aircrews knew nothing about the War and suggests that 'a more imaginative approach by the Army might have yielded different results'.[18] Notwithstanding the caution exercised in the operation of military aircraft, the RAF lost a small number to guerrilla fire. Most losses resulted from opportunist attacks on individual aircraft involved in the delivery of military mails in the south-western counties.[19] With the signing of the Treaty in London on 6 December 1921 and its subsequent ratification by the Dáil (Irish parliament) in January 1922, the RAF squadrons began withdrawing from Ireland. While 100 Squadron was withdrawn from Baldonnell in early February 1922 an Irish Flight of four Bristol Fighters was formed there in April 1922 and operated in support of the British Army withdrawal from Ireland.

CIVIL AVIATION

Meanwhile, in the field of civil aviation, Ireland was somewhat of an aeronautical backwater. There was no commercial aviation and the country had no manufacturing capacity whatsoever. Apart from the early efforts of Harry Ferguson and Lillian Brand, in what is now Northern Ireland, enthusiastic amateur constructors, that were prevalent in other countries, were not in evidence. While a significant sports 'aviation meeting' had taken place on Leopardstown Racecourse on 30 August 1910, and a fiasco of an air race (Leopardstown to Belfast and return) was staged in 1912, all participants were from outside of Ireland. As the aviation meeting had been a limited success and the air race a failure and, as Leopardstown was deemed unsuitable from a meteorological point of view, the organisers were dissuaded from attempting a third event.[20] The fact that the UK Civil Aviation Act, 1918, did not apply to Ireland reflects the paucity of aviation activity on the island of Ireland at this juncture. Even though Iona National Airways began commercial operations at Kildonan in 1931, it would be as late as 1936, coinciding with the beginning of a state sponsored air service by the newly formed Aer Lingus, before primary legislation was passed by the Dáil to provide for the regulation of civil aviation.[21] While Baldonnell was the headquarters of the RAF in Ireland it had also functioned as Dublin's civil airport from 1 May 1919. In April 1919 the RAF at Baldonnell was instructed to appoint an officer for civil aviation duties. Shortly afterwards Captain M. V. McKeon was made the Civil Aviation Officer. However, it wasn't until 25 June 1919 that the first commercial aircraft, a chartered Airco DH9B, visited Baldonnell. Other civil aircraft movements, during the remainder of the RAF stay at Baldonnell, were not recorded.[22]

BALDONNELL TAKEN OVER

As the British were withdrawing Army and RAF units from Ireland during 1922, the Provisional Government's main aviation concern was with the establishment of a civil air service and the occupation of a suitable civil airport. Colonel Emmet Dalton, in his capacity as chief liaison officer, was instructed by the Air Council to request vacant possession of Baldonnell.[23] The Irish Flight left Baldonnell on or about 1 May 1918 and operated from Collinstown until its departure for Aldergrove on 29 October 1922.[24] On 3 May 1922 Baldonnell Aerodrome was handed over to Provisional Government troops.[25] On or about 12 May the Civil Air Service, under the direction of Captain Charles F. Russell, officially assumed responsibility for the civil aerodrome, while the military camp remained under infantry control.[26]

TWO

CIVIL AVIATION DEVELOPMENTS
IN SAORSTÁT ÉIREANN

—

The Defence Forces of today, consisting of army, air and naval elements, are officially designated as Óglaigh na hÉireann or Irish Volunteers. The organisation traces its lineage and name back to the formation at a meeting held in the Mansion House on 25 November 1913.[1] Only partially quoting the aims of the Irish Volunteers, John P. Duggan described this first Irish Army as a 'volunteer force, a people's army formed to secure and maintain the rights and liberties common to all the people of Ireland without distinction of creed, class or politics'.[2] However, the Irish Republican Army (IRA) that fought a guerrilla campaign in 1919–21 with the aim of ending British rule and occupation, did not greatly reflect such lofty ideals. The Anglo-Irish War (or War of Independence) was 'characterised by guerrilla warfare, ambushes, raids on police barracks, and planned assassinations' on the part of the IRA and 'reprisals, the shooting-up and burning-up of town, executions and terrorising' on the part of the British forces.[3] The Anglo-Irish Treaty agreed on 6 December 1921, in addition to providing for British retention of the naval ports at Cobh, Berehaven and Lough Swilly, provided for the installation of military aviation facilities in their vicinity if so required by British coastal or maritime defence considerations. The Treaty also provided that 'a convention shall be made between the same governments for the regulation of civil communication by air'.[4]

The various accounts of the early days of Irish aviation cite the purchase of a passenger aircraft during the Treaty negotiations in London as the first event in the history of the State's military aviation. These accounts also suggest that the aircraft was specifically purchased in order to facilitate an expeditious departure for Michael Collins and the four other plenipotentiaries should the Treaty negotiations fail.[5] While this version of events is accepted in the mythology of Irish aviation and, though it is based on an account by an officer on the fringes of the Treaty negotiations, it will be shown that this myth falls far short of the complete story.

AIR ASPECTS OF THE PEACE AND TREATY NEGOTIATIONS

Following the Truce of 9 July 1921 that marked the cessation of hostilities between the IRA and the British forces in Ireland, the latter half of that year was dominated firstly by peace negotiations carried out at arms length, and later in the year by bilateral negotiations in London, which led to the Treaty of 6 December 1921. It was during these two negotiating phases that the initial concepts of Irish defence and aviation began to be formulated, though not well defined. An early British paper put considerable emphasis on the strategic position of Ireland:

> The position of Ireland is also of great importance for the air services, both military and civil. The Royal Air Force will need facilities for all purposes that it serves; and Ireland will form an essential link in the development of air routes between the British Isles and the North-American continent. It is therefore stipulated that Great Britain shall have all necessary facilities for the development of defence and of communication by Air.[6]

This and similar conditions, including a requirement that the new State contribute financially to the military defence of Great Britain, prefaced a negotiation process that greatly emphasised Britain's strategic defence considerations. At this early stage the Irish, particularly Robert Erskine Childers, still held out hopes of creating 'a gradually expanding, as finance allowed, modest naval force purely for coastal defence and reconnaissance'. Eventually recognition of the precedence of Britain's strategic needs combined with the financial impracticality of such a proposition put the matter of an Irish naval force on the long finger. In discussions on 'army and air' Childers suggested:

> It is no doubt agreed that we should maintain an army with a small standing force highly disciplined and well equipped, and a wider reserve; with a strategic organisation based on the idea of rapid concentration for coastal defence. A small air establishment disposed on the same principle, specialising in coast reconnaissance and perhaps in anti-submarine and commerce protection work.[7]

This modest proposal was quoted in the context of a British statement laying down a condition that Ireland contribute militarily and financially to the common defence requirements of Britain and Ireland. In their consideration of aviation matters initial British concerns were totally selfish. They claimed that Ireland's geographic position was of great importance in relation to British 'military and civil air services' and 'that Ireland will form an essential link in the development of air routes between the British Isles and the North

American Continent'.[8] Childers proposed that the condition 'that Great Britain shall have all necessary facilities [in Ireland] for the development of defence and communications by air' be opposed on the basis 'that Atlantic reconnaissance and anti-submarine work can be done by her by other means'. Thus, the proposal to retain Royal Air Force stations in Ireland was rejected on the basis that their only possible use would be against Ireland.[9] The provision of facilities for British civil and military aviation in Ireland did not feature in the Treaty eventually agreed. However, the initial civil aviation considerations, particularly that regarding future transatlantic air travel, are of interest in the context of the later development of the flying boat base at Foynes and the nearby airport at Shannon by the Irish Government. These facilities, that were to be of considerable benefit to the Allies during the Emergency, would be developed prior to and during the Second World War, many years before Irish commercial aviation had any use for them.

While the Irish peace negotiators in July 1921 were anxious to minimise, if not eliminate entirely, all aspects of the British military presence in Ireland and to negate the perception of any Irish obligation to contribute to Britain's military defence, it is doubtful if they had any defined concept of airspace or air defence. The preference in the peace negotiations, expressed as an overall defence policy, was to stand alone 'with complete independent control of our own territory, waters and forces', suggests that military aviation was not identified as a separate consideration in the defence of the country. It must be noted, however, that the statement: 'we must be clear as to what our naval and military policy would be' could be interpreted as including military aviation policy – on the basis that the term 'military' would have included not just army but also the new discipline of aviation in those early years of military aviation.[10] The Irish policy position conceded that, while naval defence was the essence of a country's defence, it would take some time to build up even a minimal capability. In the absence of air and naval defence the Irish policy position could be interpreted as suggesting that greater defence was afforded by the absence of British forces. It was in effect an early admission that the new State would not be able to defend itself in naval and air terms.

The first indication of an ideological environment conducive to the development of civil and military aviation in the new State came about during the Treaty negotiations in London in the autumn of 1921. The British appro-ached the matter of military aviation in a manner similar to the policy adopted regarding the retention of the ports and certain naval facilities. However, examination of the accounts of the informal meeting of the Defence Committee reveals that the British did not initially have a well-defined position. The RAF was not represented at the initial meeting of the sub-group dealing with defence matters which was called to discuss the naval and air aspects of concern to both sides. In fact, the opening British position was that Ireland would not be permitted either naval or air forces and that both functions would remain British responsibilities, with Britain's defence requirements as the priority.[11]

At a later meeting Marshal of the Royal Air Force, Lord Trenchard stated that 'we want bases for our aeroplanes such as would be required for the defence of Britain'. He suggested that an attack on Britain, from the continent, might be made by 'aircraft routing around by the west of Ireland' and we 'want to be in a position to put a squadron in Ireland to deal with any attack by air'. He contended that an air attack that might come through Ireland constituted a particular threat to Britain.[12] In their consideration of such matters the Irish delegation was fortunate to have Robert Erskine Childers as their secretary, though strictly speaking not a delegate. Childers's experience as a naval officer, Royal Naval Air Service/RAF observer (navigator), and his knowledge of military and naval science placed him in an excellent position to counter extreme positions adopted by the British.[13] Childers reminded Britain's main air expert (his former superior) that Ireland did not play an important part, from an air point of view, in the war with Germany. He also pointed out that aircraft with sufficient range for such an attack had still not been developed. Trenchard agreed with Childers's contention that aircraft with ranges of the order of 500 to 600 miles were then only available but considered that there was also the matter of carrier borne attack.[14] It is possible that Trenchard had his own agenda aimed at the retention of a more substantial RAF. The post-war RAF, as the third and very junior service of the British forces, was fighting for survival in the face of prejudice from the Royal Navy and the British Army.[15] It had been greatly reduced after the Great War[16] and a possible withdrawal from Ireland would lead to two more squadrons being disbanded, further undermining the cultural argument for the retention of an independent air force and parity with the Royal Navy and the British Army.

The matter was resolved by Winston Churchill who stated that any developments in air power that might be made by her enemies would be matched by similar technical advances by Britain and that it was therefore immaterial whether RAF aircraft were based in Britain or Ireland. He indicated that, in contrast to the case for naval bases and for possible naval airbases nearby, it would not be necessary for Britain to retain any RAF bases in the new Irish State. On the matter of civil aviation, while both Churchill and Trenchard emphasised the British future requirement for 'stopping places for cross-Atlantic travel', the question remained unresolved – possibly because this was still a somewhat remote concept. Collins, showing no great concern about military aviation, asked if Britain 'would give us landing places in England' for civil air services and was pleased to be reassured by Churchill that 'there would be perfect reciprocity' and that the State's future participation in civil aviation in particular would, by international convention, be on the same basis as any independent country.[17] Some alarm was later caused in the Irish camp when the British indicated a new condition that Ireland would not be allowed to develop an 'air force'. When queried on the matter the British quickly clarified that this only related to a prohibition on Irish naval aviation.[18] While most discussion on Irish defence and air matters was confined to sub-committee level the most definitive statement on defence policy was to be made in the final stages of the main negotiations. A significant amendment to Article 7 of the draft agreement, attributed by Frank Pakenham to the 'republican wing' of the negotiating team, indicated a much more positive and strident policy position on defence than had been discussed internally or previously articulated in negotiations:

> As an associated State Ireland recognises the obligation of providing for her own defence by sea, land, and air, and of repelling by force any attempt to violate the integrity of her shores and territorial waters.[19]

While this amendment, along with most of the others proposed by the Irish negotiators on 4 December 1921, was dismissed out of hand by the British and thus not reflected in the Treaty, it is not clear to what extent it would represent the actual defence ideology or doctrine that would guide the new Free State. In the negotiations the amendment did not meet with Collins's approval.[20] It could be argued that the State's 'defence by sea, land and air', as would be undertaken in the 1920s and subsequent decades, fell well short of such aspirations. The tenth amendment proposed on 4 December 1921, which stated that 'a convention shall be made between the British and Irish Governments for the regulation of civil communications by air' was ultimately included as paragraph three of the annex to 'The articles of agreement for a treaty between Great Britain and Ireland, December 6, 1921'.[21] While it is not certain at whose insistence such a provision was made, in view of his previous concerns regarding a possible future air service, it would appear that it was most likely to be Collins who insisted on such a clause.

THE PURCHASE OF AIRCRAFT

In the mythology of Irish aviation it is accepted that, while the Treaty negotiations were still in progress, a civil passenger aircraft, a Martinsyde Type A, Mk II, was bought in England in October 1921 on the authority of

Michael Collins. Based mainly on Emmet Dalton's recollections in 1951, the mythology also suggests that the machine was purchased solely to act as a ready means of escape to Ireland for Collins and the other plenipotentiaries. In 1951 Dalton responded to a query from Lieutenant Colonel W. J. Keane:

> I had discussions with Michael Collins, and together we put before the General Staff the idea formulated by me that we should purchase an air-plane in London and have it standing by in readiness to fly Collins . . . back to Dublin in the event that the negotiations broke down.[22]

While Collins's safe passage home would have been a priority it is not easy to accept that Collins, and the second Dáil, could justify the expenditure of £2,700 on a get-away aircraft.[23] Comparison of Dalton's 1951 account of the events of October 1921, that possibly relied on memory rather than a written account of how he perceived the matter in 1921, with that of Commandant Seán Dowling, indicate subtle differences.[24] It suggests that Dalton's account may not be fully accurate, that he may not have been privy to the complete plan regarding the purchase of aircraft and that his part in the events may not have been as important as he suggests. Though the identities of the principals involved in the decision to purchase aircraft are not in doubt, the particular roles played by Collins, Russell, Dalton and McSweeney require clarification.

According to Dalton he was authorised by Collins to put into effect his plan to purchase an aircraft for the purpose outlined. He had known Jack McSweeney (more formally known as William J. McSweeney), a former RAF pilot officer who was a Dublin IRA Volunteer, from their participation in a previous IRA plan and now sought his assistance in this aviation matter. Dalton was introduced to another ex-RAF pilot named Charles F. Russell, a member of the 4th Dublin Battalion IRA, by Dowling. Dalton recorded his account of the start of a very important mission:

> I called these two young men together, had a long conversation with them, became convinced of their loyalty, and sent them to England to examine the possibility of purchasing a suitable aircraft. Russell, who had spent some time in Canada, was to act as if he were making the purchase for a Canadian forestry department.[25]

Seán Dowling, formerly of 4th Battalion, Dublin Brigade, had a somewhat different view of the thinking behind the proposal to purchase aircraft. He later recalled the first meeting of Russell and Dalton:

> When I introduced Russell to Dalton they began to discuss how his special experience as a pilot could be helpful to the IRA. Russell said that, for instance, if

the negotiations for a treaty broke off and fighting began again, he could go to England and seize a plane, fly it over here and bomb enemy positions.[26]

While this proposal did not relate to that of acquiring a passenger aircraft for a hasty retreat, the idea of purchasing such an aircraft may have been the product of the discussions between Russell and Dalton, and possibly others, rather than the singular idea of one person. Subsequent action, the purchase of a military training aircraft (in addition to the passenger machine) would suggest that Russell's idea of bombing British forces in Ireland became one of the main contingency plans adopted by Collins. Russell accompanied Dalton to London at the time of the Treaty negotiations, though his function there was solely in relation to the purchase of aircraft and the making of arrangements for a possible flight to Dublin. The payment by Russell of £25 'to Lt. McSweeney, IRA' for 'expenses before the purchase of the machine' indicates that W. J. McSweeney had been in London for the aircraft evaluation phase and had returned to Dublin after suitable aircraft had been chosen. In Dublin McSweeney was to organise the personnel, equipment and arrangements for the possible arrival there of the aircraft carrying Collins.[27] That McSweeney had been on 'GHQ staff duty in London during [the] Treaty negotiation' in the autumn of 1921 is confirmed by his officer's history sheet compiled in early 1924.[28]

On arrival in London Russell, apparently accompanied by McSweeney, began a survey of aircraft manufacturing companies and an evaluation of the various aircraft on offer. He was seeking aircraft to fulfil three particular roles. Firstly he wanted a 'machine capable of direct flight [to Ireland] (a) for passengers, (b) freight'. Secondly he sought a 'machine suitable for military undertakings i.e. bombing in Ireland', and thirdly a 'machine sea-plane [*sic*] suitable for transporting freight from ship in home waters to [a] base in Ireland'. He consulted with representatives of five aircraft manufacturing companies; Avro & Co.; Martinsyde & Co.; Short Bros; Vickers Ltd and de Havilland & Co. and received quotations in respect of a variety of aircraft. The Vickers Viking and the Martinsyde Type A MK II were the machines considered for the primary role of transporting passengers or freight. The Viking, which was an amphibious aircraft with a useful load of 1250 lbs and a range of 400 miles, was rejected out of hand even though it was considered suitable for all three tasks. The aircraft 'is, we consider, out of the question at the price quoted – £4,675'. The delivery period of three months after the placing of an order, and the aircraft's handling when landing on grass, were also deemed to be unacceptable. Its greater stability on water in bad weather conditions was cited as an advantage over the Martinsyde aircraft. Russell focused on the Martinsyde:

The machine to our mind which is suitable for purpose (1) is the 'Martinsyde' Type A, MK II, 4 seater biplane. This machine is fitted with a Rolls Royce engine, and is complete with floats or land undercarriage, and is quoted to us at a price of £2,600. It has a range of 550 miles at [a] crusing [*sic*] speed of 100 miles per hour. Delivery could be made within twenty-eight days.[29]

Russell deemed that the Martinsyde would also be suitable for the ship to shore freight role. At first glance it is not obvious why Russell should have considered the Avro 504K in the context of bombing in Ireland. As noted by Russell this particular type, with its 110 hp Le Rhone engine was 'the English Army standard training machine'. With a limited load carrying capacity, over and above fuel and two aircrew, it was not configured for bombing. While a gunnery trainer version of the Avro 504K, with a 130 hp engine, had been developed by way of modifications incorporated in the basic machine there is no record of a version fitted out for conventional bombing.[30] Such was the configuration of the machine that it could only be used for bombing by the rather crude practice of having the second crew member drop bombs over the side of the rear cockpit manually – an accepted practice in earlier times. However, the major advantage of the Avro 504K would be in the fact that if questions were asked by the UK authorities it could correctly be described as a training machine. Against the 504K Russell considered the de Haviland DH 9 which he stated 'was used during the European War as a long [range] bombing machine and was found very suitable as such'. The DH 9 was considered by Russell to be suitable for the proposed bombing role but he eliminated it on the basis of cost. He stated that 'the price however, being £1,000 is exorbitant except for permanent use'.[31] This latter comment suggests a short-term bombing role for whatever aircraft was to be purchased within financial limitations.

The Martinsyde Type A Mk II and the Avro 504K therefore comprised Russell's final choice of aircraft types. In effect he selected two aircraft to carry out four distinct roles. Although the choice of the Avro 504K training aircraft in a bombing role seems to have been a compromise due to the lack of sufficient funds, what is perhaps more intriguing is why Collins felt he required an aerial bombing capability at that particular juncture. A likely explanation is that Collins, with the technical and professional assistance of Russell, was hedging his bets while awaiting the outcome of the Treaty negotiations. On the one hand he was preparing for a peaceful outcome to the negotiations by purchasing an aircraft capable of several commercial roles. On the other hand he was preparing for the possible failure of negotiations and resumption of hostilities by having the same aircraft available to get back to Ireland in a hurry while also purchasing a training aircraft that might be used for bombing purposes should hostilities be rejoined. In the event of a successful outcome to the Treaty negotiations the Avro 504K would make a very satisfactory training aircraft for either a civil or military aviation organisation, whichever was developed in the new State.

Having selected the aircraft Russell's next task was to effect the actual purchases. The financing of the aircraft purchases was, in itself, an interesting arrangement though the exact mechanics are not totally clear. While it is known that the Dáil Defence Department channelled £3,050 through the Irish Self-Determination League of Great Britain it is not certain exactly when the monies were paid to the League. The context suggested by the Dáil accounts for 1 July to 31 December 1921 and Russell's account forwarded to Collins in late February 1922 suggest that £3,050 was forwarded to the League on the basis of an estimate, by Russell, of what monies would be required to effect the purchase of aircraft.[32] This mechanism appears to have been used to obviate the necessity for Russell to pay for goods and services by means of cheques drawn on a Dublin bank, and so hide the financial transaction from the British authorities in Dublin Castle. Dublin Castle not only monitored the Dáil's Dublin bank accounts but had been engaged in 'pinching Michael Collins's "war chest" from the Munster and Leinster Bank' in October 1920 – by all accounts on dubious authority.[33]

On 19 October 1921, Russell received a cheque for £1,500 from Art O'Brien of the League.[34] About two days later Dalton and Russell reported to the COS in Dublin:

> We have succeeded in purchasing a Martynside [*sic*] Aeroplane which can carry ten passengers or 16,000 pounds weight of munitions. We intend that this shall serve several purposes – it can be used, if necessary, in a break of the present negotiations. I have the pilot over here and the machine will be ready for flight within two weeks.[35]

The content and style of his report appears to indicate that the first two paragraphs, including the above extract, were written or dictated by a non-expert (such as Dalton) who appears to have greatly exaggerated the passenger and weight carrying capacities of the aircraft. With a full fuel load, a useful load of 500 pounds and two passengers, or a maximum of about 800 pounds, would have been more correct. While the transportation of five passengers would have necessitated a reduction in the fuel load Dublin would still have been well within range.[36] The latter paragraphs, giving precise instructions covering all aspects of the arrangements to be put into effect by McSweeney in Dublin, were probably dictated by Russell. He proposed that 'it would be necessary to have six men on the approved landing ground i.e. a flat part of the race course' at Leopardstown. Equipment and materials required included two motor cars, 'sixty gallons of 1st grade Aero Petrol (this can be purchased from Lemass of [the] "LSE" Motor Company), 2 ft square of chamois cloth, five gallons of water.' The instructions went on to detail all arrangements required, including those peculiar to a possible night landing.[37]

Russell appears to have made a down payment on the aircraft on or about 20 October 1921. In the case of the Martinsyde the basic price was £2,300, with a further £300 for floats. An additional £100 was paid to have the aircraft modified in order to increase the passenger seating capacity from four to five. In the case of the Avro 504K training aircraft, originally quoted at a price of £175, a down payment of £130 was made. The eventual cost was £260, though there is no apparent explanation for the increase. On 12 December 1921 Russell drew down a second payment, this time of £1,300, and a final amount of £250 on 30 December. All payments were supposedly made 'on the instructions of the M.O.F.' (Minister of Finance – Collins).[38] The measured manner in which Russell received the monies suggests that it was paid as required to meet his purchasing obligations to a preset limit of £3,050.

The total amount finally spent on the purchase of the two aircraft and the associated expenses was £3,767. 10s. amounting to almost 40 per cent of the purchases made by the Director of Purchases of Dáil Éireann's Department of Defence.[39] In the context of the limited financial resources of the first and second Dáils, and of the then current ministerial salary of about £300 per annum,[40] the expenditure of such a sum on aircraft, for whatever purpose they were intended, represented a very substantial and risky investment. It is noted that while the two aircraft cost a total of £2,960, the associated costs increased this figure by some 27 per cent. The Martinsyde, after modification, was test flown at Brooklands on 24 November 1921 by a Captain Clarke. Clarke was paid £15 by Russell to ensure that the aircraft was kept ready to fly at short notice.

With the successful signing of the Treaty on 6 December 1921 and with no immediate necessity to use either aircraft both were put into storage and eventually delivered to Baldonnell, as freight, in June 1922.[41] One of the more interesting items of expenditure incurred by Russell in late 1921 was one of £25 paid to the Director of the London/Paris Air Service for a study of the viability, including costs, of 'the commercial possibilities of an air service between Cove and London'.[42] The acquisition of such a report in 1921–2, in conjunction with the roles specified for the Martinsyde aircraft, strongly suggests that the administration, with Collins and Russell as the prime movers, was seriously contemplating the early establishment of a civil air service.

In due course the expenditure of the monies, that had been authorised by Collins and expended by Russell, was accepted by the Provisional Government as a legitimate expense of the new Free State. On 27 February 1922 Collins wrote 'asking for [a] statement of expenses incurred' by Russell. Replying the same day from the Aviation Department of GHQ, Beggar's Bush Barracks, Russell acknowledged the receipt of £3,110 (the additional £60 was from 'other sources'), and accounted for the expenditure of £3,247.10s., suggesting that he was due to be repaid £137. 10s.[43] Subsequently Richard Mulcahy, in his capacity as Minister for Defence, wrote to the Minister for Finance:

> I desire to make application for the sum of £520, being [the] immediate financial requirements for the civil aviation Dept. An outline of the expenditure to be covered, is attached. The cash is required urgently, please.[44]

This additional expenditure was the Civil Aviation Department's estimate of the various expenses, including the storage in Britain and the delivery to Ireland of the two aircraft purchased in October 1921. The note was initialled by 'MO'C' with instructions to the effect that if 'this application is in order, please pay'.[45] Above the signature 'M. O'Coileain', five days later, Collins wrote to Defence in regard to Russell's transactions of October 1921:

> The total amount was £3,050. It was decided at a provisional government meeting that we should accept liability for that sum. It is now a matter of putting the matter formally in order, so that we can get it repaid to the Dáil. Will you please endorse this and send it forward to Mr Duggan for his endorsement, in accordance with [the] recommendation regarding Defence accounts.[46]

As the above authorisation only related to the £3,050 Russell had received from the Self-Determination League, Cabinet approval for the additional £520 was recorded on 18 April 1922.[47] It is not clear when and how the outstanding balance of the initial monies spent in London was repaid to Russell. While civil aviation was to recede into the background after the start of hostilities on 28 June, it appears to have been financed for some time thereafter. On a date between 1 November 1922 and 15 August 1923 the Army Finance Officer, Thomas O'Gorman, received some £1,364 from the Civil Aviation Account and refunded it to Dáil Éireann. It is presumed that Russell had been repaid the money due to him prior to 1 November 1922.[48]

THE CIVIL AIR SERVICE

In early 1922, while the Provisional Government was beginning to take responsibility for the administration of the new Free State and the Army was taking over a large number of military installations from the departing British

Army and Royal Air Force, Russell and McSweeney were informally appointed to positions, under Emmet Dalton, in the General Headquarters of *Óglaigh na hÉireann* in Beggar's Bush Barracks Dalton explained:

> Plans and suggestions were drawn up by McSweeney and Russell in what was known as the Aviation Department . . . This aviation dept. came into being as a subsidiary Department to my branch training [*sic*].[49]

In the initial weeks of 1922 the two officers worked in the Aviation Department, where, initially, no apparent distinction was made between military and civil matters. Subsequently a decision was made to the effect that 'the [civil] aviation service be worked as a military department'.[50] However, it is apparent from the major surviving source on the subject, (DT file S.4002), that Russell was concentrating on policy matters relating to civil aviation while McSweeney was addressing the subject of military aviation. Their appointments received the formal approval of the Air Council meeting of 23 March 1922. The minutes record that 'Mr W. J. McSweeney was appointed director of military aviation with the rank and allowance of a commandant general' and that 'Mr. Chas F. Russell was appointed director of civil aviation and secretary to the Air Council'. It was decided that the two directors would receive the same remuneration.[51] There are a number of indications that this division of responsibility was a considered decision on the part of the Provisional Government. Russell had originally been selected by Dalton and apparently confirmed by Collins, ahead of McSweeney, for the task of aircraft selection and purchase. Accounts suggest he accomplished this task to the satisfaction of Collins. Copious contemporary records suggest that Russell was the superior manager and staff officer and best suited to the development of civil aviation – the new State's priority.

It would appear that Russell, in an informal manner, had originally been appointed to the new and more important position of Director of Civil Aviation by Collins himself on Saturday 18 February 1922, when the title 'superintendent of civil aviation' was used. On that occasion Collins asked Russell for a paper exploring the manner in which commercial civil aviation might be developed in Ireland. Russell reported within two days:

> With reference to our conversation on Saturday [18 February 1922] I am sending you herewith [a] scheme for handling aeronautical affairs in Ireland, which however I have only had time to outline roughly. During the preparation of this report I have had before me reports on the management of aeronautical affairs in practically every country in the world, and while it is not an exact duplication of any one country's methods, it is more or less [on] the lines of the New Zealand government['s policy].[52]

The submission by Russell of a five-page paper on civil and military aviation, obviously based on material already in his possession, in such a brief time strongly suggests that he was, on his own initiative or on some understanding with Collins, well advanced in his research and study of military and civil aviation matters. Collins, having indicated his preference for civil aviation and for air communications during the Treaty negotiations, would have required Russell's professional background and research skills to formulate and express the State's civil aviation options.

In the preliminary remarks of his paper on a scheme for handling Irish military and civil aeronautical affairs Russell may have been simplistic when he stated that aviation could be divided into two branches – civil and military. However he was not only informing Michael Collins and his department but also educating ministers and officials of other departments as well as the General Staff, many, if not all, of whom would have little appreciation of air matters. He defined civil aviation as comprising 'aircraft construction' and 'civil air transportation which together aim at the acceleration of inter-communication and the expansion of trade by means of air transport'. He considered the ultimate objective of military aviation to be 'direct defence'. He considered that 'the duties of Air forces once war is joined' would be to 'manoeuvre to attack enemy air, land and sea forces and territory, and to defend home territory from the air' and enable 'arterial air routes to continue to operate by protecting them from hostile air attack'. He highlighted the similarities of, and the differences between, civil and military aviation but emphasised the fact that they were closely allied 'because in future the actual cadres which compose the country's service air forces in peace, can only be augmented in war by a reserve of men (who might well be placed on a volun-teer basis) by material [sic] and experience in construction and design afforded by Commercial aviation'.[53] He warned 'that commercial aviation cannot be fostered merely as a reserve for the country's military air forces' and that 'its test must be that of a commercial success'. He foresaw the potential for the development of civil aviation initially in the carrying of mails. Other roles included 'the high speed carriage of . . . small and valuable goods, passengers, and for sundry purposes such as mapping and survey work'. To support his claim for the transportation of mails he cited large savings being made in the US by a commercial airmail service operating between large and well-separated cities. However, he did not expand on how such a service might fare in Ireland or between Ireland and Great Britain.[54]

In summarising 'Ireland's position' he asserted that 'one may safely say that [as a nation] we know practically nothing about aviation either military or civil'. He stated that 'we have in Ireland about fifty ex. English army flying officers, 8 of whom served with the IRA', only one or two of whom had flown 'since the close of the European War' and that 'as far as can be ascertained we have only one commercial aviator in Ireland at the moment'. He further suggested that 'as a result of conditions in Ireland we have not yet had the necessity for a department to handle and foster commercial aviation'. Noting that the Civil Aviation Act, 1918 did not apply to Ireland, Russell listed the various regulatory duties and functions provided for in that Act that would be required to be performed by a government aviation department. These would include the testing and licensing of both pilots and aircraft and the issuing of appropriate certificates; testing the airworthiness of aircraft and the physical standards for air pilots; as well as the collection of meteorological information, wireless communication with aircraft, and much more. To carry out such an extensive range of duties Russell considered that a regulating aviation department required 'an aerodrome [that was] suitable from a commercial and military point of view', 'several aeroplanes for testing purposes' and 'suitable wireless equipment for communication within a radius of 500 miles'. In proposing that, in effect, an 'aviation department' would function as a regulatory body and as an aerodrome operator Russell was combining two functions that might later be considered incompatible. However, in a country with practically no aviation activity it probably made sense at that time. In any event this dichotomy did not arise as the approaching Civil War was to stymie early plans for the regulation of civil aviation, for state-sponsored commercial air operations and proposals for the operation of a civil aerodrome. As a result of these early setbacks and the subsequent development of military aviation during the Civil War early administrations were slow and reluctant to undertake their obligations under international civil air conventions.

Russell's proposals for the development of civil or commercial aviation included aspects that he considered would dovetail with the military requirements of the new State. Bearing in mind the current state of affairs in aviation and future military requirements as well as the necessity for Government assistance to commercial aviation he made four recommendations:

> The creation of an aviation department under the Minister for Defence. . .whose duties shall be detailed elsewhere . . . The creation of a school of aeronautics and flying at the government commercial aerodrome . . . the adoption of this school of flying and aeronautics by the military authorities . . . The appointment of a commercial air council.[55]

He further recommended that the 'Air Council' have as its president the Minister for Defence with the 'Superintendent of Aviation Dept. to be secretary to Council'. The Postmaster General, representatives from the Department of Lands, Ordnance Survey and aviation companies in Ireland as well as the officer commanding the Military Air Service were to be members. Russell detailed the duties of the Air Council as exploring commercial air possibilities, considering a scheme for a mail and passenger service from Cobh to London or between other points, and considering the possibilities of locating herring shoals from the air. He also foresaw that the small size of a military air service should preclude it being 'saddled with the expense of a School of Aeronautics' and, as a result, it would become the best customer of the 'civil aviation department of aeronautics and flying' so that the military and civil organisations would have a certain level of interdependence.

Russell proposed the appointment of a 'superintendent of commercial aviation' who would be responsible for the many regulatory duties specified by the Civil Aviation Act, 1918. These included registration, licensing and testing the airworthiness of aircraft, and the licensing and testing of the fitness of pilots. Such an officer would also be responsible for the running of the aerodrome and schools of aeronautics and flying as well as advising the Government on all aeronautical matters. The estimated annual cost of the office of the superintendent was put at £12,000, while the annual cost of the schools of aeronautics and flying was put at £13,820. He suggested that the position of superintendent of commercial aviation be announced sooner rather than later and advocated the taking over of Baldonnell which would be divided between 'the commercial people' and the military. [56] While Russell displayed considerable confidence in the future of Irish aviation, the absence of any legislation, regulation and the rudiments of commercial aviation activity – all fundamental aspects identified by him – were major obstacles to success on any level. Given these stark facts and the worsening political situation it would have been difficult for the Provisional Government to have great faith in the possible success, at that time, of his proposals as initially drawn up.

On Thursday 23 March 1922, having apparently been postponed from both the 14 and 15 March, a 'meeting of members of the Government, members of the General Staff, and officers from the Military Aviation Department' was held at Beggar's Bush Barracks to consider an agenda based on the various proposals contained in Russell's 'Scheme for handling Irish aeronautical affairs'. The following were present at the meeting:

Mr R. Mulcahy, TD, Minister for Defence – chairman.
Mr M. Collins, TD, Minister for Finance.
General O'Duffy, TD, Chief of the General Staff.
Lieutenant General J. O'Connell, Assistant Chief of Staff.
Major General J. E. Dalton, Director of Training.
Commandant General W. J. McSweeney, Director of Military Aviation.
Mr C. F. Russell, Director of Civil Aviation and Secretary to the Air Council.[57]

The status of those attending the first meeting of the Air Council bears witness to the Government's interest, at least at this juncture, in supporting both civil and military civil aviation. It is probable that Collins's sponsorship of the concept of aviation in general and civil aviation in particular, had a positive influence on the level of attendance. Many decisions, mainly of an organisational or administrative nature, were recorded. It was decided that aviation would be divided into two areas, military and civil, McSweeney and Russell were confirmed in their respective appointments and the composition of the Air Council was also decided. After the meeting the Air Council adopted the recommendation that the proposed school of aeronautics be placed under the direction of the proposed Civil Aviation Department, and that fuller information be sought on a scheme proposed by Dublin Corporation that would establish such a school within the existing technical school system. The Air Council also took on board the recommendation 'that a school of flying be started under the direction of the Civil Aviation Department' and that that school 'be adopted by the military aviation authorities for the training of their pilots.' It was decided that Baldonnell Aerodrome would be the most suitable location for all purposes. Emmet Dalton, in his capacity as Chief Liaison Officer, 'was directed to make the necessary arrangements for the taking over of this aerodrome at an early stage.' In the matter of military aviation 'it was decided that the military air authorities should aim at the organisation of one air squadron for the present'. The detailed consideration of 'air estimates' was postponed *pro tem* while 'both departments were asked for their immediate financial requirements'.[58]

Notwithstanding the fact that Collins and the administration were backing the concepts of both civil and military aviation, it is to be noted that no aviation proposal, with significant financial implications, was adopted at this stage of the planning process. There was evidence, in the first minutes, of a certain air of caution that was to become more pronounced in the weeks following. While no direct reference was made to the deteriorating political situation, as the country headed towards civil war, it was, no doubt, a major disincentive to any significant investment in personnel, aircraft, equipment or general facilities, or even a regulatory body. In this regard the small attendance at the next Air Council meeting was telling.

The meeting held on 6 April 1922 was attended only by the Minister for Defence and the two directors. The minutes of the meeting reflected a significant slowing down of the initial impetus generated for Collins by Russell. Russell's Civil Aviation Department proposal for a school of aeronautics at Baldonnell was put on hold while an aeronautical technical scheme proposed by the Dublin Corporation Technical Committee was referred to the engineers of both bodies so that a joint report could be prepared for consideration by the

Air Council. In the meantime the estimates for the Schools of Aeronautics and Flying were withheld. It was also decided that Russell would produce a memorandum on the methods and conditions to apply for entry into the schools and on a possible scholarship scheme to ensure adequate numbers of students. In his paper he was also to address the matter of entry to a military air service by means of graduation through the schools. In his proposal Russell suggested that the school of flying was the only aspect of civil aviation on which it was intended to incur expenditure for the time being, and that all purchases of equipment would have to be sanctioned by the Air Council prior to purchase. The Air Council meeting decided to forward a statement of the Civil Aviation Department's immediate financial requirements to the Government, however, no details of which were recorded in the minutes. No discussion took place on the subject of the 'purchase of machines for the school of flying', or on 'the appointment of a consulting engineer', all such matters were postponed until the next meeting. The matter of 'foreign quotations for aeroplanes' was discussed briefly but also deferred for later consideration.[59]

The Director of Civil Aviation 'enquired whether his department or the military department would be responsible' for taking over Baldonnell Aerodrome. It was decided in the Air Council meeting 'that the military authorities should take over the aerodrome and arrange with the Civil Department to let them have the required number of sheds'.[60] This particular decision was somewhat ambiguous in that both nascent air services came under military control and 'military authorities' was not defined. In any event, on 3 May 1922, the complete installation was taken over by the Army as just another military barracks.[61] Subsequently, at some date between 3 and 12 May 1922, responsibility for the aerodrome and hangars and the civil aviation functions of the facility were taken over by Russell in his new role as Director of Civil Aviation.[62] The composition of the small staff of the Civil Aviation Department that moved into Baldonnell in May 1922 confirms that it was initially administered as a civil rather than as military aerodrome.[63]

On 25 April 1922 Russell had reported that he was in receipt of queries from two UK based air service companies who had expressed an interest in running air services between Dublin and such cities as Manchester and London. One company was preparing to commence operations between London and Dublin via Manchester on 1 June 1922. The second company wanted to operate between London and Dublin and were requesting support from the Irish Government in the form of a subsidy. These overtures brought into focus a number of problems that the Government had not addressed. Firstly Baldonnell, still occupied by the RAF in April 1922, was not designated (by the new administration) as an aerodrome for civil aircraft entering the State nor was it designated a customs aerodrome as required by the Convention for the Regulation of Aerial Navigation. There were five other important areas, including the formulation of local flying regulations at Baldonnell, where the absence of legislation and regulation was hindering the opening up of the country to civil aviation. The situation regarding commercial activity was similarly difficult as the Government had not adopted a policy in respect of subsidies to be offered to companies interested in serving Baldonnell or elsewhere.[64]

Russell attempted to convene a meeting of the Air Council for 5 May 1922 and distributed an agenda of seven items, including two matters postponed from the poorly attended meeting on 6 April, in addition to consideration of the proposals made by the two British companies. However, no minutes of a meeting on 5 May 1922 appear to have been recorded. A further meeting of the Air Council was requested by Russell for 15 May 1922 with an agenda of sixteen items that included those carried over from the two previous agendas. The new items for discussion included a small number concerning the infrastructure and services at the recently taken over aerodrome at Baldonnell and some items relating to the schools of flying and aeronautics. Once more the absence of minutes indicates that the meeting did not take place. To overcome the absence of legislation and regulation an undated 'notice of motion by Mr C. F. Russell' proposed that selected articles of the International Air Navigation Act, 1919 be adopted 'to form as a temporary measure, an arrangement whereby foreign aircraft can immediately undertake commercial air services to and from Ireland'. He also proposed the adoption of the appropriate articles of the same Act 'to form as a temporary measure an arrangement whereby the necessary control over foreign aircraft arriving in Ireland, may be obtained', and included appropriate draft regulations. As with much of the previous correspondence on the file the total absence of comment or annotation suggests the attention of the President of the Executive Council was elsewhere.

In the meanwhile Russell had prepared an exhaustive study of the policy and practice internationally in the matter of 'subsidies for civil aviation'. In an 11-page memorandum, dated 2 May 1922, he examined the direct and indirect assistance provided to air service companies by the governments of some 11 countries, mostly European but including the United States and New Zealand. He defined indirect assistance as the provision of 'aerodromes and ground mechanics, light houses, pilot training schools, meteorological information, and technical and medical testing of pilots'. Direct assistance was supplied 'by means of subsidies generally based on the number of flights' or 'of passengers or so much per pound of freight'. The study was intended as an aid to Collins and the administration to formulate a policy on subsidies that might be used to foster commercial aviation. The material, indicating the definition of subsidies and exploring current international practice, would have left the administration in no doubt as to the range of options that might be considered.[65]

It appears that Russell elicited no response to any of his correspondence after 6 April 1922, nor did he succeed in assembling the Air Council again. With the inexorable approach of hostilities the Government had no time to consider aviation matters, in general, and civil or commercial aviation, in particular. In effect the file, 'Civil aviation – developments in Saorstát Éireann', in the office of the President of the Executive Council, and presumably similar files in other departments were to remain closed until January 1924.

Another item that confirms the Provisional Government's considerable interest in civil aviation was the dispatch of Captain E. A. K. Mills to London to examine the matter of medical examination of civil pilots. Mills subsequently reported:

> I visited the Air Ministry, Kingsway, W. C., where I interviewed Colonel Heald who explained and demonstrated the medical tests. I also saw a pilot under examination. He facilitated me in every way and answered my queries to my entire satisfaction, and has given me complete insight to the various tests and scientific instruments necessary for the working of this department.[66]

Mills was supplied with a complete list of the instruments and equipment required for the medical examination of pilots. He was brought to Oxford University and met specialists who were developing more searching medical tests for pilots. They advised him further on the general subject and offered assistance if required in the future. Mills also acquired a full bibliography of the English, American, French and Italian major works and articles relating to aviation medicine and got advanced notice of material about to be published.[67] Although medical examination was originally intended specifically for civilian pilots operating the proposed new air service, starting in late 1922, the Army Medical Corps would first use the expertise for the benefit of military aviation.[68]

CONTRARY OPINION

To what extent Michael Collins and his fellow ministers on the Air Council accepted Russell's staff papers, minutes of meetings and financial projections relating to the new State's aviation policy and plans for the development of commercial aviation is not at all clear. The main source for the period, the file 'Civil aviation – developments in Saorstát Éireann' (DT, S.4002) covering the period 1922 to 1932 includes documents supposedly forwarded to the President of the Executive Council for information only, but reflect no interest or action in response on his part. Without the benefit of the handwritten notes, queries and comments normally found on working documents it is not easy to judge how the matters may have been viewed by GHQ, by the Minister for Trade and his department or indeed by Michael Collins himself in his capacity as Minister for Finance. It is considered that while Russell's standing as an aviation specialist resulted from his considerable expertise in such matters, the unrecorded mandate received from Collins may also have protected him from possible detractors. Although the Air Council, on Russell's recommendations made various recommendations for the expenditure of substantial sums the Provisional Government, through the early Department of Finance, approved no capital investment. Nor did they sanction any significant recurring expenditure. The expense incurred involved no more than relatively nominal figures on, for example, the taking over of Baldonnell as a civil aerodrome and the putting in place of a small civilian air staff. Some civil aviation staff had been recruited as early as April 1922 and, notwithstanding waning interest caused by the worsening political and military situation, a total of 16 personnel, including Russell, was eventually on the Army payroll by 20 July 1922.[69]

This is not to state that there was no dissenting voice to the various aviation proposals. Criticism came from a Captain M. Dunphy, a member of the staff of GHQ. In April 1922 a brief report was made on 'the financial estimates presented by the aerial directors showing the initial outlay and general expenditure for one year' as presented to the GHQ. Severe criticism of Russell and McSweeney and of their plans for the expenditure of £47,550 on civil aviation and £137,846 on military aviation was detailed:

> I understand that the expenditure of this amount will be for all practical purposes in the hands of the aerial directors. The directors did not furnish me with proofs of their competency to act as expert purchasers of machines and stocks or of their experience in selecting men for what may be regarded as lucrative appointments.[70]

Dunphy suggested that the fact that Russell and McSweeney were pilots was insufficient to justify the possible investment of £180,000 for 'the establishment of an air force'. He was similarly critical of the estimated costs of raising a military squadron and warned that a squadron costing £350,000 annually in Britain could not be raised in Ireland for a sum between £95,000 and £115,000. He observed that for the pilots to establish an air service was analogous to an engine driver organising and establishing a railway company and that employing two from the ranks of hundreds of unemployed airmen would not further the aims of the Air Council. He was also critical of the fact that the aviation proposals had not 'been discussed with the trading community, or chambers of commerce'. He recommended that a 'scientific organiser competent to advise the Air Council on all matters relating to military and civil aviation be secured on loan from a foreign government' and that, taking into consideration the amount to be expended, the approximate cost of £2,000 for the services of a consultant expert for one year would be money well spent.[71]

While one could not question the advisability of having the civil and military aviation proposals critically examined from a financial point of view an impartial observer might consider that there was more to Captain Dunphy's assessment of the aviation plans and estimates and to his implied personal criticism of the individuals responsible for formulating them. Given the personal relationship between Collins, Russell and McSweeney, and the confidence the latter pair obviously enjoyed, it is possible that Dunphy may have perceived the two young directors in their former guise as enemies of the IRA and possibly owing too much allegiance to their former service – the RAF. It is a recognised fact that certain factions of the Army could not countenance the concept of ex-British personnel, particularly those without pre-Truce service, serving in the emerging Free State or National Army.[72] In the context of the odium applying to ex-British personnel two very young officers, having the rank or status of commandant general and enjoying considerable influence at the highest level of government, were bound to become the objects of prejudice and professional jealousy. In any event it is unlikely that Russell and McSweeney appreciated the analogy between their profession as pilots and that of engine drivers. No doubt it would not be the last time that such sentiments would be expressed, though possibly not recorded.

It must be noted that the aviation plans of early April 1922 were still underdeveloped and had not been subjected to the scrutiny of the Department of Finance. At the same time no substantial expenditure had been sanctioned while the personnel of the Civil Department were on the Army payroll. Russell himself had recommended 'the appointment of a consulting engineer' who would no doubt have evaluated the proposals before they were forwarded to the Department of Finance.[73] Captain Dunphy's report made no allowance for the immature nature of the plans and, in this regard, his criticism may have been premature.

CONCLUSION

The peace and Treaty negotiations of the latter part of 1921, apart from confirming that the new State could raise an army that included an aviation element, did not identify air defence as a national priority. It is clear, however, that from his relatively brief contribution to the discussions on civil aviation,

Michael Collins had a greater preference for advancing the cause of civil air transportation services between Ireland and the UK. It is probable that it was at Collins's instigation that provision was made, as an annex to the Treaty articles of agreement, for an Anglo-Irish convention on civil aviation. The purchase of the Martinsyde aircraft in October 1921 was an indication of Collins's confidence in the outcome of the negotiations and of his clear intention to facilitate the commencement of commercial aviation at an early date. It does not seem at all reasonable that such an aircraft would be purchased solely as a means of escape from Britain. The attention, by Russell, to the specification of the machine and to the three distinct commercial roles that it could fulfil, very strongly suggests that the aircraft's commercial potential was paramount and that the escape function was a secondary consideration. The priority of attaining the Avro 504K seems to have been in preparation for a possible resumption of hostilities in Ireland.

The working relationship between Michael Collins and C. F. Russell appears an intriguing one in the context of the efforts made, in the first half of 1922, to develop plans for the development of civil aviation, in general, and the establishing of a civil air service, in particular. In Russell Collins found an equally enthusiastic and energetic individual who had the professional aviation expertise, a great facility for research and staff work and a broad vision that allowed him to identify and articulate the State's obligations in civil aviation regulation and its options in terms of developing and subsidising a civil air service. It is not easy to identify which of the two was making the running. On balance it was probably an equal partnership with Collins having the broader vision of air communications as a symbol of Irish independence and being in a position to endorse and authorise those ideas projected by Russell which he considered best suited his, and the nation's, purpose. It is not an adverse reflection on either that political and military circumstances diminished the priority of air services and that civil aviation had to be abandoned for the time being.

THREE

MICHAEL COLLINS, THE MILITARY AIR SERVICE AND THE CIVIL WAR

—

THE START OF MILITARY AVIATION

With matters relating to both military and civil aviation being administered under the aegis of the nascent National Army and Russell continuing to elicit support for various aspects of civil aviation, W. J. McSweeney was carrying out a similar though less visible exercise in support of military aviation. An early proposal, for an 'air service department separate and distinct from any other department', included recommendations concerning the status and functions of the 'Chief Executive Officer of the [Military] Air Service'. While they were not expressed in Russell's articulate manner the recommendations were to the effect that the officer commanding military air services should be a member of the General Staff and have equal rank with the heads of other headquarter departments and Army corps. This was proposed by McSweeney on the basis that the Army headquarters should have the benefit of professional expertise on aviation matters, to ensure that the Air Service would have the appropriate status and so that the officer commanding the Air Service would be *au fait* with the overall military operational situation and make decisions accordingly.[1] The record does not reflect how this particular matter was received by the Army leadership.

McSweeney's first submission to the Air Council was sent via the Chief of Staff (COS). The document, apparently his first to be considered by the council, made an opening, slightly inaccurate, statement to the effect that the aviation department, under the Director of Training, had 'one 5 seater aeroplane purchased at a cost of £2,600 and one dual control Avro machine purchased at a cost of £130', and that the total expenditure to date was £3,000. The paper went on to detail three options for the development of military aviation, all involving the disposal of the Martinsyde passenger aircraft (that was still in storage in London) and the purchase of varying numbers of single seat Martinsyde F.4 (Buzzard) scout or reconnaissance type aircraft, and of the two seat version, the Martinsyde F.4A.

A detailed and priced proposal for the establishment of a military air service and for the purchase of aircraft and other equipment had the establishment of a school of flying as its immediate objective. This school, which would cost £23,595 to set up and run for six months, would train the personnel for a squadron of sixteen aircraft in six months. However, an 'Air Squadron', consisting of only eight officers and forty other ranks, would cost four times as much to set up and run for six months. The 'grand summary air estimate' of £150,026, providing for an air squadron – £95,346; school of aeronautics – £8,583; school of flying – £23,597 and air reserve – £22,500, was put to the Air Council meeting of 23 March 1922. The schools of flying and aeronautics were considered to be part of the civil department while the air reserve idea was lost. In the context of the time the expenditure of such sums of money appears to have received little consideration. The only decision relating to a military air service to be approved was that 'the military air authorities should aim at the organisation of one air squadron for the present.'[2] As with the proposals on civil aviation, no financial sanction for the proposed air squadron was sought, or provided. The establishment of such a squadron remained in limbo awaiting a significant change in the military situation. It was of no help to McSweeney that the Government, even as civil war was threatened, tended to favour civil aviation almost to the exclusion of military. On the outbreak of civil war, therefore, a somewhat notional Military Air Service consisted of no more than eight personnel including 'Miss M. Kiernan, typist'.[3] While the headquarters appears to have remained at Beggar's Bush Barracks for a short time after the takeover of Baldonnell, on 3 May 1922, the first non-commissioned personnel of the Air Service initially reported to Captain W. Stapleton of the Baldonnell garrison.[4]

The general circumstances surrounding the lead up to civil war are possibly best summarised in the words of the acknowledged authority on the subject and period:

> It took six highly confused and tense six months for the divisions over the Anglo-Irish Treaty to result in civil war. During that period sundry attempts to settle the political and military divisions, or at least to postpone them, failed. On all sides, however, there was reluctance, right up to the Four Courts attack, to concede that war was inevitable. The Treaty left many issues open to debate and interpretation; such ambiguity made it less likely that divisions would quickly come to a final test.[5]

Such was the pace of the evacuation by the British Army that military installations were being taken over by the local IRA with the Provisional Government powerless to prevent anti-Treaty forces occupying many such posts. In terms of territory the anti-Treaty forces occupied about two thirds of the country.[6] In Dublin many elements of the No 1 Brigade had adopted an anti-Treaty stance though all barracks in the city remained under the control of the pro-Treaty IRA.[7] The military takeover of the Four Courts by anti-Treaty forces of the Dublin No 1 Brigade on 13 April 1922 was a symbolic act that illustrated how tenuous was the authority and control of the Provisional Government. The failure to prevent the occupation, and the delay in ending it, was to emphasise the new Government's military weakness as well as to try the patience of the British Government as they awaited decisive action.[8] On 28 June 1922, using artillery pieces borrowed from the British, the Free State forces had commenced the shelling of the Four Courts; nonetheless, initial efforts had not succeeded in dislodging the rebels. The following day, in his capacity as Chairman of the CID Sub-Committee on Ireland, Winston Churchill indicated that he was particularly anxious that adequate supplies of artillery ammunition should be available to the British forces not yet withdrawn from Ireland should the Provisional Government relent on its opposition to British military assistance and agree to the offer of heavy artillery. Similarly Churchill proposed that the RAF should lend aircraft, painted in Free State colours, and pilots, to the Free State forces. The Chief of the Air Staff was opposed to painting RAF aircraft in Free State colours on the grounds of the possible adverse effect on the morale of aircrew; however, he would do so if ordered by higher authority. He suggested that the RAF could bomb the Four Courts with 112lb bombs containing delayed action fuses, which would burst inside the buildings.[9] To get ready for such an eventuality the Royal Air Force either prepared for dispatch to Dublin, or actually sent, a number of aircraft capable of dropping 112, 200, 250 and 550lb bombs. Over 700 bombs of various types were prepared or dispatched.[10]

Notwithstanding the Irish caveat regarding British military assistance, the Irish Flight, RAF had prepared for the possibility that it would be ordered to bomb the Four Courts. On the evening of 29 June, even as Provisional Government forces were bombarding the Four Courts, RAF crews were practising their bombing techniques. Between 18.30 hours and 20.15 hours that evening at least four RAF crews carried out bombing practices on their aerodrome at Collinstown using Bristol Fighter H.1485 (and probably more of the flight's four aircraft). This particular aircraft was to be handed over to the National Army within the week.[11] While aviation folklore suggests that British aircraft did in fact bomb the Four Courts, no evidence has yet been found to support the contention that the RAF carried out bombing on behalf of the Provisional Government.

THE FIRST MILITARY AIRCRAFT

In the meanwhile GHQ at Beggar's Bush Barracks had taken steps to initiate the purchase of at least one military aircraft. On 20 June 1922 the COS, General Eoin O'Duffy, received a receipt from McSweeney recording that the latter had 'received from Chief of Staff the sum of one thousand three hundred pounds [for the] purchase of [an] aeroplane'.[12] Before the end of the month O'Duffy apparently issued a cheque for £2,500, bringing the total for which McSweeney would subsequently account to £3,800.[13] It is unlikely that O'Duffy had a leading role in the matter other than that of supplying the funds, and, later, a letter authorising the purchase. As Collins did not assume the functions of Commander-in-chief until 12 July 1922 McSweeney's actions had to be authorised by his military superiors. The subsequent involvement of Collins in military aviation matters would strongly indicate that the initial decision to purchase a reconnaissance aircraft was his – probably advised by Russell and under pressure from the British to do something about the occupation of the Four Courts.[14]

Accounts suggest that McSweeney travelled to London on 21 June 1922, apparently with only verbal orders to purchase a military aircraft. On 24 June he visited the Aircraft Disposal Company, the firm charged with the disposal of British war surplus aircraft. He handed over a cheque for £400, drawn on one of two accounts he held in Dublin, as a deposit on a Bristol Fighter F2B. However, without written authorisation, he was not allowed to take delivery of the machine. Examination of McSweeney's expenses for late June 1922 suggests that he spent some time shopping in London. In Gamages department store he spent two guineas for unidentified goods and a further three pounds seven shilling was paid to 'C. Baker'. For reasons that are not obvious, apart from improving his sartorial appearance and wasting time spending money on himself in retail outlets, he stayed in London until 29 June, even though it was obviously urgent that the purchase and delivery of an aircraft be completed.[15] On arriving back in Dublin on the morning of 30 June McSweeney proceeded to GHQ and collected a letter of authorisation signed by the COS. This was addressed, incorrectly, to 'Martinsyde & Co., Woking':

> The bearer, Commandant General McSweeney, has authority to purchase one two seater reconnaissance machine which he will fly back to Ireland. The account will be settled on being furnished to me.[16]

As O'Duffy had already given McSweeney two cheques totalling £3,800 it is not clear why he should indicate that he would settle the account. In any event, he did not. Accompanied by Volunteer Thomas Nolan, who was to act as observer or navigator for the delivery flight of the aircraft, McSweeney returned to Britain on the evening mail boat on 30 June 1922. On arriving at the Croydon offices of the Aircraft Disposal Company on 1 July McSweeney signed an undertaking, typed on ADC notepaper, confirming the purchase of a Bristol Fighter aircraft:

> At the direction of the Chief of Staff, General O'Duffy, I hereby place with you a firm order for one new Bristol Fighter fitted with [a] new 300 hp Hispano Suiza engine, at a price of £875 delivered to me in flying condition at your Croydon works. The machine to be fitted with one Vickers gun and one Lewis gun, at an additional price of £225 . . . The above price to include one dual instruction flight and one solo flight on your stock machine.[17]

The provision in the contract that McSweeney would undergo one instructional flight and one solo flight was most likely to refresh McSweeney's flying skills in view of the fact that he had probably not flown any aircraft since being discharged from the RAF in July 1919.[18] On the same day he paid a further £400 to the ADC. Two weeks later (15 July) he was to make a further payment of £1,100 that brought the total paid to the ADC to £1,900 though the contract price for the Bristol Fighter was £1,100 – compared with an original new price in the region of £2,561.[19] It is not clear why McSweeney paid so much over and above the agreed price for the single aircraft. It is probable that the additional £800 represented the price of two RAF aircraft that had been handed over in Dublin while he was in London.

Despite the fact that McSweeney had informally ordered the aircraft a week earlier and confirmed it in writing on 1 July the aircraft, 'Machine No H.1251', was not ready to be test flown until 3 July. This delay was probably caused by the removal of the 200 hp Arab E.3534 engine and the fitting of a 300 hp Hispano-Suiza engine, the most powerful of no less than 11 engine options available for the Bristol Fighter.[20] The machine was test flown by a Captain Stocken for fifteen minutes on 3 July. Even at this point, however, the delivery flight could not commence, as McSweeney had not completed his familiarisation flights. These supposedly occurred on 4 July. On that day McSweeney signed an undertaking or indemnity agreeing 'to exonerate unconditionally The Aircraft Disposal Company, Limited, from any responsibility whatever for any accident that may occur to me while flying machines the property of the said company'. He also agreed to pay for any damage done to Avro 504K D.9358.[21]

At this point the key records relating to the events of early July 1922 suggest certain ambiguities. The signing of the undertaking and the completion of two familiarisation flights on 4 July indicate that the aircraft could not have left Croydon until the early afternoon of that day. The aircraft log book records that the 'machine B.F.2B [H.] 1251 arrived at Baldonnell [on] 4/7/22. Time in air three hours', without citing any particulars of the crew, the route taken or the dates and times of individual legs of the journey. Such a chronology was possible within the limits of the daylight available. [22] However, McSweeney's expense account, not tendered until August 1923, indicates that he and Nolan had departed Croydon on 3 July, landed at Shotwich, presumably late in the afternoon, and spent that night, and a second one, at a hotel in Chester and therefore could not have arrived at Baldonnell until 5 July 1922 at the earliest. This chronology does not allow for the familiarisation flights on 4 July and therefore may well indicate a slight error in McSweeney's expense account. The expense account entry that reads 'flew to Shotwick – arrived 3/7/22, Hotel Chester 3-4-5/7/22' might more correctly relate to a departure on 4 July followed by two overnights (4–5 and 5–6 July) in Chester and a departure to, and arrival at, Baldonnell on 6 July 1922. This latter scenario is also suggested by a brief telephone message, recorded at GHQ on 7 July, stating that 'Comdt General McSweeney rings from Baldonnel to say he has arrived with plane and awaits instructions'.[23] This message indicates that McSweeney had arrived at Baldonnell late on 6 July but did not inform GHQ until the following morning. In the circumstances, while it cannot be stated categorically on what date Bristol Fighter F2B 'B II' was delivered, it is strongly suggested that it was not on 4 July 1922 and that, on balance, it is probable that the aircraft arrived in Baldonnell on 6 July 1922. The log book entry recording 4 July 1922 most likely relates to the departure from Croydon rather than the arrival at Baldonnell.

TWO RAF AIRCRAFT ACQUIRED

In the meantime, with the retaking of the Four Courts by Provisional Government forces the Civil War was under way in Dublin and surrounding counties. At an early stage Michael Collins had identified an urgent requirement for air reconnaissance and took decisive action. On the morning of 4 July he made representations to Dublin Castle. As a result of these Alfred W. Cope, Under-Secretary at Dublin Castle, sent an urgent, yet long and detailed telegram addressed to Lionel Curtis in the Irish Office at the Colonial Office, London, for the attention of Winston Churchill:

Collins wants two aeroplanes one with undercarriage for bombing and one without. . . . Telegraph and telephone communication is interrupted and particulars of the surrounding country are not available. Reports come in of concentrations of Irregulars in Dublin county and neighbouring counties. Troops . . . fail to find them and men and time are wasted. Collins is satisfied he could clean up the country districts if he could get early information of concentrations and keep up communications . . .[24]

The telegram suggested, incorrectly, that the aircraft purchased in London had not yet arrived due to inclement weather and that aircraft were urgently needed to make up the lack of communications caused by the destruction of telephone lines. Cope endorsed the request for the supply of a single aircraft at once and stating that 'the P. G.' had 'one or two efficient airmen'.[25]

The appeal from Collins, which was received in the Colonial Office at 11.39am the same day, in addition to constituting an urgent request for reconnaissance aircraft, explains much about the military situation as it was developing in the aftermath of the re-taking of the Four Courts. Collins recognised immediately that the absence of adequate communications rendered it very difficult, if not impossible, to counter the activities of the Irregulars who were, of course, responsible for the destruction. Even in the areas close to the city Provisional Government forces were operating at a considerable disadvantage, a situation that required the type of intelligence that aircraft operating from Baldonnell could provide. With Churchill and the Colonial Office well disposed to Collins and the Provisional Government, the British Government, in keeping with the policy of granting whatever military help might be requested, responded quickly. At 14.20 hours on 4 July 1922 the War Office sent a secret dispatch to the British GHQ in Dublin stating that 'two Bristol aeroplanes from [the Irish Flight at] Collinstown will be handed over at once to provisional government' and that the aircraft 'should be equipped as provisional government may desire'.[26] Later that day the head of the RAF in Ireland, Group Captain Bonham-Carter received a secret dispatch:

> Orders have been issued through War Office to supply provisional government with two service aeroplanes. You will hand over two serviceable Bristol Fighters armed and equipped as they may require . . . Offer any technical advice and ensure that machines are efficient in every way. British marking to be removed.[27]

With McSweeney still taking delivery of the first aircraft C. F. Russell, the Director of the Civil Aviation Service that was also under military command, was the only pilot available to take delivery of aircraft. Russell proceeded to Collinstown the following afternoon. There he took possession of one of the only three serviceable Bristol Fighters of the Irish Flight, RAF. Taking off at 15.00 hours and, allegedly wearing a bowler hat, he flew Bristol Fighter E.2411 the ten-mile journey to Baldonnell in fifteen minutes. The fact that Russell's aircraft was given the Air Service serial number 'B I' would suggest that this aircraft actually arrived at Baldonnell before the one flown from London by McSweeney As discussed above it is unlikely that McSweeney's aircraft arrived at Baldonnell before 6 July – and therefore was given the number 'B II'. While log book entries should be the most reliable historical record manifest inconsistency in the keeping of log books, by an inexperienced and as yet poorly organised ground staff, was to make it difficult to detail the chronology of aircraft flights, to identify individual missions and assess the overall operational use of aircraft throughout the Civil War period.

AIR OPERATIONS

As the instigator of efforts to establish military aviation Michael Collins controlled and directed the operational use of aircraft during the months of July and August 1922. Telephone messages and other correspondence in the Mulcahy Papers indicate that the designation of particular missions was done by Collins, in consultation with Russell or McSweeney or both, mainly by telephone but sometimes in person with the flying officers. There is no evidence that the staff of GHQ had any active role in the matter though, being in receipt of reconnaissance reports received by telephone or in writing, they would have been well aware that an air operation was in hand. At eight of ten meetings of the General Staff or War Council held in July and August, the subject of 'air' or 'aviation' featured on the agenda. Despite this the relevant minutes reflect no discussion of such matters. This possibly reflected a policy, dictated by Collins, which restricted the air intelligence to those who needed to know. While it is also probable that GHQ staff had very little interest in the activities of two ex-RAF pilot officers and had little faith in the use of aircraft for intelligence purposes it is possible that Collins's reports to the War Council on aviation matters went unrecorded on his instructions.[28]

Although there was no apparent policy or overall plan for air recon-naissance the air operation fell in to three fairly distinct phases – the month of July in the Leinster area, the month of August mainly in the Munster area and, from October 1922, in the Munster area with missions conducted from bases at Fermoy and Tralee. During the month of July about 12 recon-naissance missions were flown in the Dublin area and the south Leinster counties of Wicklow, Wexford, Carlow and Kilkenny, against Irregulars who were being driven south by Free State Army troops. C. F. Russell, now second-in-command to McSweeney, flew the majority of the reconnaissance patrols during July and August 1922. Given the standard of log book keeping and the paucity of reconnaissance reports it is not possible to be definitive about the extent of the air operation, either in July or later. In many cases the aircraft log books do not identify the mission area while flights are not recorded in chronological order. Some entries appear to have been written up days in arrears without accuracy with regard to date. Accounts suggest that an unknown, though probably small number of entries were omitted entirely.[29] No reconnaissance reports appear to have been made prior to 16 July while only three were available for that month.

It will be appreciated that, in early July 1922, the Military Air Service had only two pilots and two (untrained) observers, and, with the delivery of a second RAF Bristol Fighter from Collinstown on 10 July, a total of three aircraft.[30] W. J. McSweeney, who had been on the Army payroll since at least April 1922 had, as his observer or navigator, Lieutenant Tom Nolan who was still of volunteer rank when hurriedly pressed into service on 30 June for the delivery flight of B II. Nolan is recorded as being granted a 'commission as 2nd Lieutenant in the Aviation Section IRA' by McSweeney on 7 July 1922, the day after his return from London, 'subject to ratification by the Chief of Staff.[31] C. F. Russell, the only other pilot, had as his observer Captain W. Stapleton who flew for the first time on 10 July 1922.[32] Stapleton, who had been in charge of the garrison at Baldonnell and who had 'put in a temporary transfer to aviation as an observer' is recorded as actually joining the Air Service on 11 July.[33] Like Nolan, Stapleton appears to have had no aviation training or experience before commencing flying with the Air Service.

The first reconnaissance mission for which a report survives took place on Sunday 16 July 1922. That afternoon McSweeney and Nolan left Baldonnell at about 15.00 hours in Bristol Fighter B III to carry out reconnaissance in the Tullow and Baltinglass areas, apparently in preparation for an attack by Government troops on Irregulars that was planned for the following morning. They observed nothing unusual at Baltinglass and observed at Tullow that 'the town was full of men' but that 'they were only standing around and there appeared to be no activity of a military nature'. They also reported that the roads into the town were partially blocked but that there were no sentries at the barriers. En route to Newtownbarry (Bunclody, County Wexford) the engine of McSweeney's aircraft began to cut out forcing him to turn for home and to use the handpump to maintain fuel pressure. Eventually the engine failed from fuel starvation resulting in a forced landing in a field at Ballycane, Naas, County Kildare. The aircraft was badly damaged when it hit an open ditch and Nolan was injured. The aircraft was to remain out of service until February 1923.[34]

A number of significant factors may have contributed to this first accident for the Military Air Service. Firstly, there was the matter of McSweeney's relative inexperience and lack of recent flying practice. Secondly, his lack of familiarity with the operation of the systems of the particular aircraft type, a type he had not flown during his service with the RAF, put him at a severe disadvantage. It appears that he did not have sufficient time to receive technical instruction on the Bristol Fighter at Croydon earlier that month. Even if he had received instruction his lack of experience with the fuel system would have shown through.[35] This unfamiliarity probably led McSweeney to mismanage the somewhat complex fuel system of the Bristol Fighter resulting in the fuel starvation that caused the accident.[36]

Subsequent events suggest that air reconnaissance missions, and the intelligence they afforded, were already important aspects of the fight against Irregulars after only one week of such operations. With the reconnaissance mission of 16 July to the Wicklow/Carlow/Wexford area not completed C. F. Russell and his observer, Captain Stapleton, most likely on direct orders from Collins, were detailed to undertake the same mission early the following morning. In very poor weather conditions Russell and Stapleton left Baldonnell at 07.00 hours in an unidentified Bristol Fighter (probably BI). They carried out reconnaissance of the towns of Tullow and Baltinglass in particular. On their return to Baldonnell at 08.45 hours Russell reported the presence of barricades on the roads to the north and south of Tullow, at about 07.40 hours, though he 'found the town asleep' and saw neither Irregulars nor Free State troops. When he arrived over Baltinglass he found 'a good number of people about the streets' and all roads and bridges intact. He reported that his aircraft had been hit by one of a total of eight shots, fired from three separate locations in Baltinglass, but that he was unable to return fire because of the poor visibility and mist at 200 feet.

On his return to Baldonnell Russell's report was relayed by phone to Army Headquarters where it was recorded as being received at 09.15 hours.[37] What affect this mission, and the information it provided, had on the ground operation against the Irregulars would be a matter of conjecture but by 22.40 hours that night GHQ had issued a press statement to the effect that at 14.00 hours that day troops had captured Baltinglass and that they occupied Baltinglass and Newtownbarry. Twenty-five men and a significant amount of arms and ammunition had also been seized.[38]

ORGANISATION AND PERSONNEL

By the outbreak of the Civil War on 28 June 1922 there were three separate organisations, paid off the Army payroll, based at Baldonnell. The aerodrome at Baldonnell was actually under the military control of a garrison which had moved in after the aerodrome had been taken over by Captain O'Grady from the Irish Flight of the RAF at midday on 3 May 1922. The newspaper reports of the matter suggest that the initial garrison consisted of troops from Clonskeagh Castle. On 28 June 1922 the garrison numbered about 90 all ranks and its main function and preoccupation was the security of the Camp.[39]

In April 1922 the Civil Air Service, that was later given Air Council approval to assume the functions of the civil aerodrome authority at Baldonnell, had seven staff members, including Russell. In addition to its intended task of managing and running the civil aerodrome Russell's department was notionally preparing to facilitate commercial aviation if and when such activity received Government approval. By 20 July the Civil Aviation Department had a total of sixteen personnel, fourteen civilians including 'Chas. F. Russell, the Director of Civil Aviation and secretary to the Air Council' and two Army volunteers. Five of the sixteen were recruited after the commencement of the Civil War. Two engineers were included. Charles J. O'Toole was employed as an aero ground engineer since 14 June 1922. William J. Guilfoyle, a civil works maintenance engineer was on loan from Irish Lights since 30 April 1922 on the understanding that he would be reinstated to his previous position should his services not be required permanently by the Civil Air Service.[40] The most recent member of staff, Eamon Broy, accountant and clerk, was employed with effect from 19 July at a salary of £5 per week. Broy's employment in the Civil Department was apparently as a result of representations he had made to Collins who in turn arranged employment with the new air service.[41] Arthur J. Russell, younger brother of C. F. Russell, was paid one pound ten shillings per week as a junior clerk.[42]

In contrast, the Military Air Service, which aspired to be equipped and to function as the effective military air element of the Army, consisted of only five personnel, including 'Commandant General' W. J. McSweeney, at the end of April and had only risen to a total of eight by 28 June. The Civil War was to bring about significant, though contrasting, changes in fortune for the two aviation departments though the function of GHQ in these developments would appear reactive rather than proactive. By 22 July 1922 the strength of the personnel in military aviation had increased to a total of 36 and included eight civilians.[43] Much of this increase can be put down to the immediate effect of the general recruitment call of the Minister for Defence in early July. With continuing recruitment, one hundred officers and men had joined the Air Service by mid-November 1922 by which time the garrison unit, now known as the Air Service Infantry, had reached a similar figure.[44]

However, on or about 22 July 1922, for all practical purposes the Civil Department ceased to exist. This decision was taken in light of a review of organisation and personnel requested by the Minister. On 17 July Richard Mulcahy had written to the Air Service, and sent a reminder four days later, requesting a statement of the number of men employed in that Department, plus a diagram indicating the organisation. 'Will you also let me know their ranks, duties, date from which they have been employed and their pay' and 'let me have the same particulars regarding the men employed in Civil Aviation under Russell'.[45] McSweeney had Russell supply the required details of the 16 personnel of his department and subsequently forwarded nominal rolls for the two air elements indicating the required detail. The nominal roll of McSweeney's 35 subordinates in the military department included: five officers, 22 other ranks and eight civilians. The names of five volunteers, posted to the unit on the 22 July, had been appended in manuscript. McSweeney also forwarded a covering letter that confirmed the amalgamation of the civil and military aviation resources at Baldonnell:

> During the present hostilities, Civil and Military Aviation have been combined and the various Civil and Military Departments are working together as one unit, all under military control with the Director [of Civil Aviation] as 2nd in command.[46]

The organisational functions or departments were defined as discipline; aeroplane repair and maintenance; electric power, water supply and sewage; buildings, repairs and technical stores; aerodrome labourers (carpenters, labourers, etc.) and quartermaster stores. McSweeney explained that 'each department was under a responsible man' but that the organisation was 'only at an embryonic stage of development'. He stated that the goal of maximum efficiency depended on approval for increased pay for qualified mechanics, the supply of more aeroplanes and new transport, all of which he indicated had been previously requested by him. While the Military Air Service had no transport of its own it had a transport depot that maintained the vehicles on charge to the garrison. McSweeney included a copy of a newly drafted standing order that detailed the daily routine to be observed by all military personnel, as well as standard practices and safety precautions to be observed, by military and civilian alike, in the hangars and workshops.[47]

While the bulk of the Civil Department employees merely transferred to the military payroll as civilians a small number, including C. F. Russell and his brother Arthur, joined the Military Air Service.[48] On 19 July, McSweeney had already requested the COS to approve the appointment of Russell, Director of Civil Aviation to the rank of Commandant.[49] Eamon Broy also made the transfer from civilian to military. In light of his Dublin Metropolitan Police background and his distinguished record in Republican intelligence during the War of Independence his employment, initially in civil and later in military aviation, might appear unusual. However, he appears to have had a certain interest in aviation from at least October 1921 when he was in London with Collins and had been on the periphery of events associated with the purchase of the two aircraft. As a result of research carried out by the late Colonel Billy Keane in the 1950s we know that Broy, on 28 October 1921, had purchased a book entitled *The German Air Force in the Great War* and, much later, a German Air Force handbook for the year 1936 suggesting an interest at least in German military aviation.[50] While Eamonn Broy's level of interest in, and knowledge of, aviation may be uncertain it appears that his drift towards civil aviation, in the first instance, and later into military aviation, may more correctly have been a drift away from intelligence work. A compelling reason for Broy's change of career is recorded in his military personal file:

> Broy was in Oriel House up to some time in April 1922, when he left as a result of a disagreement with Tobin, who was at that time in charge . . . He later joined the Army, sometime in July or August 1922 I believe.[51]

While Broy had been employed in the Civil Air Service from 19 July 1922 ten days later he was given the commissioned rank of Commandant and appointed Adjutant, Air Service by order of Michael Collins and subsequently functioned as Second-in-Command to the Director of Military Aviation until May 1923. Accounts record that he did so with considerable dedication to the Air Service and loyalty to McSweeney.[52] The suggestion by Padraic O'Farrell that Eamonn Broy adopted an anti-Treaty stance during the Civil War is quite erroneous.[53]

Two days after announcing the amalgamation of the civil and military aviation resources at Baldonnell McSweeney made a submission to the COGS which implied he considered that he should have overall command of all Army elements at Baldonnell, including the garrison, with himself holding the appointments of both DMA and Station Commander. His submission was imprecise suggesting that the garrison's strength was in the region of 100 all ranks, including five officers and 16 NCOs, but that the personnel were liable to be posted elsewhere, for civil war action, without being replaced. Indicating that a barrack staff of 30 men, a daily guard of 21 men and a weekly emergency guard of 21 men could not be maintained from existing resources, he recommended that the garrison be increased to 250 men or two companies with a barrack staff of 50. He considered that the ratification of certain appointments would effect a considerable improvement in organisation. He recommended that Captain Conry be made permanent Officer Commanding the camp and that Lieutenant Wilson be promoted to Captain and appointed Camp Adjutant.[54] The inference in McSweeney's submission regarding the command of the various elements at Baldonnell did not escape the notice of the COGS, Richard Mulcahy, though his instructions to Emmet Dalton, GOC 1st Eastern District, were brief and somewhat ambiguous. Mulcahy directed Dalton to make arrangements, sooner rather than later, 'to bring the Garrison at Baldonnell up to the numbers necessary to carry out garrison duties.' In a manuscript postscript he indicated that 'the division of authority and responsibility between McSweeney and OC Garrison require clear definition and understanding'.[55] The inference in this direction appears to be that Mulcahy considered the Air Service and the garrison to be of equal status under officers of similar standing. Dalton confirmed that he would increase the numbers in the garrison up to 100 (plus 50 of a barrack staff) and that he was 'also arranging the division of responsibility'.[56] However, there is no evidence, then or subsequently, that Dalton issued any directive clarifying the matter of command.

At this early juncture in the formation of the Air Service it appears that the infantry and air personnel had already divided along cultural lines undermining whatever authority McSweeney may have considered he had over the garrison on the basis of his rank. When an unknown GHQ officer visited the aerodrome on 30 July 1922 he recorded that he had met Russell and 'visited both camps, lower camp and upper camp, generally speaking careless, number of men at present much too small for effective care and control'. While he observed that 'it is proposed to put the entire air establishment under army control' – something that had in fact been done a week earlier – he made no reference to the command status of the various formations at Baldonnell.[57] A further indication of the apparent divide was the existence, in August 1922, of two guardrooms as well as separate sleeping quarters and officers' messes for infantry and air personnel.[58] That Dalton failed to act on the matter of the division of authority and responsibility can be inferred from a later communication. McSweeney listed the various aspects of command and control, in particular of military aviation operations, which at that late stage, still remained to be clarified.[59] The failure of GHQ to approve a formal establishment and to clarify the command status of the Air Service was to leave McSweeney in a nebulous position, in effect, up to his eventual dismissal in March 1924. Records also show that Dalton also failed to increase garrison numbers before he was transferred south to Cork in early August. At the time that the Army census was taken, on 22–3 November 1922, the total of the garrison and barrack staff combined was still only about 100.

RECONNAISSANCE MISSIONS IN MUNSTER

Before the end of July the general line of contact between the Army and the Irregulars had cleared south Leinster as the latter group retreated southwards and westwards and the hostilities were becoming somewhat concentrated in

Munster.[60] The radius of action of a Bristol Fighter operating from Baldonnell, as the RAF had found previously, extended only as far as Fermoy.[61] As a result aerial patrols became less practical and productive due to the constraints of aircraft range and endurance. The report by Russell and Stapleton on their 'reconnaissance and propaganda-dropping flight' of 22 July illustrated the difficulty. Having been directed to drop newspapers in addition to carrying out reconnaissance Russell reported successfully distributing '600 copies of the *Weekly Freeman's Journal (War Edition)* and 9,000 copies of An tOglach (War Special)' among the main towns in the county of Kilkenny. However, having insufficient endurance to pursue the required reconnaissance objectives he had to curtail the flight though he did report that one of the bridges in Clonmel had been blown up.[62] While, according to a garbled report recorded in GHQ, a propaganda-dropping mission was flown into County Tipperary on 2 August 1922, the evolving military situation dictated that aircraft would need to be operated from airfields further south and eventually be based closer to the ground action. On 4 August Michael Collins recorded a brief note in relation to the Air Services:

> Interviewed Commandant Russell, 4th August, 11am, arranged that as soon as practicable reconnaissance will be made of Youghal and Cork. This will be carried out probably from Waterford. The Waterford station is in the process of being fixed.[63]

This requirement was already known to McSweeney who had gone down to Kilkenny by road on 29 July and, with Commandant General Prout, had identified two landing grounds. He considered that one of the grounds, which had been used by the British during the War of Independence, was only safe for landing 'in the directions NW & SE'. He picked a second field 600 yards away 'for landing in the directions NE & SW'. The location of this former RAF landing ground was not specified but was apparently one of three landing grounds in Kilkenny, that is Callan, Jenkinstown or Johnstown, designated as landing grounds during the earlier hostilities. He also visited Waterford and inspected and approved a landing ground at an unspecified race course, presumably that at Tramore, about seven miles south of the city.[64] McSweeney prepared detailed instructions for the use of the forward 'aero bases' for the guidance of the troops that would have to support Air Service use of the selected fields. His instructions included provision for security, marking of landing grounds and the cutting of meadow grass to allow aircraft to operate safely. In particular, he made arrangements for fuel and oil that the Air Service would supply, to be forwarded to Kilkenny military barracks in advance of a mission involving any of the Kilkenny landing grounds.[65]

As McSweeney had departed for London on 30 July to purchase more aircraft it fell to Russell and Stapleton to carry out the reconnaissance mission to the Cork and Youghal area, though, for reasons that are not clear this did not get under way until 7 August. On 6 August the Air Service confirmed to Collins that 'arrangements have been made for the establishment of operational bases at Kilkenny and Waterford' and that supplies of petrol and oil supplies had been dispatched and had been received at both locations.[66] The log book for B I indicates that Russell and Stapleton left Baldonnell at 11.35 hours on 7 August 1922 to commence the mission to Cork which had four specific objectives. The main task was to meet the Cork Commander and inform him that various requisites were being dispatched from Dublin. Russell was also to carry out reconnaissance of the whole area, give armed support to ground troops where required and distribute copies of a special air edition of An tOglach.[67]

With McSweeney and Russell away from base Commandant Broy was in charge at Baldonnell and was to keep GHQ informed as to the positions, as he knew them, with regard to the Cork mission and McSweeney's trip to London. On 9 August Broy reported to GHQ that he had been unable to communicate with Kilkenny or Waterford by telephone or wireless since the aircraft had left. 'I therefore had a private motorcar commandeered yesterday and sent Lieutenant Nolan and a man in mufti to get through to Kilkenny and Waterford'. Having left Baldonnell at 20.45 hours on 8 August Lieutenant Nolan made contact with Russell in Waterford at 05.45 hours on the 9 August. Russell reported that 'they had made a successful landing at Kilkenny and Waterford' on the 7 August. Russell later reported to Broy that he had tried to send a message to Collins to inform him that he had not been able to fly on 8 August due to the fact that it had been raining all day.[68] Accounts suggest that the actual reconnaissance of Youghal and Cork took place on 9 August, though neither the aircraft log book nor Russell's report of 10 August 1922 make this clear. The reconnaissance mission coincided with the latter stages of the ship-borne attack by Government troops on Cork and its environs.[69] It was a relatively short mission, leaving Waterford at 16.10 hours and returning by 18.10 hours. Russell subsequently reported that 'the message was delivered to the CO at his HQ, Rochestown'. He also stated that Thomastown was very quiet while Youghal was similar with Free State troops moving freely about the town. He observed that two ships were tied up at Passage West and, while a few troops were about, there was no fighting and that civilians were moving freely through the streets. His report on Cork suggests that he had arrived over the city at or close to the termination of the military operation that had successfully cleared the Irregulars from the area.

One would imagine to see Cork city from the air that the whole town was enveloped in flames. Closer examination revealed the fact that all barracks, police and military, were on fire. Also what appeared to be a private house, half a mile north of Victoria Barracks. Victoria Barracks was, in spite of smoke and flame, a scene of great activity. Large numbers of men were moving about in a very excited manner.[70]

Russell confirmed in his report that a total of 4,000 copies of *An tOglach* had been dropped over Youghal and Cork and that while no opportunity had presented itself to use machine gunfire in co-operation with friendly troops, he had returned fire after coming under attack at Midleton on the return flight to Waterford.[71]

After the taking of Cork Collins concentrated aerial reconnaissance on the general area of the counties of Limerick, Tipperary and Cork, with some missions coinciding with his own visit to the area. Russell and Stapleton had been due to go to the Limerick area on 12 August but were again delayed by bad weather.[72] At noon the following day Russell received a message from Collins indicating that he should report to Limerick Junction at any time after eleven and that all arrangements, including a car, were in place.[73] After some difficulty with damaged telephone lines Broy replied, at 14.50 hours, stating that the aircraft had left at 13.00 hours for Limerick Junction.[74] It is not at all clear whether Russell landed at Limerick Junction or not. However, it appears that he did land at the Fair Green in Limerick on 13 August in B 1 and operated from there on 13 and 14 August. Late on the first afternoon Russell carried out a patrol of a large part of north Cork. He subsequently reported that the railway bridges at Mallow and near Buttevant had been blown up while the latter was still burning fiercely. He also reported that three small road bridges near Buttevant had been destroyed and that his aircraft had come under heavy fire from a machine gun post a half mile north of Mallow.[75] On the morning of 14 August Russell patrolled the areas of Bansha, Ballylanders and Kilfinnane in Tipperary and reported little of interest other than being shot at from Ballylanders.[76] In the afternoon he patrolled Charleville, Buttevant and Liscarroll in the county of Cork, subsequently reporting to have seen only Free State troops in Charleville and after having 20 shots fired at the aircraft in Buttevant.[77]

Some technical aspects relating to the operation from Limerick give rise to a degree of ambiguity at this time. In preparation for the Limerick operations the Air Service had to have fuel and oil dispatched to Limerick, but as a result of there still being no indigenous road transport, apparently arranged for the convoy bringing Collins southwards to convey the necessary supplies. Collins had directed that 'only the best cars' were required for the journey south.[78]

Difficulties with the transport arranged for the journey of 12 August resulted in the fuel and oil not reaching their destination. The matter is explained by a letter of complaint from Collins to the QMG:

> I have to report to you that on Saturday morning, 12th Inst., there paraded as follows; one touring car, one Crossley tender, one Lancia car. The Lancia went on fire at Naas, was restarted, had difficulty getting to Roscrea. Near Roscrea the Crossley ran out of petrol – no spare petrol in any of the cars – and both the tender and Lancia would, therefore, have been left on the road were it not for a supply of aviation spirit which was being taken, at the request of the Air Service, to Limerick.[79]

The folklore suggests that, as a result of not having aviation fuel at Limerick, Russell's aircraft was supplied with motor petrol and that engine failure, a forced landing and damage to the undercarriage had ensued. Indeed Collins's own diary and other records of the time confirm that the aircraft sustained a certain level of damage, probably on Sunday 13 August. At 10.43 hours on 14 August Broy received a wireless message, via 'O/C Troops Limerick' and Portobello Barracks, requesting the dispatch of enough fuel and oil for ten hours of flying and also a 'spare wheel and two mechanics for duty here: send direct to Limerick City'.[80] Within the hour Broy had reported that Russell's requirements had been dispatched at 11.30 hours.[81] While it is not clear how and where Broy acquired the necessary transport at such short notice it is probable that he negotiated the use of a vehicle on charge to the Baldonnell garrison. On the following day Collins made a cryptic note in his diary confirming that the 'aeroplane wheel and strut smashed. Mechanics arrived for repairing on Monday 14th' at about midnight and that he had visited the Fair Green at 11.30 hours on Tuesday to find that the work had not started.[82]

Although all the indications are to the effect that Russell's aircraft had sustained, at some time on Sunday 13 August 1922 or early on the Monday morning, a level of damage that might have precluded it being flown, the reconnaissance reports indicate that one mission had been flow on Sunday afternoon and a further two on Monday 14 August. While the circumstances appear contradictory it must be assumed that the damage, while requiring a replacement wheel, was not sufficient to render the aircraft totally unserviceable. The fact that the repair work had not been initiated on the Monday afternoon and completed by Tuesday morning gives further weight to this supposition. A more complete understanding of the matter is hindered by the fact that the aircraft log book of B I not only fails to record the reconnaissance flights of 13 and 14 August but also contains no reference to repairs to the undercarriage of the aircraft about that time. Strangely, the record of the completion of other repairs, carried out during August and September 1922, appears to be duplicated.[83]

MODEST EXPANSION AND SUPPORT SERVICES

At the general staff meeting of Friday 28 July 1922 Michael Collins reported on the 'position with regard to the air force', based on a meeting he had with McSweeney and Russell earlier that day. He announced his decision to approve the purchase of two SE 5A aircraft which he noted had a range of 200 miles and were fitted with two Vickers and one Lewis guns each as well as bomb racks. He indicated that 'these will be wired for this evening and one way or another will be across [here] on Monday evenings'.[84] While Collins gave no rationale for the decision to purchase more aircraft it was probably due to the poor serviceability of the existing machines and to the fact that two new pilots, F. S. Crossley and T. J. Maloney, were about to be appointed.[85] Notwithstanding the urgent tone of his briefing to the General Staff the objective of having two new aircraft delivered by the following Monday evening was to prove very optimistic.

McSweeney and Russell had apparently used their meeting with Collins to highlight matters on which they hoped he could be of assistance. While Collins subsequently took up in writing several aspects of the support services required by military aviation, he noted, in particular two subjects for mention at the general staff meeting. He informed the meeting, for the specific attention of the QMG, that the Air Service required 20,000 rounds of selected aerial ammunition and 2,000 rounds of tracer rounds as well as a Crossley tender and a three-ton lorry. On this matter Collins subsequently wrote to the QMG reminding him that on 28 July he had taken 'certain notes for the above, namely 20,000 rounds of selected aerial ammunition' and 2,000 tracer bullets in addition', and inquired as to whether he had received delivery yet. Collins emphasised that he had been informed by Russell that aircraft guns were jamming after only five, 10 or 15 rounds while using ammunition intended for infantry weapons.[86]

In the first two weeks of August, while Russell and Stapleton were carrying out such reconnaissance flights as the weather would allow, McSweeney was in London acquiring the aircraft recently approved by Collins. He left for London on 30 July, remained there for eight days that included certain delays, and completed the purchase of a Martinsyde F.4 (Buzzard) and a SE 5A – not two SE 5A aircraft as had been indicated by Collins. Due to the urgency indicated by Collins McSweeney endeavoured to keep him informed at every stage. On Wednesday 2 August, writing from the Imperial Hotel, Russell

Square, he informed Collins that he intended inspecting the aircraft on the following day with a view to carrying out test flights on the Friday and leaving for home the same day.[87] However, on Monday 7 August McSweeney again reported to Collins indicating that, although the aircraft were due to be ready for delivery on Friday 4 August, one had a leaking radiator while the other had faulty guns. As a result, the aircraft were not ready until 17.00 hours on Saturday. The departure was further delayed by bad weather and the bank holiday.[88]

Eventually the two aircraft, the SE 5A being flown by McSweeney, and the Martinsyde F.4 by a Mr Perry, left Waddon on Tuesday 8 August routing to Shotwick in Chester. En route the aircraft ran into a thunderstorm – the same or similar weather to that experienced by Russell on his mission to Cork about the same time. Writing from the Lamb Hotel, Nantwich, on the following day McSweeney reported that 'the rain tore the fabric off one blade of my propeller and owing to the vibration and running out of petrol I had to make a forced landing in the nearest field' near Nantwich, Cheshire. The second aircraft, in a similar condition, had landed four miles away. The circumstances suggest that it is probable that the damage was caused by hailstones rather than by rain. McSweeney indicated that he had made arrangements to have new propellers fitted, hopefully by 10 August, with the intention of reaching Baldonnell the same day.[89] Having first engaged the services of six men to picket and cover the two aircraft, McSweeney verbally contracted the services of L. B. Fitch and an assistant to effect the replacement of the propellers. These two men stayed with the aircraft and completed the repairs on 10 August after which the two machines were flown to Shotwick. There the S.E. 5A was found to be unserviceable and, after a further delay of five days that is not explained, the Martinsyde was flown to Baldonnell by Perry on 15 August while McSweeney had supposedly returned by the mail boat on 14 August.[90] An account tendered by ADC suggests that the Martinsyde F.4 cost £2,297. 10s.[91] The SE 5A was eventually delivered in early September. This new aircraft apparently cost £2,532 in total.[92] In effect the urgently required delivery of two aircraft had, due to technical and weather difficulties, taken about five weeks. The difficulties did not end there. Within days the SE 5A was lost. On 8 September 1922 the machine was being flown to Limerick by Lieutenant F. S. Crosseley when he got lost in low cloud in north-east Cork. He reported that the engine lost oil pressure due to the failure of the big end bearing and that he had been forced to land in the vicinity of Macroom. The damaged aircraft was subsequently burned by Irregulars.[93]

The next, and final, consultation with McSweeney and Russell was to result in Collins making policy decisions regarding aircraft, personnel and support services. These decisions and actions were to have major ramifications for the future of air operations and, ultimately, for the survival of the nascent air organisation itself. With Collins and Russell returning from the Munster area and McSweeney back from London a meeting was arranged for Baldonnell at 22.30 hours on 15 August 1922.[94] Earlier that day McSweeney had carried out a reconnaissance mission in the Dundalk area. This apparently was the only such mission north of Dublin and the first by McSweeney since his accident on 16 July.[95]

Collins, in summarising the major decisions taken at the meeting stated that 'we discussed the question of air services generally and came to the conclusion that it is essential that we order a flight of, say S.E. 5s, also two Avro [504K]s and a spare engine for an Avro [504K]', for which McSweeney was to submit an estimate of costs. He also approved the decision that 'up to six pilots are to be taken on immediately' confirming that 'these will be admitted on approval and if not satisfactory will be dispensed with at once'. The third major decision authorised the occupation of an air base in Munster. Arrangements were made for Russell to fly to Limerick and on to Fermoy to report on the condition of the ex-RAF aerodrome there. If Fermoy was found to be satisfactory Collins wanted McSweeney to fly down a second aircraft, and 'if the condition [was] not satisfactory the Board of Works would be requested to provide some temporary covering'. The ultimate intention was to conduct air operations from a base such as Fermoy as the ground war was well outside the range of aircraft operating from Baldonnell. Of lesser concern to Collins was 'the fact that a previous decision, to have pilots paid two and a half times the rate of pay appropriate to 'ordinary volunteers', had not been acted upon.[96]

Accounts attest to the fact that McSweeney and Russell used the occasion to renew their representations regarding such matters as communications, ordnance, transport and meteorology though actions already initiated by Collins, mainly referring such matters to GHQ staff, would suggest that they were pushing an open door. On 17 August 1922 he wrote to the Quartermaster General on the matter of a 'car for Baldonnel aerodrome' pointing out that the 'old ford' was not reliable enough to bring reconnaissance reports in to GHQ after flights, and requested that arrangements be made to 'exchange it for some really reliable serviceable car'.[97] In the absence of a wireless and reliable road transport, the telephone was the only means of communication available to the Air Service. While Collins noted the matter of air communications at the aerodrome it is not known if he made any firm direction on the matter of a wireless for Baldonnell.[98] He had earlier contacted the Postmaster General with regard to the 'telephones at Baldonnel Aerodrome', claiming that it was difficult to be heard on a defective system which he described as being extremely unsatisfactory.[99] The system was reported to be under repair on or about 21 August and was expected to be restored to service before the end of the month.[100]

Indicating the continuing urgency of the matter of aviation ammunition Collins sent the QMG a second reminder, eight days after the previous, asking curtly 'I should like to know if anything has yet been done about the matter'.[101] The QMG eventually replied on 22 August 1922 and begged to inform Collins that he had 'succeeded in procuring 4,800 rounds of special ammunition', which he understood was all that was available at the time. The response was too late to receive Collins's attention, while the related matter of the supply of bombs would later be taken up by McSweeney with Richard Mulcahy, who replaced Collins as CIC.

Although how much Collins was involved in the matter is not totally clear, problems relating to meteorology were the subject of notes between GHQ, the Air Service and the Department of Agriculture. The basic problem probably related to the absence of meteorological forecasts for aviation purposes. In the context of Air Ministry's continuing responsibility for the management of the country's meteorological stations (until 1936), any short-term solution would do little to improve the situation. On 10 August the Department of Agriculture, replying to Collins's minute of the previous day, reported to him that it had arranged for the 'Meteorological Office of the Ordnance [Survey] Office' to supply observations of 'barometer, rain, sunshine, approximate wind speed and direction, humidity, atmospheric temperature and fog' at 10.00 hours each morning. The Air Service at Baldonnell did not take observations and had no expertise to generate forecasts. In the circumstances, the availability of the observations recorded in the Phoenix Park would have been of little benefit.

Collins's last dealings with aviation matters occurred on 22 August 1922, the day of his assassination. At 22.35 hours that night the wireless station at Portobello Barracks received a message, 'from Commander in Chief in Cork', intended for Russell or McSweeney – whichever of them was available to fly:

> Tell Russell that Fermoy [is] suitable for landing. Ask him if he will fly over . . . Macroom to Ballyvourney to Inchegeelagh, Bandon to Dunmanway . . . most anxious to have these places reconnoitred.[102]

About midday on 23 August General Dalton requested confirmation from GHQ that the requested mission was to proceed.[103] There is no record of a response. Similarly there is no record of any air reconnaissance in the weeks immediately following Collins's death.

CONCLUSIONS

While Russell's Civil Air Service expanded modestly after the start of hostilities the Civil War caused the administration to turn its attention to military aviation. On or about 20 June 1922 records attest to the fact that Collins authorised the purchase of the first aircraft with a view to bombing the Four Courts. While his subsequent request to London for aircraft indicated air reconnaissance as the preferred role Collins did not rule out bombing or other armed roles. Throughout July and August 1922, Collins, in addition to directing operations, actively supported McSweeney and Russell in their endeavours to get various requisites and support services.

By any objective yardstick the output for the period 6 July to 22 August 1922, in terms of the number of missions flown, was very modest, amounting to about one mission every other day. The question remains as to how timely and valuable the air intelligence became in the overall context of the Civil War. In the absence of evidence that such evaluation was conducted at the time inferences must be drawn from the fact that the aborted patrol to the Wicklow/Wexford area, on 16 July 1922, had to be completed early the following morning in preparation for an operation by Free State troops planned for that day. Similarly, Collins saw the necessity for an extended patrol into Waterford and Cork in early August 1922 and subsequently authorised the posting of aircraft to Fermoy and the expansion of the reconnaissance operation into Munster.

Many factors conspired to frustrate Collins's intent. Not least of these was what must have been seen as inordinate delays in procuring aircraft. Equally, it could be said that he might have authorised the purchase of more aircraft and hiring of more pilots at an earlier juncture. An underlying prejudice amongst Collins's Army subordinates towards ex-British servicemen was probably a major deterrent to early expansion.

While Michael Collins was the superior authority, the air effort during July and August was conducted by a veritable triumvirate – Collins, McSweeney and Russell. Under Collins's overall direction McSweeney was very much occupied with matters of administration, organisation and the purchase of aircraft. Russell did the bulk of the flying and possibly liaised with Collins on operational matters to a greater extent than McSweeney. Notwithstanding McSweeney's relative inexperience, and his nebulous position in the Army command structure, air reconnaissance assumed an important role during July and August 1922 due to Collins's direction and sponsorship. Despite increased numbers of aircraft and pilots, the positions and roles of Russell and McSweeney in the overall scheme of things were to change significantly with Collins's untimely demise.

FOUR

FROM CIVIL WAR TO ARMY MUTINY

—

THE IMPLEMENTATION OF EXPANSION PLANS

On the day Michael Collins was killed it was reported that, while the Irregulars were being driven from their bases, they were still not at all vulnerable in Cork and Kerry; and that, although the Free State Army was capturing towns, they were not capturing Irregulars or their weapons in significant numbers. In effect the guerrilla phase of the War had commenced and, initially at least, Government troops were not getting the upper hand. In the *Irish Times* of 15 August 1922, it was also suggested that if the pro-Treaty forces had moved quickly enough, the War could have been over in three weeks.[1] It was for this phase of the War, apparently, that Collins had authorised the expansion of the Air Service with the intention of conducting reconnaissance operations in the south west. Despite the urgency attached to Collins's expansion the taking on of additional pilots was the only matter that was initiated with little delay but even that process was to prove tediously slow. A full month was to pass before McSweeney set off for London to buy more aircraft. The last of these aircraft would not be delivered until 22 November, three full months after Collins's death while Fermoy would not be occupied by an air detachment until 1 October 1922, a delay of almost six weeks.

The initial recruitment of additional pilots had started about 11 July with the signing of Lieutenant J. McCormac, who, for reasons that are not stated, was dismissed the same week. A Lieutenant A. J. Gogarty is also recorded as being recruited at about the same time but did not report for duty.[2] On 29 July Lieutenant Frederick S. Crossley was put on the payroll followed, on 19 August, by Lieutenant Thomas J. Maloney.[3] Like McSweeney and Russell the new pilots had previous service with the RAF. Crossley had initially served with No 41 Squadron and later with No 1 Squadron in France where he was wounded on 6 July 1918. Maloney had served with No 57 Squadron.[4] Unlike the two senior pilots the new pair had no IRA service. Probably greatly influenced by his own embarrassing experience at Naas on 16 July 1922, having returned to flying after a three-year break, McSweeney sent Crossley and Maloney to the Aircraft Disposal Company in Croydon in early September 1922 to avail of instructional and solo flying as their re-introduction to flying duties and as preparation for active service. While the number of hours flown by each is not known, it is probable that they only received a small amount of dual instruction on the Avro 504K and some solo flying on Bristol Fighters.[5]

In September four more pilots were taken on. Delamere's later account of the early days of the Air Service, that reflects the adverse effect of the passage of 50 years on the memory, gives an insight into the informal manner in which individual pilots were recruited. In his own case he recalled that while at home in Dublin, on holiday from his engineering employment in England, he was approached by William 'Bird' Flanagan, an individual with connections to W. T. Cosgrave, President of the Executive Council. He was asked if he would like to fly again and, having answered in the affirmative was subsequently called for an interview by General McSweeney. Following the interview he 'was accepted for an appointment as a pilot in the new Army Air Service'. He suggests that the manner in which other pilots were recruited was most likely similar.[6] It is strongly implied that the matter of recruiting ex-British service personnel, particularly those without pre-Truce experience in *Óglaigh na hÉireann*, was such a sensitive matter that the Army resorted to word-of-mouth rather than advertising the positions for pilots – which perhaps explains the somewhat surreptitious method of recruitment described by Delamere. It was to be early December 1922 before the last of the six pilots, as authorised by Collins in August, were taken on. With the departure of C. F. Russell to the Railway Maintenance and Protection Corps about the middle of September, the dismissal of Captain John Arnott in mid-December, and the recruitment of an additional two pilots earlier that month the total number in service at the end of 1922 was still only ten.[7] With no operational missions that can be identified from log books or elsewhere being flown after 16 August, the latter part of August and much of September 1922 appears to have been taken up with the conduct of a series of instructional and re-familiarisation flights for newly recruited pilots.[8]

Correspondence between the new CIC and the Air Service in early September on aviation matters suggests a period of adjustment and review on the part of General Richard Mulcahy. It also explains the delay in ordering and taking delivery of aircraft. A file memo, of *c.*5 September, illustrates an appraisal of the aviation resources required in the context of air operations in the west and south west. While the considerations listed were very similar to those matters on which Collins had made decisions on 16 August the fact that such decisions had already been taken was not mentioned. Mulcahy appears to have generally endorsed Collins's decisions though in some respects the new proposals went further. Where Collins had decided to acquire a flight of aircraft (of unspecified number) Mulcahy, indicating that he had mislaid McSweeney's documentation on aircraft estimates, gave formal authority to proceed with the purchases as previously agreed, that is 'the instructional machines, three Bristol Fighters [and] three [Martinsyde] Scouts'.[9] Very shortly afterwards Mulcahy indicated that, while he proposed to discuss aviation matters in general at a General Staff meeting, he was disposed to authorise the purchase of two more Bristol Fighters at a cost of £1,160 each. He proposed that Fermoy would be the main airfield for the whole south west and had intended that Russell would take charge of the area. He expressed his intention to attach four machines to Fermoy and two to Limerick. He further proposed to have enquiries made by General Seán McEoin, GOC Western Command, with a view to finding the best location for basing two aircraft in the Clare area.[10]

A few days later Mulcahy wrote to the Air Service indicating that, in view of the loss of the SE 5A a few days previously, McSweeney should make arrangements to order two additional machines for the Western Area.[11] The authorisations of 5 and 11 September translated into an order for two Avro 504K training aircraft, five Bristol F2b Fighters and three single-seat Martinsyde F.4 scout aircraft – a total of ten aircraft. As he later recorded McSweeney commenced his journey to London to purchase ten aircraft on 22 September 1922 and stayed there for five days. While there he confirmed an order for five Bristol Fighters and two Avro 504K aircraft with the ADC at Croydon. He travelled out to Woking in Surrey and there he confirmed an order for three Martinsyde F.4 machines with the ADC acting as agents.[12]

While he had been authorised to purchase ten aircraft, in fact 11 machines were delivered between 16 September and 22 November 1922. The additional one was a third Avro 504K. The first aircraft of this order, B IV, was taken on charge at Baldonnell on 16 September, a week prior to McSweeney's latest trip to London.[13] This information suggests that the ADC accepted a written or telegraphed order from McSweeney on or about 14 September, fitted a 300 hp Hispano Suiza engine to Bristol Fighter E.1958 on 15 September and had had it delivered by a company pilot, to Baldonnell the following day. This aircraft, B IV, which was delivered with Lewis and Vickers guns, and bomb racks for 20lb bombs, was damaged on landing at Baldonnell on its delivery flight and did not enter service in Air Service colours.[14] It appears that the additional Avro 504K (a training aircraft) was an 11th aircraft delivered by the ADC by way of compensation for failing to deliver a serviceable machine on 16 September. Unlike the circumstances around purchasing the Bristol Fighter in July McSweeney did not have the funds or the authority to settle the account. The account, for a total of £15,000, was forwarded for payment to the AFO via the QMG.[15]

THE MOVE TO FERMOY

While Collins had wanted to have aircraft based at Fermoy before the end of August, in reality, the first aircraft did not arrive there until 1 October 1922.[16] It is not totally clear why this was so. Despite the fact that – at four – the number of pilots was very low, a single aircraft and crew could have been posted in the south west before the end of August had the new CIC wished to put Collins's decisions into immediate affect. Similarly the non-availability of aircraft appears not to have been a serious concern as the service aircraft ordered on 14 September only began to enter service from 13 October, well after the actual occupation of the aerodrome.

Although the Army leadership may initially have been waiting for the delivery of new aircraft before occupying Fermoy, it is possible that their hands were forced by circumstances in Baldonnell. Accounts suggest that J. C. Fitzmaurice and T. J. Maloney were, in fact, sent to Fermoy at short notice after a disagreement with McSweeney. The point of conflict appears to have been around the promotion to the rank of captain of John Arnott. Arnott had joined the Air Service (on 15 September) after both Maloney and Fitzmaurice and, in theory, should have been junior to both. Not only was he promoted to captain but he was also designated as 'acting 2nd in command of flying'. The general tone of Fitzmaurice's unpublished memoir suggests that he had no doubts about his own worth as an officer and pilot and so it would have been out of character for him not to have objected to such action. It is suggested by Teddy Fennelly that Maloney and Fitzmaurice were contemplating resigning over this matter when they were ordered, at short notice, to fly to Fermoy. Three pilots are reported to have flown to Fermoy on 1 October 1922. In addition to Fitzmaurice in Martinsyde F4 No 1, Lieutenants Maloney and Crossley are reported to have travelled in a formation of three unspecified aircraft.[17] The assertion that the three available service aircraft had moved to Fermoy on 1 October 1922, however, cannot be confirmed from the respective log books. The log book for Martinsyde F4 No 1 confirms that Fitzmaurice flew to Fermoy on 1 October 1922. The log book for B II shows that Crossley, with Airman Spittel arrived there the following day. The log book for BF I records no flying, or servicing, between 24 September and 10 November 1922, and only indicates that the aircraft had been flown from Baldonnell to Fermoy on 13 November 1922 by Lieutenant W. P. Delamere. The Army census shows that Fitzmaurice, Crossley and Maloney were at Fermoy Aerodrome on the night of 12–13 November 1922. Eventually four aircraft and crews were posted to Fermoy and a single machine, with air and ground crew was based at Tralee.[18] As the senior officer Lieutenant F. S. Crossley was the first commanding officer at Fermoy. However, it appears that Fitzmaurice and Maloney were not happy to have him as CO. With Maloney's connivance, Fitzmaurice apparently brow-beat Crossley into vacating the position and Maloney was appointed to the post. Subsequently, after Maloney had been transferred back to Baldonnell, Fitzmaurice took over.[19]

Notwithstanding Collins's suggestion that the Board of Works should make good any damage to the aerodrome in preparation for aircraft, the sight that greeted the Air Service was less than wholesome. The departing Irregulars had left the aerodrome buildings in a ruinous state. Fitzmaurice subsequently provided a colourful description:

> The aerodrome presented the appearance of having suffered an attack by a flock of locusts possessing a voracious appetite for galvanised iron sheets, wood, glass and everything that went to make up the aerodrome buildings. . . . We discovered that the various buildings and station equipment had been dismantled and sold by auction and the materials scattered about in the numerous farmsteads for miles around.[20]

Fitzmaurice opined that the proceeds of the auction had gone to 'Mr de Valera's war coffers'. However, he claimed that all the material could be traced through the auctioneer's books.[21] In the context of the aerodrome's immediate use it was of particular note that the steel-framed Robin-type hangars built by the British had been stripped of the galvanised sheeting leaving a bare metal frame. The repair of this damage led to industrial unrest and to a question in the Dáil. On 18 October the *Voice of Labour* was reported as carrying an article suggesting that soldiers were acting as scabs in the matter of the rebuilding of Fermoy aerodrome.[22] On 25 October 1922 McSweeney, based mainly on information he had received from the officer in charge of the detached flight at Fermoy, replied to a query from the COS. He confirmed that the corrugated sheeting of the hangars had been removed and that it was essential that aircraft be covered against the elements. McSweeney reported that he had initially gone to Fermoy and had bought and supplied a number of tarpaulins to provide temporary cover for aircraft. The quartermaster had been instructed to engage the services of a local contractor to carry out more permanent repairs but it was alleged that the contractor's labourers would not work due to high winds. Due to the urgency of the situation the Air Service engaged the services of unemployed locals at soldiers' rates of pay – £1. 4s. 6d. per week, plus overtime, until the job of covering one shed had been completed.[23] In reply to the Dáil question of Tomas de Nolga, Eamonn Duggan, on behalf of the Minister for Defence, put a slightly different slant on the matter:

The aerodrome at Fermoy is not being rebuilt. It was decided to cover one shed with corrugated iron sheets, and the work entailing the employment of 15 or 20 men at most was given to Mr Mahoney. His men gave trouble by refusing to hold down the sheets on the shed in a gale, and soldiers had to be put on the job in order to get it done. Except in this case soldiers were not employed on the same work as civilians.[24]

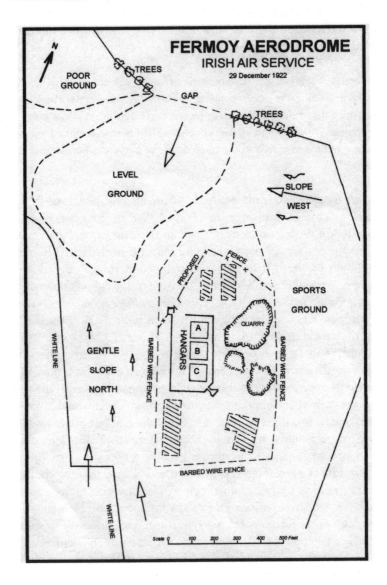

Fermoy Aerodrome, 29 December 1922

FERMOY OPERATIONS

While the record of the Air Service detachment in Fermoy is not well documented the air support role appears to have differed somewhat from the armed reconnaissance missions carried out under the direction of Collins. J. C. Fitzmaurice recalled the *modus operandi*:

> Our duties consisted of providing air escorts to military convoys moving through difficult mountain countryside. These convoys were engaged in cleaning up operations which called for the establishment of a garrison in every town and village. They were subject to ambush only in difficult country where the terrain was suitable to the Irregulars, that is to say, stretches of country allowing them a safe commanding fire position from which a river or deep, wide rivulet prevented pursuit by the ambushed party. Trees were felled across the roads to contain the convoys during the period of these ambushes. Our arrival over such scenes brought an abrupt end to these capers.[25]

This general description suggests a change of emphasis rather than a change of role. Nonetheless, it must be noted that the first missions flown by Fitzmaurice, and many others carried out over the winter and spring of 1922–3, were reconnaissance patrols – initially to west Cork on 9 October and to east Cork on 10 October 1922. Other missions involved the dropping of propaganda material in areas held by Irregulars. Despite the fact that the Fermoy detachment consisted of four aircraft and four crew little is known of the day-to-day operation. This is mainly due to the fact that the central control of the Collins era had been dispensed with and, with the Air Service Fermoy detachment under the direct command of General Emmet Dalton, GOC Cork Command, the dispatch of reconnaissance reports to GHQ was discontinued. It must be presumed, therefore, that the type of mission to be flown was directed by the local commander on a day-to-day basis.[26]

TRALEE OPERATIONS

Lieutenant W. P. Delamere, who reported to Baldonnell on 21 September 1922 and was posted to Fermoy on 13 October, subsequently operated out of Tralee from 1 December 1922 to 12 October 1923, initially under General W. R. E. Murphy, GOC Kerry Command and, from January 1923, under Major General Paddy O'Daly.[27] The Tralee detachment consisted of Delamere, who, along with his observer Lieutenant Charles (Tiny) Flanagan flew a Bristol Fighter, and two unnamed technicians. The Tralee landing ground consisted of what appears to have been a marginally suitable field of about 22 acres that adjoined the Militia Barracks at Cloon More on the south-east side of Tralee. It had a small hangar associated with it. This was located in an adjacent field and had been part of the original RAF Class B aerodrome of the period 1919 to 1921. The departing RAF had apparently left, pinned to the hangar wall, a hand-drawn sketch of the airfield showing the best approaches.[28]

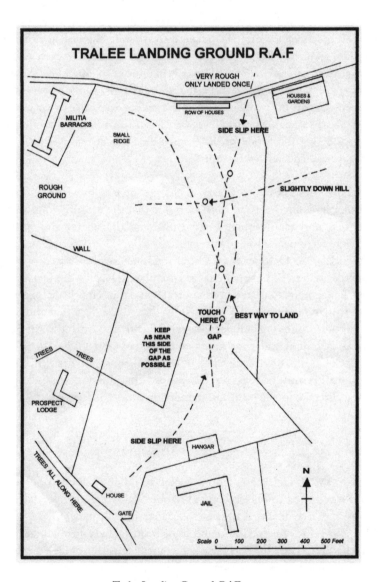

Tralee Landing Ground, RAF, *c.*1921

Unlike the practice at Fermoy most, if not all, of Delamere's operational flights appear to have been reconnaissance patrols, as distinct from escorts, over the 'very wild country' of the mountains of Kerry. Between 4 and 21 December 1922 Delamere and Flanagan flew nine flights, mainly reconnaissance patrols that were somewhat curtailed by the mountainous terrain and adverse winter weather. On 14 December he reported being fired on at Ballyheige while patrolling in the Listowel/Ballybunion area. He dropped propaganda pamphlets in the Ballyheige area on 19 December, and in the Farranfore area two days later. Of a total of just six patrols carried out by Delamere and Flanagan in January 1923 only one was of note. On the afternoon of 16 January 1923 Delamere recalled being fired on by Irregulars while patrolling in the Brennan's Glen area. He dropped two bombs and 'held the Pajoes in houses until dark – troops approaching and attacked the Pajoes', killing one and wounding two.[29] The Kerry Command's Report of the event read slightly differently:

> Army aeroplane flying over Brennan's Glen fired on by party of 30 Irregulars. One bomb dropped and machine gun fire opened from aeroplane. Simultaneously, Dublin Guards from Killarney arrived on [the] scene and attempted encircling movement. Irregulars retreated and in running fight Irregulars suffered six casualties.[30]

From an operational point of view February 1923 was even quieter. Ten flights were flown, to Fermoy and Baldonnell, all for technical or administrative purposes. On 10 February, Delamere and Flanagan flew to Baldonnell in B I and, due to technical difficulties with the replacement aircraft, B VIII, did not get back to Tralee until 9 March 1923. The rest of March 1923 was similar to the previous month, with only three flights out of 13 being of an operational nature. A further three flights, all reconnaissance patrols, were carried out during April 1923.

In May 1923 Delamere carried out nine operational flights. The first of these he recorded as the first day of a 'big round-up operation' in the Kenmare area. From an air reconnaissance point of view he recorded that there was little to report. Operating under new orders, they patrolled the areas of Killarney, Caragh Lake and Castleisland, on 8, 17 and 18 May, respectively but recorded that 'nothing of importance was observed'. With the Irregular leadership declaring a ceasefire on 24 May 1923, Delamere was to fly his last operational mission on 20 June when he observed the 'railway troops [being] withdrawn from the lines' in the Killarney area. On 28 June 1923 he flew back to Fermoy and, on the following day, commenced two weeks leave.[31] In the absence of an explicit air operations policy, at GHQ or Command level, and of patrol reports it is not at all easy to assess the effectiveness or otherwise of the Tralee detachment. Even allowing for the difficult terrain and adverse weather, the completion of only 31 operational missions in seven months of civil war operations appears to represent a modest return while the military intelligence value must remain a matter of conjecture.

On 16 July 1923, still flying B VIII, Delamere and Flanagan, returned to Tralee and flew on the occasion of the parade and march-past for W. T. Cosgrave's visit to Ballymullen Barracks on 22 July. After the Civil War, between July and October 1923, the detachment completed only nine flights, mainly between the bases of Tralee, Fermoy and Baldonnell. Tralee closed on 12 October 1923 and the small detachment moved back to Fermoy. Similarly, aimless flying continued at Fermoy until it too was eventually closed on 14 April 1924.[32] It is not clear why Tralee was kept open until October 1923 and Fermoy until April the next year. While it is possible that GHQ wanted to have aircraft in the south west in case of any minor hostilities it is more probable that the detachment was simply forgotten and only came to the attention of GHQ again in the context of the administration of the demobilisation and the reorganisation processes of 1923–4.

PERSONNEL AND SUPPORT SERVICES

While, in August 1922, Collins had been convinced of the necessity of providing appropriate support services, such as transport, communications, ordnance and meteorology to military aviation, and had initiated staff action, these matters were not to receive similar priority from the new leadership. The matter of communications, probably the most fundamental and essential support service, was a case in point. With no sign of a wireless station being established at Baldonnell by 26 September 1922 McSweeney made requests to Mulcahy:

> In view of the establishment of an aero base at Fermoy two wireless sets are extremely urgently needed, one at Fermoy and one at Baldonnel to communicate with each other. It would be very economical if we could arrange the handing over of eight aeroplane wireless sets by the British Government for reconnaissance work . . . worth £40 each . . .[33]

This was not an unreasonable request given that wireless telegraphy sets had been standard equipment on operational aircraft since the Great War and that training in wireless telegraphy was a fundamental aspect of pilot training. Similarly, communication between Fermoy and Baldonnell, using sets of appropriate frequency and power, would be considered essential to the management and operation of air resources. As had been demonstrated during Russell's mission to Waterford and Cork in early August, up to this time aircraft had no wireless communications with Baldonnell or elsewhere while all longer range communication with Baldonnell had to be relayed, via GHQ, by telephone or by road. Having been requested by Mulcahy to address McSweeney's requests for ground stations and aircraft wireless sets, Liam Archer, OC Communications Department, quoting from a report on the distribution of 30 watt wireless sets, indicated that one would be installed at Fermoy Aerodrome, apparently for communication with GHQ, but that no set was available for Baldonnell.[34] In response Mulcahy asked for clarification on the matter of aircraft W/T sets, which Archer had pointedly ignored. Archer's response was short and to the point and indicated that his report of 29 September 'had covered all queries raised by the memo of the director of aviation'.[35] Mulcahy did not pursue this line in the matter of aircraft radios further.

Subsequently, he approached the question of installing a wireless set at Baldonnell from a different direction. He suggested that the principal function of such a radio would be to receive meteorological reports. 'I would like to know if the 30-watt receiver will do their work effectively and whether there is a set to spare.' Archer replied succinctly that 'no 30-watt set was available for Baldonnel'.[36] In effect Archer was bluntly ignoring the request of his CIC. Archer's approach to dealing with aviation communication requests could be interpreted as a reflection of an anti-British stance which resulted in a decision prejudicial to the ex-RAF staff of the Air Service. Mulcahy's failure to resolve this matter confirmed the fact that the air communications was to be an independent signals function which would be exercised without reference to the actual needs of military aviation. Unresolved difficulties in respect of responsibility for aviation communications policy was to have a major influence on the relationship between the Signal Corps and the Air Corps in later years; in particular, it was to adversely affect the standard of the air-to-ground and ground-to-air communications during the Emergency.

The position in relation to ordnance was not unlike that pertaining to communications, though in this case Mulcahy himself was responsible for the prevarication. On 22 August 1922 the QMG had reported to Mulcahy that he had succeeded in acquiring 4,800 rounds of aerial ammunition, presumably from the departing RAF, in response to Collins's requests on the matter during the early weeks of August. One must question the high level of interest in having the Air Service develop a capacity to drop bombs and paradoxically, standard hand grenades, from aircraft on patrol duties.[37] It is not clear if armament was carried on all operational missions. Most aircraft appear to have carried Vickers or Lewis machine guns, or both, in addition to bombs.

In September 1922 McSweeney had a limited stock of nine-pound bombs that had most probably been acquired from the departing RAF and which he tested for their potential against Irregulars. He reported that he had used two aircraft on Sunday 24 September to test the bombs at Baldonnell. He stated that four bombs had been dropped from 500 feet and that holes, two feet wide and one foot deep, had been made in the selected grass area, while pieces of shrapnel had been scattered 20 feet around. He suggested that such bombs should be very effective when used on roads. The nine-pound bombs were not compatible with the bomb racks (supplied for 20 lb bombs) purchased with many aircraft which made it necessary to manufacture launching tubes to be used by the observer in two-seat aircraft. He suggested that single-seat aircraft, such as the Martinsyde F.4, required a compatible bomb and bomb rack combination that could be operated by cable. He also reported sending a Cooper-type bomb, which he had ordered to be stripped, to the Director of Munitions in the hope that similar bombs could be manufactured locally.[38] Having received no response to his report on the testing of bombs, McSweeney reminded Mulcahy that he had attempted to order over 300 bombs and 105,000 rounds of assorted aerial ammunition, when he was last in London. The War and Colonial Offices required Mulcahy's authorisation before supplying the munitions. Mulcahy replied, stating that he was not taking any steps in relation to such munitions until a conference was held with representatives of the British War Office. By February 1923 Mulcahy had not met with officials of the War Office and did not perceive a genuine demand for such armament. He recorded a file memo to the effect that the bombs were not required at once and that the question of obtaining them need not be considered until Army representatives visited the War Office and, in effect, continued to prevaricate on the matter.[39]

PILOT TRAINING

By the end of 1922 a total of 17 aircraft had been acquired by the Air Service, though at least two of these had been badly damaged as a result of accidents. At the same time ten pilots, of 12 recruited that year, remained in the service. C. F. Russell's last recorded flight during the Civil War had taken place on 6 September. He was transferred from the Air Service in mid-September to take charge of the new Railway Maintenance and Protection Corps.[40] This Corps was needed as a result of the wanton destruction of the railway infrastructure perpetrated by the Irregulars. Air Corps folklore reflects the popular belief that Captain John Arnott, the seventh ex-RAF pilot recruited, (on 15 September 1922), was lucky to escape with his life when 'dismissed' on or about 21 December the same year. Allegedly he had been identified as a former Auxiliary and was requested at gunpoint to take the mail boat to Britain – and did so!

With five pilots and five aircraft stationed in the south west the remaining resources at Baldonnell were directed to be used in training pilot. While the commencement of pilot training was not formally announced by GHQ until 20 December 1922, aircraft log books suggest that training had commenced, on an *ad hoc* basis, as early as October 1922. About that time a number of officers, including Lieutenants A. Russell, Tom Nolan and Ned Fogarty had commenced flying training on Avro 504K.[41] On 20 December 1922 the Adjutant General advertised a limited number of vacancies for pupils in the 'aviation department of the Army'. Officers, between the ages of 18 and 23 wishing to transfer to the 'department of aviation' were invited to apply through their command HQ. After interview successful applicants were to be attached to the 'Flying Corps' for an unspecified time and if found satisfactory as pilots would be transferred on a permanent basis.[42] In the context of the involvement of a rudimentary Military Air Service in the Civil War, its ill-defined functions and poorly organised nature, the training of new pilots drawn from the officer body of an untrained army made little sense. Its only logic was in the perception of the army leadership that it was necessary to replace ex-RAF pilots with officers of an acceptable nationalist background as quickly as possible. However, this pilot training effort would not produce the desired effect.

DEMOBILISATION

As soon as the Civil War had come to an end in May 1923 it was inevitable that action had to be taken to reduce Army strength, and to reorganise roles and functions more compatible with the new peace situation and the impecunious state of the country's finances.[43] The initial demobilisation proposed was a reduction in total Army numbers from the May 1923 figure of 55,000 to 31,300 by January 1924.[44] Even before demobilisation had taken place, the secretary of the Department of Finance questioned the necessity for the Air Service in terms that put its survival in severe doubt:

As the Minister of Defence is aware the position of the public finances is such as to render it imperative that drastic economies be effected in all services which are not immediately essential in the public interest. In this connection the Minister of Finance would be glad to learn whether the Minister of Defence sees any urgent reason for the maintenance at the present time of an air branch in the Army. The

Minister . . . finds it difficult to believe that the upkeep of an Air Service in this country at the present time can be justified. . . . The minister would also be glad to be supplied with details showing the strength, distribution and equipment of the Air Service at the present time. Pending the further consideration of this matter the proposals put forward on 31st ultimo by the army finance officer for the grant of additional and extra pay to Air Service personnel in certain cases are being held over.[45]

Nothing on the Department of Defence file indicates on what criterion Finance based their rather blunt opinion. It is generally acknowledged that military expenditure soared out of control during and immediately after the Civil War.[46] However, it is not generally appreciated that the actual expenditure on the Army vote for the Civil War period was of the order of £7,459,104 for 1922–3 and £10,461,401 for 1923–4.[47] On a purely financial basis it was clear that an army of over 55,000 simply could not be supported. Nonetheless, when one considers the total numbers in uniform during the War and the rather minute size of the air component the cost of the aviation element could not be considered disproportionate. The Military Air Service, including the troops known as the Air Service Infantry, had peaked at a total of 540 all ranks, 298 infantry and 242 aviation, in June 1923.[48] The entire Air Service represented less than 1 per cent of the whole Army at its maximum strength while the specific aviation element amounted to 0.44 per cent. The cost of 0.44 per cent of the Army for the two years comes to about £79,000. To this must be added the cost of purchasing and operating aircraft that would have been over and above the cost of infantry soldiers. The total cost of buying 22 aircraft, plus the operating cost of spare parts, fuel and oil during the Civil War was put at £29,000 by McSweeney, a figure that was not contested by the Department of Defence or the Department of Finance.[49] With no additional aircraft being purchased in 1923–4 a notional £500 would probably have more than covered any additional costs in the financial year 1923–4 – a total of £29,500. In effect the Air Service had cost the state less than £110,000 out of the total of £17.91 million expended on the Army in the two-year period. It is possible, however, the Department of Finance considered that the £29,500 spent on aircraft did not represent good value for money.

Despite several reminders and requests, Department of Defence files do not indicate that the Minister gave the matter due attention or that he had sought an input from either GHQ or the Air Service on the matter. Thus, the minister was again reminded that approval of a flying pay regime for some Air Service personnel, decided by the Army Pay Commission in May 1923, was being withheld pending a reply.[50] In November he did forward the strength return requested for mid-August. As McSweeney had, within two days, provided the required information, on the numbers and distribution of personnel and details of the aircraft on charge, there is no obvious reason for GHQ's delay in forwarding this information on to Finance.[51] As late as 24 December 1923 the Department of Finance was still endeavouring to elicit the case required to justify the retention of the Air Service from the Minister or from his department.[52] Not only did Mulcahy fail to supply the required case but failed to even acknowledge that such a case was required. In the absence of such essential information it is not known what Mulcahy's policy on aviation may have been or why he was seemingly so ambivalent about military aviation. Based on the indifference to military aviation he displayed during the Civil War it might be considered that Mulcahy was reluctant to support the continuation or demise of the Air Service and, in effect, seemed content to leave the judgement and decision to others.

In the meanwhile McSweeney, GOC Air Service, had submitted a proposal for an Air Service consisting of a headquarters and two squadrons to be included in a reorganised force. He cited the necessity of having sufficient, but undefined, striking power available to counter potential enemies. He also indicated the necessity of being able to patrol fishing grounds to identify and monitor foreign trawlers. While McSweeney considered two squadrons to be the minimum size of aviation unit that would be effective and viable the reasons he cited may not have been judged sufficient to GHQ staff who had previously displayed little appreciation of air power and the operational application of aircraft.[53] McSweeney was informed that an establishment for two squadrons was being recommended but that financial considerations might not permit this. It was stated that 'in all probability it may be decided by the Executive Council to abolish the air force [*sic*] entirely'.[54] In any event the officer establishment, eventually published in February 1924 and intended for activation in the following April, provided for a headquarters and two squadrons with a total of 43 officers. With the addition of an appropriate establishment of NCOs and men this establishment, had it been proceeded with, might have been expected to bring total Air Service numbers to about 500 of all ranks. In the reorganised Army McSweeney was to have been reduced to the rank of colonel.[55] However, the scheme eventually proposed for the Air Service only provided for the reduced total of 287 all ranks – 43 officers, 60 NCOs and 184 privates.[56]

As was the case throughout the Army, the reorganisation and demobilisation process in the Air Service was interrupted and complicated by the Army mutiny or crisis of March–April 1924. The difficulty arose due to the manner of the demobilisation process aimed at reducing officer numbers from 3,300 to 1,800 and from the manner in which officers of War of Independence repute were allegedly being targeted for discharge while many ex-British officers were to be retained. Three categories of officers were identified as being liable to dismissal: unsuitable officers, post-Truce officers who had no special qualifications and pre-Truce officers who were surplus to requirements. In Baldonnell some eight officers fell into these categories. On 7 March 1924 some 900 officers, including four officers of the Air Service Infantry and four of the Military Air Service, were demobilised.[57] The latter group included Lieutenant William McCullagh who had been injured in a flying accident on 25 June 1923 and was classified as 'long term sick'. Also demobilised was Second Lieutenant J. V. Norton, one of the trainee pilots taken on in 1922–3.[58]

Brigadier General Liam Tobin and Colonel C. F. Dalton who had assumed the leadership of the pre-Truce officer group, took exception to freedom fighters being discarded while ex-British officers and soldiers were being retained. In fact only 157 technical officers, that would have included 11 former RAF pilots in the Air Service, were so retained.[59] After written and verbal confrontation with the government the 'mutiny' was successfully contained and turned into what would nowadays be termed a redundancy scheme.

Thirteen Air Service officers, including McSweeney, who had been nominated for appointments in the reorganised Army, are recorded as having 'resigned due to the crisis'.[60] While in general the particular circumstances surrounding the mutiny and the resignation, discharge or demobilisation of individual officers are not detailed in the surviving records, circumstances applying to some of the flying officers in Baldonnell can be elucidated. This information is available due to access gained to the Military Archives files of a number of pilot officers. General W. J. McSweeney was one of the most senior officers to be listed as a mutineer and the only officer, other than Tobin and Dalton who were actually named, who can be identified from the Dáil debates of March 1924. It is most curious that Colonel C. F. Dalton, one of the founder members of the Irish Republican Army Organisation in January 1923 and later one of the two officers who challenged the Government on 6 March 1924, was McSweeney's adjutant at Baldonnell from 30 June 1923 to 29 March 1924.[61]

While there is no evidence to confirm any ulterior motive in the matter it might be considered extremely odd that a disaffected officer like Dalton would be appointed adjutant to any military formation. His co-conspirator, Major General Liam Tobin had been appointed aide-de-camp to the Governor General. The thinking behind the latter appointment was possibly that Tobin might not be in a position to spread the rot of dissent throughout the Army barrel. The appointment of C. F. Dalton as adjutant of the Air Service could be viewed in the context of the prejudice of the old IRA against ex-British officers holding commissioned rank in the National Army of 1923–4:

The old IRA men in the army generally objected to the presence of those who had never participated in the national movement, and particularly to those who were regarded as enemies prior to the Truce with England.[62]

As the Air Service had a significant concentration of such personnel it might be considered that it made little sense to appoint a disaffected officer such as Dalton to the position of adjutant. Considering the disciplinary aspects of the attendant duties the appointment of C. F. Dalton as successor to Broy could be viewed, at best, as careless and at worst, as being deliberately seditious. It cannot be ruled out that Dalton had been moved to Baldonnell in the full knowledge that he might foment dissent amongst both the ex-IRA and ex-RAF officers. In the circumstances his relationship with the latter group was bound to be particularly fractious.[63]

When C. F. Dalton was appointed to the position of Adjutant, Army Air Service in July 1923 he succeeded Colonel Broy who had retired in June 1923 after less than a year in uniform.[64] It is not recorded why Broy retired but his marriage in July 1923 may have been an influence.[65] There are a number of other possible reasons, however, that will be considered. Broy had originally been specifically appointed to his position in the Air Service by Michael Collins having just previously been briefly on the staff of the Civil Aviation Department.[66] In September 1922 the Adjutant General had reported that Broy 'had brought the standard of discipline to a very high pitch and that he is a person well suited to accept the responsibility of carrying on in the absence of the director [of military aviation]'.[67] In May 1923 GHQ eventually got around to formally endorsing those temporary commissions and appointments of Air Service officers originally authorised by McSweeney and, in Broy's case, by Michael Collins. Thus, Broy was the first Air Service officer so endorsed by means of publication in General Routine Orders. On 4 May 1923 'Colonel E. Broy' was confirmed in the appointment of 'adjutant, Air Service, and O/C ground organisation' in the COS's Department. Over three weeks later, on 28 May 1923 'Major General John [*sic*] McSweeney' was confirmed as 'officer commanding, Air Service', in the same department. Without the definition of the term 'ground organisation' it is unclear what Broy's precise status and responsibilities were. In publishing Broy's appointment as 'O/C Ground Organisation' before that of McSweeney GHQ may have been trying to make a clear distinction between the Air Service *per se*, that would be under McSweeney, and the Air Service Infantry and the garrison troops under Broy. This distinction had not been made clear in 1922 or 1923 and was still no clearer in March 1924 when McSweeney and others were being dismissed as alleged parties to the mutiny.

Notwithstanding confirmation on 4 May 1923 of his original appointment of 29 July 1922, Eamonn Broy decided to resign with effect from 22 June 1923.[68] It is likely that Broy, who was not a pilot and who was 11 years older than his immediate superior McSweeney, did not see a future for himself in military aviation thus precipitating a move back to his earlier calling as a policeman. While his retirement may have been influenced by his forthcoming marriage, or indeed, his age, it is also cannot be ruled out that his resignation was related to the appointment of C. F. Dalton. The fact that Broy departed Baldonnell and the Air Service rather abruptly may be a further indicator of this. Shortly after the mutiny, Commandant Mason, who took over the duties of camp commandant at Baldonnell, found two trunks of property belonging to Eamonn Broy in a room more recently vacated by McSweeney. Apart from personal items one trunk contained some 896 rounds of .303 service ammunition.[69] Such a discovery is curious, as it would have been very much out of character for Eamonn Broy to abandon service ammunition in such a manner. The fact that McSweeney was best man when Broy married Elizabeth Dooly, on 16 July 1923, strongly suggests that Broy left the Air Service on the best of terms.[70] It is likely, therefore, that it had been intimated to Broy, by higher authority, that he accept a change of appointment to make way for Dalton; in response Broy probably chose to resign rather than move elsewhere, and leaving Baldonnell in somewhat of a huff, he neglected to tidy up his affairs.

While the effect of Dalton's influence in Baldonnell cannot be accurately judged it is significant that 12 infantry and air officers of an IRA background are recorded as having been dismissed as a result of the Army crisis of 1924. Only one ex-RAF officer, McSweeney, is similarly listed, though subsequently two more would follow. As these pilots were not in any of the three categories of officers designated for demobilisation the discharge of three pilots, particularly that of McSweeney, requires explanation. On an undated list of 'resignations, dismissals and absenters' McSweeney was recorded, along with C. F. Dalton as a deserter.[71] On a list dated 19 March, apparently later than the first, McSweeney and Dalton are recorded under the heading 'Additional resignations due to crisis'.[72] While his adjutant, Colonel Dalton, was one of the ringleaders there is no direct evidence that McSweeney took an active part in the mutiny. In fact they were unlikely bedfellows – if bedfellows they were. McSweeney was one of the 157 ex-British officers whose proposed retention in the Army so antagonised Dalton and others.

The circumstances surrounding McSweeney's dismissal are not clear and, in some respects, appear contradictory. J. C. Fitzmaurice, in his unpublished memoir, states that McSweeney, on some unspecified date about the time of the mutiny (March–April 1924), had travelled down to Fermoy in his own car allegedly in possession of a significant quantity of misappropriated arms. Fitzmaurice states that he was amazed to find that McSweeney had taken the side of the mutineers. In stating that Fitzmaurice was questioned by McSweeney as to his attitude and that of his officers in the matter, he implies that the reason for McSweeney's visit was to persuade Air Service officers to join the mutiny. Fitzmaurice, who had no time for McSweeney, claims to have affirmed his own allegiance to the State and to have persuaded McSweeney to leave Fermoy.[73] In the absence of confirmatory account it is not possible to judge the veracity of this serious allegation. The alleged incident does not sit well with known aspects of the mutiny period insofar as McSweeney was concerned. Had the incident happened it probably would not have gone unnoticed. Similarly an assertion by Commandant J. J. Flynn is difficult to understand. In the course of contesting his own dismissal Flynn states that he found that his GOC was absent from Baldonnell on Monday 10 March 1924. However, the circumstances of McSweeney's resignation would appear to contradict this opinion.

On Saturday 8 March 1924, two days after the ultimatum to the Government that had initiated the crisis, three line officers of the Air Service had absconded from Baldonnell with three Lewis guns and a Crossley tender. On Monday 10 March 1924 McSweeney made a phone call to General Mulcahy, CIC and Minister for Defence. The telephone call was mentioned on 11 March when the matter of the mutiny, including the taking of arms from Baldonnell and elsewhere, was being reported to the Dáil:

> In connection with the Baldonnel incident the OC of the aerodrome yesterday tendered his resignation on the phone. He was told his resignation would not be accepted in that way, and he said that if that was so he would have to be regarded as a deserter.[74]

The clear inference in the Minister's statement is that McSweeny's resignation was directly related to the taking of weapons by absconding officers. It might be inferred that the Minister had demanded the GOC's resignation holding him responsible for the actions of his subordinates. In the heat of the moment McSweeney may have tendered his resignation verbally, indicating that he was unlikely to put it in writing. In any event McSweeney submitted not one but two letters of resignation. After the telephone call to the Minister he wrote:

> Sir, I have the honour to tender my resignation from the Army. I rang you up on the phone this evening and you accused me of breaking my word of honour. I assert now that I kept my word to the letter, also my Oath.[75]

Following the telephone conversation a GHQ officer, with unknown instructions, was sent out to Baldonnell. The visit caused McSweeney to write again, this time, without due deference, to the COS:

> I desire to tender, from today, my resignation from the Army, and in doing so I wish to state that I faithfully kept my word I gave to you & the M.D. last night. Judging from Col. O'Connor's arrival in Baldonnel, and the document he carried, you do not appreciate the word of honour of an officer.[76]

The arrival of an officer from GHQ and the writing of a second letter of resignation add little to our understanding of the complex details surrounding McSweeney's resignation or dismissal. While he may have contacted the Minister on his own initiative it is more likely that he was responding to a query from the Minister about the misappropriation of arms, in the context of an inevitable Dáil debate. The tone of the letters of resignation suggests a difference of opinion on some unknown matter. As McSweeney was one of a number of ex-British officers with no pre-Truce service, whose appointment and proposed retention was cited as a factor in the mutiny, it cannot be ruled out that it possibly suited Mulcahy to hold him responsible for the mutinous actions of three line officers of the air station. Indeed, the misappropriation of arms by his subordinates was probably more than adequate reason for requiring McSweeney's resignation.

In spite of McSweeney's early resignation and the fact that the Army Enquiry Committee generally avoided giving reasons for individual 'resignations' the enquiry found that 'Major General William J. McSweeney' had absented himself in such a manner as to show wilful defiance of authority'.[77] That he was absent is not supported by the fact that he had telephoned the Minister on the 10 March, was available to receive an emissary from the COS and had resigned with immediate effect. Perhaps McSweeney was deemed to have been absent on the basis that he left his post before his resignation had been formally accepted. Unfortunately the records fail to reflect the precise circumstances of his resignation or dismissal, so there can only be conjecture, rather than conclusive fact.

The question still arises as to the exact nature of McSweeney's position vis-à-vis the mutiny. While, for obvious reasons, he might not have seen eye to eye with C. F. Dalton, as an officer who owed his rank and career to Michael Collins and who was due to be reduced to the rank of colonel in the proposed reorganisation, McSweeney may well have held a grudge against the current leadership. Some evidence to support this is provided by Captain Patrick (Joe) Mulloy, a former IRA officer and later of the Air Service Infantry, who was an Air Service observer at the time of the mutiny. In a pamphlet published some fifty years after the events he observed on the ease with which the infantry at Baldonnell might have contributed to a general *coup d'état* initiated by Tobin and Dalton:

> In Baldonnell the headquarters of the newly formed Air Corps [*sic*] the GOC, General McSweeney: the adjutant Colonel Dalton, one of the signatories on the ultimatum presented to the government, and the bulk of the officers of the garrison were involved [in the mutiny], and it would be a comparatively simple matter for the GOC to issue instructions that orders from him only were valid. The flying personnel were not involved as they were largely ex-RAF and would take their orders from the GOC. Thus the whole camp, with the Air Corps [*sic*], could be taken overnight, without a shot being fired.[78]

These first-hand observations by Pat Mulloy seem to suggest that while McSweeney was in sympathy with the mutiny his position was a personal one and that he did not attempt to influence either the rest of the ex-RAF group of officers or others. Equally, it could be argued that those infantry officers at Baldonnell, and the line and air officers of the Air Service, who supported the 'mutiny' and who would have had good republican records, were more likely to have been rallied to the cause by Dalton rather than by McSweeney. Fitzmaurice intimates that McSweeney did not enjoy the confidence of the ex-RAF pilot group though we only have his seemingly jaundiced views on this matter. He described McSweeney as a 'youth who bore the exalted rank of major general' and who was 'an ex-cadet of the Royal Air Force whose flying experience was practically nil'.[79] On the basis of his background and military culture McSweeney would have been held in similar odium by the former IRA officers of the garrison. It is of note that McSweeney, who appears to have had no direct involvement in the mutiny, resigned on 10 March 1924 while C. F. Dalton did not resign until 25 March.[80] Their respective personal files in Military Archives indicate that McSweeney was paid off with £100 while Dalton was paid £225 for 'excellent service prior to the Truce'.[81] There is a certain irony in the fact that one of the main functions performed by Dalton, as Air Service adjutant during the demobilisation and reorganisation process, was to bear witness to, and certify, the satisfactory nature of the service of the individual ex-RAF pilot officers.[82]

In addition to McSweeney, only Commandant J. J. Flynn is recorded as having resigned due to the crisis. Flynn, who had been in charge of pilot training, was found to have absented himself in the same manner as McSweeney. The reason cited is the fact that he was declared to have been absent on or about 10 March 1924. Details of the circumstances surrounding his resignation are contained in Flynn's correspondence with higher authority in April 1924. At the time of the Army mutiny Flynn was second in command to McSweeney. It appears that he had received, on 8 March 1924, McSweeney's verbal permission to be absent from Baldonnell for 24 hours so that he could attend to private business in Sligo. On his return he found Baldonnell had been taken over by troops from the Dublin command and claimed that McSweeney had absconded. Flynn was arrested and spent ten days in Arbour Hill Detention Barracks. On 21 March 1924 he was released having given his 'parole to come up for trial when duly summoned'.[83] About four weeks after his release he wrote to the Minister for Home Affairs, Kevin O'Higgins, giving an understandably biased account of his travails over the previous six weeks and seeking the redress that he could not get elsewhere. He explained how he had been on leave and how he had been arrested on his return and spent ten days in Arbour Hill. He indicated that he had made his loyalty to the state known to the COS.

Although he was considered to have been absent without leave he was not charged. While in Arbour Hill he had been referred to as a mutineer though he was kept apart from that group of officers. On his release he was initially not allowed back into Baldonnell but was summoned there on 5 April and had an interview with the COS to whom he explained his situation. As a result he was to be allowed remain at the aerodrome, however, he retained his rank, was removed from his position as squadron commander and made subordinate to Captain T. J. Maloney (the new commanding officer). Dissatisfied with the situation he felt compelled to tender his resignation and did so on 10 April 1924. On 13 April he was incorrectly informed, by Maloney, that his resignation had been accepted by the COS. He was directed to report to the Staff Duties Office in GHQ and was instructed to proceed on leave while awaiting the decision of the Army Council as to whether his resignation would be accepted or not. Citing the fact that those at the top of the Air Service were being given greater recognition for their service with foreign armies than he, as an old IRA man, received, Flynn pleaded that, as the only remaining flying officer at Baldonnell with an IRA record, he continued to support the Treaty. He requested that Kevin O'Higgins have the whole matter investigated and that he should be allowed to withdraw his resignation.[84]

Kevin O'Higgins's office referred the case to the Office of the President of the Executive Council. From there the matter was sent to General O'Duffy, the recently appointed GOC Forces, for his opinion. O'Higgins had suggested that there might have been a mistake made and that the treatment of Flynn, therefore, could have been too harsh.[85] Nonetheless, O'Duffy recommended that Flynn not be allowed to withdraw his resignation. He did so on many grounds. Not least of these was that Flynn could not prove that he was not absent on 10 March 1924. O'Duffy put great emphasis on the fact that Flynn had been 'absent from his post at the time practically the entire staff absconded from Baldonnel' stating that it was more than a coincidence. Without doubt, the most damning comment was that in his letter of resignation Flynn had expressed opinions that echoed those of Tobin and Dalton the chief mutineers. Citing, also, Flynn's 'mutinous and indisciplined remarks', his intemperate language in his letter of resignation and his intemperate behaviour when dealing with the COS, O'Duffy suggested that he 'could not reaccept him in the army in any capacity' and recommended rejection accordingly.[86]

O'Duffy's assessment of J. J. Flynn and his recommendation suggest that Flynn was unfortunate to be considered absent at an important juncture and overreacted and thus contributed greatly to his own misfortune. The circumstances of the McSweeney and Flynn cases suggest that many officers like them must have been discharged or dismissed on the flimsiest of hearsay evidence as no formal charges were brought against alleged mutineers, not even Tobin and Dalton, the ringleaders.

The last of the four ex-RAF pilots to be let go was Wilfred D. Hardy who was discharged on 27 June 1924.[87] His commanding officer, the newly promoted Major T. J. Maloney, argued strongly in favour of his retention and protested 'at the proposed demobilisation of a good officer who was in line to be appointed flight commander'. Maloney suggested that Hardy had been selected for discharge on the simple basis that he was non-Catholic.[88] However, Maloney had apparently not been told that, in his capacity as GOCF, Eoin O'Duffy had indicated to the Executive Council on 29 May 1924 that Hardy had two brothers in the 'Six Counties Special Constabulary', and was being dismissed on those grounds.[89]

As the ex-RAF pilot group at Baldonnell were notionally exempt from demobilisation it could be concluded that GHQ made maximum use of the confused circumstances surrounding the mutiny to dismiss as many of that group as possible. In doing so they were, in effect, pandering to the prejudices of those, both serving and demobilised, whose main objection was to the proposed retention of ex-British officers who had no pre-Truce service. The pilot group, even though all were of Irish birth or origin, represented a considerable concentration of such officers and was a ready target for demobilisation.[90] In effect no distinction was made between the RAF who left Ireland in 1922 and a small number of ex-RAF Irishmen who served the State well during the Civil War. One of the ex-RAF pilots, J. C. Fitzmaurice, writing some years later described, in very strong terms, the position in which that group had found themselves in 1924:

Unfortunately the stinking evils of patronage, nepotism and corruption, now rife in my native country and slowly bringing it to ruin, commenced raising their ugly

heads about this [time] and we Irishmen who had held His Majesty's commissions were treated with grave distrust by the politicians and the majority of the
Old IRA officers who always referred to us as the 'Exers' – delightful term![91]

A similar, though less trenchant, opinion was expressed by another officer:

> I am, of course, well aware of the prejudice against British officers which is openly
> exhibited by some officers of the National Army and under the circumstances that
> prejudice is inevitable, but I make no apology for the part I played in the Great
> War.[92]

It must be noted, however, that not all ex-RAF officers had this experience.
Colonel C. F. Russell, when asked by the mutiny inquiry committee if he had
experienced hostility or jealousy on the grounds of being a former British
officer, indicated that he had not experienced such treatment from either
colleagues or from higher authority.[93]

CONCLUSIONS

Although Michael Collins had taken some time to authorise an expansion of
the Air Service, his ideological commitment to military aviation was not in
doubt. His belief in air reconnaissance was not taken on board with any
enthusiasm by Richard Mulcahy. This is shown in particular by the length of
time taken to review Collins's decisions and to buy aircraft and recruit pilots.
The delay in the expansion and movement of air operations into Munster was
symptomatic of the general failure to execute a rapid push into the south west,
a movement that might have foreshortened the War.

    At Tralee, with the aircraft under local control there was a modest return
in terms of operational missions flown. With the air operation out of Fermoy
under the command of General Dalton and with control exercised at local
level it is probable that the overall return, in terms of the missions flown by
four aircraft and crews, was similarly modest. The effectiveness or otherwise
of these escorts and patrols carried out cannot be properly gauged due to lack
of reconnaissance reports. After the Civil War it is doubtful if there was a
good military reason for maintaining the Fermoy and Tralee detachments in
place until well into 1924.

    While the demobilisation and mutiny processes had a significant affect on
the Air Service the number of IRA officers who resigned or were dismissed
during the mutiny would imply that C. F. Dalton had a greater malign
influence on that group than on the ex-RAF pilots. Although, theoretically,

Air Service pilots were in a special category that was not subject to demobil-isation, the authorities appear to have taken advantage of the confused circumstances of the mutiny to discharge a significant proportion, four out of 11, of the remaining ex-RAF pilots. As was the case of many officers discharged as a result of the crises, the discharge of the pilots appears to have been based on ill-informed perception rather than hard evidence and due process. Though McSweeney may have felt aggrieved at the prospect of being reduced in rank, there is scant evidence of him being absent on 10 March 1924. As suggested by Mulcahy in the Dáil he was held responsible for the loss of arms that occurred at Baldonnell. His resignation of 10 March was accepted with effect from 29 March 1924. While Commandant J. J. Flynn had the misfortune to be missing from Baldonnell at a crucial time and had reason to feel aggrieved with the way he was treated, he contributed to his own downfall by intemperate behaviour. Lieutenant Hardy's only sin was to have two brothers in Northern Ireland's Special Constabulary. It is not possible to assess the rights or wrongs of Lieutenant W. A. McCullagh's demobilisation on medical grounds, however.

In the wake of the Civil War it is not at all clear on what precise financial basis the Department of Finance proposed to abolish the Air Service. The department supposedly perceived aviation as hugely expensive though, in fact, it had cost less than a half of one per cent of total Army expenditure for the period in question. Subsequently, between May 1923 and March 1924, because of the ambivalence of the Army leadership, the Air Service appeared to survive more by accident than design. It was to take General O'Duffy's scheme of reorganisation to place its successor, the Army Air Corps, on a slightly firmer footing.

FIVE

ORGANISATION, POLICY AND COMMAND, 1924–36

—

THE 1924 REORGANISATION

In March 1924, as a consequence of the mutiny, General Eoin O'Duffy had been appointed Inspector General and General Officer Commanding Forces. His main function was to oversee the completion of the demobilisation and reorganisation processes that had eluded General Mulcahy. O'Duffy's reorganisation proposal provided for an army of 18,966 all ranks of which 155 all ranks (or less than 1 per cent) was to be the establishment of Army Air Corps.[1] The rationale behind O'Duffy's scheme of reorganisation was to place the aviation element in a precarious position in an infantry dominated army. As the scheme was basically one for the reorganisation of the Army in the absence of a Government statement on defence policy, O'Duffy had to make assumptions in terms of what threat was to be guarded against. He decided to couch his proposals not in terms of national defence against external aggression but rather in terms of the threat to national security still posed by the IRA:

> The question to be now considered is whether the Saorstát has greater reason to be apprehensive of an attack by forces from outside the state or an attack by Forces within its boundaries. The experience of the past two years combined with present day knowledge would go to show that internal disorder is more imminent and more to be apprehensive of. We must next decide as to the most effective arm of the service to cope with internal disorder. Again our experience has shown that the highly trained and mobile Infantry man was the most effective weapon used against the Irregulars while the practical utility of the Air Service was not considerable.[2]

Though the above statement reads like a preamble outlining defence philosophy or doctrine it was not cited in order to elucidate such matters but rather to minimise the value and potential of army aviation in an internal security context of the State's defence requirements, and to justify the establishment of what was to be a token Air Corps. O'Duffy's scheme of reorganisation was very heavily weighted towards 'the highly trained and mobile infantry man', and in effect established the precedence of an 'infantry arm' that was to dictate the military doctrine of the Defence Forces. This precedence was reflected in the essentially infantry nature of the establishment of GHQ and of its three military departments (Chief of Staff, Adjutant General and Quartermaster General), the three territorial Commands and the Curragh Training Camp, nine brigades and 27 infantry battalions. The 'infantry arm', in effect, comprised approximately 75 per cent of the reorganised Army.[3]

While the case stated for the leadership structures, demonstrating the precedence of the infantry ethos, were detailed and cohesive, some cases made in support of the inclusion of individual corps and services were perhaps too general and vague, demonstrating a lack of understanding of the nuances of the roles and functions of standard military corps. In particular, the weak cases made for such corps as the artillery, armoured car, air, cavalry and transport suggest that those drafting them lacked conviction as to the military value of, and necessity for, some of the more technical elements of a modern army. In contrast, the case for Army Corps of Engineers indicated a more professional approach and, in presenting a cogent argument for military engineering, reflects the superior staff work of its first director, Colonel C. F. Russell.[4] In general terms the explanatory notes portrayed the infantry arm as being indispensable and the other corps and services as optional extras, endorsing the belief that only the infantry soldier was capable of mounting a defence against an internal threat.

The case made for an Army Air Corps in the new organisation was unstructured, vague and somewhat aspirational in tone. It indicated that any aviation element established would, at best, have but a very minor and peripheral function in the overall scheme of defence. Shockingly, in an Army structured for internal security, and dispersed around the country largely in the manner of the garrison units of the previous force, no operational function was envisaged for military aviation. The main argument for the inclusion of an Air Corps could be seen to be somewhat contradictory:

> The question as to whether our financial resources would permit making the Army more complete and efficient by means of an adequate air service was to a certain extent answered by the actual existence of such a unit containing personnel, plant and machinery and machines. . . . The necessity for the inclusion of an Air Service in the organisation of a modern Army is scarcely necessary to demonstrate. . . . Having regard to our limited finances it is not possible to build up an air force of adequate strength to afford protection against external aggression.[5]

Although it was indicated that it was not necessary to state an ideological case for an air element in a modern army, the main argument in favour of including this air element in the new establishment was simply that such an organisation already existed. Though the case suggests that the inclusion of an air element in a predominantly infantry Army would make it more complete and efficient, it was contended elsewhere that the effectiveness of military aviation in an internal security situation was not significant. In effect the case stated for an Army Air Corps was so lacking in conviction that it provided greater scope for higher authority to exclude, rather than to retain such an element.

It must also be noted that no review had taken place to establish the effectiveness, or otherwise, of the Air Service during the Civil War. It could be argued that such a review could not have taken place, anyway, as General O'Duffy and Colonel Hogan did not have any record available of the operational use of aircraft for the period July 1922 to May 1923 on which to make such a judgement. The principal repository of material relating to armed aerial reconnaissance patrols and other missions during the Civil War, the files of the Commanders-in-Chief (Collins and Mulcahy) – which now constitute the Mulcahy Papers in the archives of UCD – had been commandeered and retained by Mulcahy when he resigned in March 1924. The following instance is indicative of the GHQ's poor appreciation, and inadequate record, of military aviation activity in the early years. In early 1925, while still trying to decide McSweeney's severance pay, GHQ found it necessary to write to the retired General Mulcahy to inquire about the former officer's service as GOC Air Service in 1922. Mulcahy's rather vague and non-committal reply suggests that he had little recollection or appreciation of McSweeney's involvement in military aviation during the Civil War despite the fact that he, himself, had the key records for the period in his possession.[6]

In such circumstances it is not easy to understand how O'Duffy and Hogan, neither of whom were on the staff of GHQ for the full duration of the Civil War or had expertise in air matters, could have made a valid appraisal of the effectiveness or otherwise of military aviation. Appropriate reflection might have indicated to them that Collins had demonstrated considerable faith in the intelligence value of military aviation in an internal security situation. Thus, the appraisal of the effectiveness of military aviation, apparently drafted by Hogan and endorsed by O'Duffy, was most probably based on inadequately informed perception.

The Army Air Corps proposal was not without progressive elements. It recognised that the scheme for pilot recruitment and training instigated in late 1922 had been singularly unsuccessful and, therefore, outlined a scheme for the recruitment of civilians that would become the standard cadet intake system, firstly for the Air Corps and eventually for the Army generally. The case for the future Army Air Corps was summarised in simple terms:

> There is therefore, no alternative but to decide what is the smallest aerial unit which would be sufficient to keep progressive thought stimulated . . . to give our troops a knowledge of the value of aerial co-operation, to train a small number of Infantry as Pilots, and for the purposes of research and watching the progress of other countries. After due consideration it was decided that one squadron consisting of 155 all ranks would meet these requirements. An annual purchase of one or two aeroplanes of the latest design would keep the unit conversant with modern developments.[7]

Despite this forward thinking in relation to recruitment and training, in the context of Army structures, and of command and control, the Army Air Corps was put in a uniquely disadvantageous position. The explanatory notes on the reorganisation scheme had extolled the merits of having the various corps disciplines represented in the GHQ staffs and of having corps staffs in the three command headquarters (Eastern, Southern and Western Commands) and the Curragh Training Camp and, similarly, of having corps units in the brigades in each command.[8] In sharp contrast the Air Corps, while designated as an army corps for the purposes of the 1923 Defence Act, was outside the GHQ/command/brigade chain of command. There were to be no air staff officers in GHQ or at the territorial command level. In addition, the minute organisation proposed could not be dispersed throughout the commands or brigades in the manner of other service corps units. Neither did it have the status as an independent service. The corps's only tenuous connection to the Army chain of command was that it was to be subject to the inspection of the 'first assistant chief staff officer of the Chief of Staff's department'. This was, in effect, the Assistant Chief of Staff, a mainly administrative functionary who would not have possessed aviation expertise.[9] In terms of strength, organisation and structure the proposed Army Air Corps was, therefore, to be a corps in name only. For some years after the implementation of the establishment under Orders No 3, which came into effect on 1 October 1924, the Air Corps was, in effect, a tenant on an inadequately staffed aerodrome in a military camp garrisoned by various infantry detachments in the territorial command of GOC, Eastern Command.[10]

Although it had been proposed that a single squadron of 155 all ranks, acquiring one or two new aircraft a year to keep up to date technically, would fulfil the aviation requirements of the Army there is good reason to believe that the instigators of the scheme did not believe this themselves. In 1925, while making a case for substantial improvements to the October 1924 establishment, Major T. J. Maloney (OC Air Corps) recalled how, in April

1924, he had originally been directed to draw up a scheme of reorganisation based on one squadron:

> I received specific instructions that the new organisation of the Corps was to consist merely of a maintenance party sufficient to keep aircraft and equipment in a serviceable condition while the existence and the future of the Corps were being considered.[11]

At the time Maloney reluctantly recommended the establishment of an organisation comprising about 33 officers, 58 NCOs and 141 privates, a total of 232 all ranks, and remarkably close to the maximum number (242) for the Air Service as recorded in June 1923 when total Army numbers were over 50,000. In General O'Duffy's reorganisation scheme Maloney's proposal was reduced by about one third. Maloney considered that the establishment numbers eventually decided upon were totally inadequate to maintain the aircraft and equipment to a standard commensurate with military efficiency.[12] While O'Duffy purported to establish a viable air unit, Maloney understood otherwise, and considered that the final size and shape of the Air Corps had yet to be decided and that the proposal for an establishment of 155 personnel was, in effect, a care and maintenance organisation.

Not surprisingly there were many organisational and structural inadequacies in the proposed Army Air Corps. The most glaring of these was the fact that it had no provision for communications personnel – no signals officer, wireless operators or even a switchboard operator for the telephone system in Baldonnell. In the explanatory notes the matters of support services at Baldonnell were referred to only in the context of international obligations in respect of civil aviation.[13] Elsewhere in the notes it was recognised that provision for a signal or communications facility was most essential to the Army Air Corps and suggested that the matter was being carefully considered.[14] Similarly, no specific provision was made for such aerodrome staff as would be standard on military aerodromes elsewhere, nor was there provision made for essential elements such as: stores, messes, canteen, security and administration, not to mention transport and meteorology. While such functions may notionally have been included in the 155 all ranks, the said establishment was to prove totally inadequate for a corps headquarters, a flying unit and an aerodrome, garrison and support services of a military aerodrome.

Notwithstanding the contradictory aspects of the case put forth by O'Duffy, fortunately, his AAC proposal was sufficiently coherent to ensure the retention of military aviation, albeit in a rather tenuous condition and in seemingly token numbers. After it was approved by the Executive Council, and before it was put into effect on 1 October 1924, the Army Air Corps establishment, by sleight-of-hand, was amended down to 151 all ranks (83 in the headquarters and in the single squadron of 68).[15] It may be no coincidence that 151 was the Air Service strength recorded in August 1924 having been reduced from the 242 of June 1923 by demobilisation, the mutiny and natural wastage.[16]

At this juncture the newly designated Army Air Corps, with a token establishment and some 22 mainly obsolete aircraft had, at best, aspirations to perform viable aviation functions. During the rest of the 1920s the small air unit was, as Air Corps folklore would suggest, little better than a publicly funded aero club. The totally informal manner in which flying was initiated each morning on the say-so of a junior officer illustrates this point. In a similarly informal manner Colonel C. F. Russell allowed Senator Oliver St John Gogarty to fly military aircraft as and when he was so inclined during the 1925–7 period.[17]

The perilous position regarding the future of the Air Service in 1923–4 is further illustrated by the fact that higher authority had considered that the circumstances relating to the period 1 April 1923 to 31 March 1924 had rendered it unwise to spend monies authorised for the purchase of spares in the financial year. However, notwithstanding these circumstances, and with tentative arrangements made for the purchase and delivery of spares, the AFO sought sanction from Finance:

> For some time the Army Air Service has suffered neglect as regards the maintenance of the necessary air craft [sic] and at a recent meeting of the Council of Defence it was decided that efforts should be made to render this branch more efficient.[18]

Stating that the COS considered that it would not be practicable to obtain delivery of more than half of a consignment costing almost £6,000 before the end of the financial year sanction was sought for 'Air craft spares and fittings up to a sum of £3,000 out of the monies provided for the current year'.[19] On 10 March 1924, the day McSweeney had resigned, Finance replied stating that the Minister had no objections to spending £3,000 out of existing Army funds.[20] Despite receiving authority to spend the money, the spares were not purchased before the end of the financial year. This matter was discussed, subsequently, at an Army finance meeting:

> The army finance officer referred to the fact that authority had been obtained in the last financial year for a sum of £3,000 to be expended on these spares, but owing to the circumstances at Baldonnel at the time at which the authority was secured action could not be taken towards the purchase in the last financial year.[21]

The meeting was informed that Commandant Maloney, who had recently taken charge of the Air Service, had requested that authority be granted for spending – in 1924–5 – £3,000 that had been withheld the previous financial year.[22] Referring to the previous approval the AFO explained that when the time had came to inform the Air Service that sanction had been granted for the purchase of spares it was found that General McSweeney had left the service as a result of the Army mutiny and no Air Service officer had been appointed to replace him. Stating that the 'air force was now in the charge of a responsible officer' the AFO requested that Maloney be given the necessary authority.[23]

The spending of the £3,000 in the 1924–5 financial year was, thus, approved.[24] Following negotiations with the Aircraft Disposal Company Maloney ordered £1931 worth of aircraft and engine spares. On the direction of the Department of Finance 'J. F. Crowley and Partners, Consulting Engineers, 16 Victoria St, London' functioned in the capacity of an agency, with Maloney 'being associated . . . as technical advisor with knowledge of what actually was required'. Given that the firm supposedly had no aviation expertise it is not clear what its exact function was. A cynical observer might conclude that some form of commission was being paid by the Department of Defence to consultants who were not expert in the field.[25] Though £1,000 was paid to the suppliers in advance, to ensure delivery before the August bank holiday weekend, the full consignment was not completed until December 1924.[26] Before the end of the financial year the Department of Finance sanctioned the expenditure of a further £1,118 on airframe and engine spares. Goods to the value of £1,283 were ordered from various companies, and eventually delivered and paid for.[27]

THE ARMY'S AIR CORPS POLICY, 1924–35

In April 1925 Maloney was instructed by a GHQ staff officer to forward a proposal for improvements to the October 1924 establishment. Due to the fact that his previous proposal had been reduced by one third, Maloney forwarded a 'new scheme of Organisation and Establishment' for the Air Corps by return of post. He proposed an increase in personnel, from 151 to 223 all ranks, to provide for a self contained fighting unit capable of co-operating with the infantry and other special services. He also recommended that provision be made for the training of ten cadets as pilots, and for observers and technicians as well as provision for unspecified civil aviation commitments. An increase in the squadron establishment, from the existing 68 all ranks, to a new figure of 139, for three flights with eight aircraft each, was suggested.[28]

Despite being instructed to make the submission Maloney's proposal appears to have been ignored. In the context of the retrenchment in Army numbers being imposed by Finance it is not surprising that an increase of one third in Air Corps numbers was unlikely to be approved at that time. Notwithstanding the rejection of Maloney's April 1925 proposal the General Staff claimed to have effected improvements in policy and organisation:

> During the period under review [1923 to 1927] all endeavours were directed towards perfecting the organisation of the corps and train suitable personnel to fill vacancies in future military and civilian developments. The army crisis of 1924 gave a very serious setback to the development of the [Air] Corps. In addition, the organisation allowed in Orders No 3 was found to be absolutely inadequate. Very little progress was made until 1926.[29]

The above only serves to disguise the antipathy of the Army and Department of Defence to the future of military aviation and the fact that few positive attempts had been made in the period to improve on a care and maintenance organisation. The mention of progress in 1926 was an oblique reference to the fact that in April 1926 seventeen infantry officers had been attached to the corps for a course of flying training. In the context of the review it was suggested that a new organisation 'which had been passed by the organisation board' would 'allow for the efficient organisation of the corps'. The Army's main aims for the Air Corps were as follows:

> To train a sufficient number of flying officers and mechanics to man the proposed peace-time coast defence and army co-operation units. To create a reserve of flying officers and mechanics capable of filling appointments in future civil aviation concerns.[30]

On achieving the above it was proposed to develop other aviation aspects for the benefit of the country. In brief these were the setting up a meteorological service at Baldonnell, the conduct of aerial photography for survey and archaeological purposes and co-operation with the Ministry of Fisheries. The carrying of American mails from Cobh to England and the continent, and the setting up of a passenger service between Dublin and London – in effect the civil aviation policies supported by Collins in 1921–2 – were also foreseen.[31] In reality little if any progress was actually made.

The few instances of interaction with the Army in the 1920s were in exercises in September 1925 in the Curragh area and, in September 1926, during manoeuvres involving the Eastern Command and the Curragh Camp. In the latter exercise a flight of three aircraft from No 1 Squadron operated from the Phoenix Park in support of the Red Army while a second flight supported the Blue Army of the Curragh. The main functions of the pilots and observers were to provide aerial observation of the opposing armies, to take oblique photographs of their dispositions as observed and to keep a complete record of all messages and reconnaissance activities.[32]

Although the General Staff seemed to be proposing an Air Corps primarily capable of coastal defence and army co-operation roles, there was great indecision and confusion as to how to proceed. Early in 1926 the COS reminded C. F. Russell (who was OC Air Corps since June 1925) that the Minister had adopted a three-year expansion programme for military aviation. This was to consist mainly of 'the completion of one complete fighter squadron by the year 1928–9'.[33] In fact six new Bristol Fighters had already been purchased (at a cost of over £15,366), and had been delivered in October and November 1925.[34] It is not clear how the two, quite separate and differing plans were to be reconciled by GHQ. In the end no dichotomy arose, as neither plan was pursued to completion. The fighter squadron did not materialise and the new aircraft, the primary role of which, in RAF service, was army co-operation, were initially used as the advanced training aircraft for the 1926–8 wings' course and, much later, in army co-operation training while coastal defence was, at least for the time being, abandoned.[35]

The consideration of more substantive roles for the Air Corps by the General Staff, and the later dispatch of pilots on courses with the RAF possibly stems from General Hugo McNeill's appreciation of the increasing importance of military aviation in defence generally. During his time in the US with the Military Mission in 1926–7 McNeill had informal discussions with US Army Air Corps officers. He was particularly interested in the range of courses that might be availed of by Irish Air Corps officers. On his return he reported on the benefits of the courses available in the Tactical Flying School. In particular he considered that courses dealing with observation, attack, pursuit and bombardment and with 'cooperation with ground forces and independent air missions would be of particular value to Irish Air Corps officers'.[36] Despite his recommendations no such courses materialised. Nonetheless, as a result of these discussions it is probable that McNeill initiated the adoption of an army co-operation philosophy and influenced the abandonment of the Minister's three-year fighter squadron programme. The abandonment of the fighter squadron option and the informal establishment of an army co-operation squadron, in 1930, would support this theory.

The 18,000 plus establishment of the 1924 Army would be down to 6,545 by 1931–2 though the Air Corps establishment would increase marginally to 160 by 1 December 1928, and to 214 by 1931–2.[37] The initial increase was a number of appointments that were specifically required to facilitate the commissioning of the seven cadets of the 1926–8 wings' class. The later increase, that introduced a workshops' branch in Air Corps Headquarters, may have been in response to the maintenance requirements of the eight Vickers Vespa army co-operation aircraft bought in 1930 and 1931. The purchase of these aircraft confirmed army co-operation to be the Air Corps's main army support role, and the period 1930 to 1935 was to be dominated by training for that task.

In the years 1929–30 and 1930–1 a substantial investment in such aircraft was authorised. The Council of Defence meeting of 4 November 1929 noted that OC Air Corps had made a final recommendation as to the types of aircraft to be purchased:

1. 4 Army Co-operation Vickers Vespa aircraft c/w (Geared Jaguar) engine @ £4,500 = £18,000-0-0.

2. Equipment, wireless, camera, navigation lights, observers' instruments, armament and other service equipment @ £442 per machine = £1,768-0-0.

3. One workshop tool kit Jaguar = £30-0-0.

4. Three Avro Type 621 Training aircraft @ £1,700 = £5,100-0-0.[38]

The meeting approved the expenditure of £24,898 and specified that the seven aircraft should be supplied before 31 March 1930.[39] Though only one aircraft, an Avro 621, was delivered before 31 March a total of £20,905 was expended before that date. Capital expenditure, on aircraft and armament, to a total of £23,957 was incurred during 1930–1. The major part of this was the £19,768 for the purchase of four more Vickers Vespa aircraft.[40]

An important aspect of the increasing emphasis on army co-operation was the participation of two pilots, W. P. Delamere and L. T. Kennelly, as students on an Army Cooperation Course at Old Sarum, Wiltshire, from 5 May 1930 to 25 July 1930.[41] Participation in the Army's combined exercises in the autumn of 1933 was a practical and useful training exercise. A detachment of the 1st Co-operation Squadron was placed under command to the Eastern Command brigade that constituted the Yellow Forces who were based in Phoenix Park from 9 to 25 September 1933. The main emphasis was on the production of oblique and vertical photographs to accompany reconnaissance reports.[42]

Without [a] specific provision for a dedicated squadron establishment the Vickers Vespas were initially operated by 'B' Flight of No 1 Training Squadron of Air Corps Schools, initially within the 160 all ranks limit of 1928, and later within the 214 limit set by the peace establishments 1931–2. By October 1934, an increased Air Corps establishment of 284 all ranks provided for the '1st Co-operation Squadron Cadre' of 51 all ranks, for the operation and maintenance of the Vespa aircraft.[43]

At a time of financial retrenchment, in other ways the GHQ policy for the Air Corps in the period 1929 to 1935 was more enlightened than might have been expected.[44] At a time when pupil intake was very modest the older training aircraft, the Avro 504Ks and DH Moths were replaced with a total of 17 Avro machines (three Avro 621s, four Avro 626s, six Avro 631 Cadets and four Avro 636s). More importantly perhaps, a decision was taken to send students on an RAF flying instructors' course. With no evidence of prior consultation with the Air Corps, GHQ initially indicated that it intended to send two officers, Captain O. A. P. Heron and Lieutenant A. Russell, to the Central Flying School, RAF Wittering, in February 1932. Within days it was stated that the second officer was to be replaced by Lieutenant D. J. McKeown.[45] Early in January 1932 the officer commanding, Major J. J. Liston, was made aware that a further change was being directed by the COS:

> It has been decided to send one officer of the Air Corps to attend the Central Flying School instructors' course, RAF, Wittering, which will commence on 2nd February, 1932 and end on 16th April, 1932. The Officer selected to attend the course is Lieutenant W. Keane.[46]

The context of the decision suggests that the Air Ministry had invited the Irish Army to avail of a student placement on the particular RAF course. In the matter of student selection it is not certain what influences were brought to bear that might have effected changes in the final decision, or what appreciation GHQ might have had of the abilities and potential of individual pilots. After informal consultation with the newly established Office of the Director of Military Aviation in GHQ it is possible that Captain Oscar Heron, an ex-RAF pilot, was not considered acceptable to the Army leadership. Similarly, Lieutenants A. Russell and D. J. McKeown (graduates of the 1922–3 wings' course) may have been considered to have been inadequately qualified, while it may have been decided that all three were simply too old. The eventual selection of Lieutenant W. J. Keane, the senior graduate from the cadet class of 1926–8 could be viewed as an inspired decision. If made with such considerations in mind, it represented faith in the more highly motivated youth of the Corps who had neither RAF nor IRA baggage. After the course Keane reported that he had become a 'B' category flying instructor, the highest qualification available to him:

> I got third place in the examination on ground subjects, qualified as an instrument or 'blind' flying instructor and competed in the final of the aerobatic and inverted flying competition for the Clarkson trophy.[47]

In the seven months after his return Keane ran two instructors' courses and helped a total of 14 pilots to qualify as flying instructors. He subsequently requested authorisation to return to Wittering for re-categorisation.[48] While Major Liston supported his case and requested the appropriate approval, re-categorisation did not take place until June–July 1935 where Keane attended a refresher course and was then graded as an 'A1' category flying instructor.[49]

CHANGES IN COMMAND

W. J. McSweeney was discharged with effect from 29 March 1924, and, thus, Captain (later major) T. J. Maloney was appointed OC Air Corps at the beginning of April. On 24 July 1925, after less than sixteen months in the appointment, however, he was replaced by Colonel C. F. Russell as OC. The reasons and circumstances are not obvious. On or about 25 July Maloney received a brief directive from the COS:

> Colonel Charles F. Russell is appointed officer commanding, Army Air Corps as from this date. You will on receipt of this communication hand over to him all the duties of corps commander. Pending further instructions you will act as squadron commander.[50]

As there was no evidence of dissatisfaction with the Maloney's effectiveness in the appointment Russell may well have been appointed simply due to his own very satisfactory record in various posts. The necessity to perform functions rising from his position in civil aviation may have been another factor – technically he was still Director of Civil Aviation. Early in 1926 Russell was the Department of Defence representative on an 'Interdepartmental committee on civil aviation' where his function was to support the Minister's desire that his Department should control all aviation, civil and military within the Saorstát and to have the Government 'appoint a director of civil aviation, a civilian, answerable to Defence'.[51]

At a time when the Air Corps had no defined function in a peacetime Army and when its future was not assured, Russell was only to serve about two years as officer commanding. His major achievement was the production of a syllabus for the flying training of pupil officers and cadets. Though he was probably under pressure to produce a syllabus in time for the start of the course in the summer of 1926, Russell took his time, and probably took advice from RAF sources before completing an instrument that would set a very satisfactory standard for *ab initio* flying training and establish a very satisfactory basis for future syllabi.[52]

In a manner similar to the termination of Maloney's service as officer commanding Russell's military service ended abruptly and without satisfactory explanation. He is recorded as having been appointed OC 3 Brigade, Cork, with effect from 1 February, and as having reported there on 8 February 1927. He was then appointed to the GHQ Inspection Staff with effect from 25 April 1927, and finally retired on 30 April 1927. It is possible that his removal from the appointment of OC Air Corps and his sudden subsequent retirement stemmed from the fact that he no longer met the medical requirements for military flying.[53]

Commandant J. C. Fitzmaurice took over the duties of OC Air Corps on 7 April 1927. For reasons that are not apparent from the records he had been performing the duties in an acting capacity from 11 October 1926. He went on a general course for senior officers at the Army School of Instruction in October–December 1927 and achieved a mark of 81.7 per cent. Despite this result there is little in the official record to suggest that Fitzmaurice was enthusiastic about the administrative responsibilities of his appointment. For example, in February 1928 he received a missive from the Chief Staff Officer to the General Staff reminding him that a number of files, requiring his attention, had not been returned to GHQ.[54] It has been noted also that before and after the senior officers' course much of his energies were centred on his ambition to achieve the first East–West, non-stop crossing of the North Atlantic – an interest that no doubt distracted him from the more mundane duties of command. In September 1927 he was part of the crew of the Princess Zenia that made an unsuccessful attempt at the Atlantic crossing.[55] In April 1928 he was the second pilot (with Hermann Koehl and Baron Von Huenefeld) on the Junkers W33 (Bremen) that made the first successful crossing of the Atlantic from east to west.[56] Subsequently, he spent much time on leave of absence but was back on duty in time to present commissions to the successful cadets of the 1926–8 wings' class, on 5 November 1928.[57]

With ambitions to capitalise on his fame Colonel Fitzmaurice submitted his application to retire on 29 January 1929, which was accepted with effect from 15 February; he vacated his quarters by 15 March.[58] Records suggest he did not leave the service on the best of terms with higher authority. At the time of resigning he cited the poor state of the Air Corps and the fact that little progress had been made in the previous year. Early in 1929 he is reputed to have submitted a copy of his 1927 report, for 1928, on the basis that so little had changed it made no difference.[59] It did not seem to occur to him that he had spent much of 1928 pursuing his own ambitions and business, and thus had done little to improve the actual state of the Corps. About the same time it was reported that the Air Corps had come to such a point of stagnation that Fitzmaurice had 'informed the Minister for Defence that the Air Corps as then organised was a useless organisation, costing £100,000 to maintain. Its equipment was a collection of junk and its mechanical personnel was [*sic*] inadequate'.[60] With or without the prompting of Fitzmaurice's derogatory comments GHQ was, in fact, actively pursuing a more progressive stance in terms of equipping the Air Corps with Vickers Vespa aircraft for army co-operation functions in support of ground troops.

DIRECTOR OF MILITARY AVIATION

On 15 February 1929, the effective date of Fitzmaurice's retirement, Commandant G. J. Carroll was appointed officer commanding and served in that appointment for a largely unrecorded 33 months at the end of which he appears to have been replaced on a veritable whim. Accounts suggest that Carroll fell into disrepute with GHQ mainly as a result of circumstances highlighted by the fallout following a flying accident at the Curragh on 9 April 1931. It must be noted, however, that the convening of a number of courts of inquiry in 1930, and his subsequent observations on findings, had already placed his judgement in such matters under examination. A memorandum on the subject, presented to a COD meeting on 23 March 1931, found that three courts of inquiry had not been convened in the proper manner with the appropriate personnel and that, as a result of the investigations being conducted by Air Corps officers only, the relevant factors were not thoroughly examined and reported upon.[61] Matters were finally brought to a head as a result of the proceedings and findings of the court of inquiry into the accident on 9 April 1931, and of the particular circumstances in which it had taken place. On 23 June 1931 the COD considered a memorandum that commented upon the proceeding and findings of the court of inquiry. Without examining the precise circumstances of the accident (supposedly Vespa No 4 was destroyed) the memorandum highlighted the fact that OC Air Corps needed no authority other than his own to send an aircraft to the Curragh to give pilots flying practice and to give air experience flights to officers of the Curragh Camp. It was suggested that the QMG might have some unspecified function in the matter.[62] Examination of Air Corps standing orders, nonetheless, indicates that the important function of the granting of authorisation to undertake a flight in an aircraft was not specifically provided for prior to June 1931. As late as 1937, one of the more important aspects of such a function, deciding whether the weather conditions were suitable or not, was still being exercised by the aerodrome duty officer.[63] The inference of the comments on the inquiry into the accident on 9 April 1931 was that the officer commanding informally authorised flights on the basis of an assessment of weather conditions by the aerodrome duty officer. On 29 June 1931, apparently as a result of adverse comments from GHQ on the manner in which the contentious flight had been authorised and administered, a new standing order, providing for the 'Flying Detail', was drafted and issued. The order specified the manner in which the corps commander, squadron commander or chief instructor could detail, in writing, flying for training, service or test purposes as well as the manner in which all flights would be recorded and administered.[64]

The incident of 9 April 1931 was further considered at the COD meeting of 16 November 1931 when it was concluded that none of the officers immediately involved in the accident could be held responsible:

> but in this case . . . it was clear that the administration of the Army Air Corps leaves a great deal to be desired. While the present officer commanding is possibly as suitable for his position as any other officer within the corps it will be necessary to seek outside an officer who will administer the corps. The chief of staff will nominate a suitable officer.[65]

Within a week, on 23 November 1931, Commandant J. J. Liston, an infantry officer who had spent much of his career in GHQ, was appointed OC Air Corps (and promoted to major). Coinciding with this appointment the Peace Establishment of 1931–2 had re-introduced the previously unofficial appellation 'Director of Military Aviation' (DMA) which had before that been attached to the appointment of OC Air Corps. It also established the 'Office of director of military aviation' in GHQ with a single appointment for a captain. The same instrument specified that 'officer commanding, Air Corps' (as well as other corps commanders) would 'act on the general staff as directors when required'. Carroll was appointed to the office of the DMA on 14 November nine days before Liston was made OC Air Corps. The establishment tables specified that the OC Air Corps also 'acts as director of military aviation'.[66] However, in the absence of a regulation or other explanatory instrument, the authority and functions of the DMA, and of the GHQ office, were not obvious. Although Corps directors were, by custom of the service, suitably qualified officers with particular expertise in the techniques and technical nuances of a particular army corps, this was not the case with Liston. There was nothing in his training or experience that fitted him to be OC Air Corps or DMA. He was, thus, unqualified and unsuited except for the fact that the rank and appointment authorised an infantry officer to exercise command over the pilots and other officers of the Air Corps.

While this cleared the way for an inexpert officer to act as OC Air Corps it does not explain how Liston could act as Director of Military Aviation, in a capacity which required appropriate aviation expertise. In any event Liston only held the two titles for less than eight months, from 23 November 1931 to 30 June 1932. On 1 July 1932, in accordance with a minor amendment to the 1931–2 peace establishment, the staff appointment in the office of the DMA was raised to the status of commandant and J. G. Carroll was made director of military aviation while the said appellation was removed from Major J. J. Liston. Carroll was to continue as DMA until 1 April 1935.[67]

Though there is little material on the subject the resolution of contentious flying issues appears to have been the main function of DMA. During May and June 1933 night flying exercises were being conducted under the direction of Captain P. J. Hassett, OC No 1 Squadron. The aircraft were being flown in conditions that were clear of cloud but very dark. Four pilots reported difficulty in performing normal turning manoeuvres while some had entered inadvertent spins. Recovery from such spins, at night and with very poor ambient light, proved difficult and dangerous. The reported incidents highlighted the fact that pilots had not been instructed, and were insufficiently practised, in instrument flying. In addition Bristol Fighters (and other) aircraft were not adequately equipped for instrument or night flying while cockpit and navigation lights were generally inadequate. Landing incidents also indicated that the aerodrome was poorly equipped for night flying, mainly in terms of the lighting of obstructions. Captain P. J. Hassett, OC No 1 Squadron, argued that the aircraft were no less equipped than the RAF aircraft of the period since 1918. He indicated that turn indicators had only recently been installed in RAF aircraft – implying that Air Corps machines were not so equipped. He stated that he had 'never had any difficulty in flying the machine by [the visible] horizon or by occasional ground lights' and that, in the past, 'night flying was carried out under far worse conditions'.[68] The tone of the complaints, and of the Squadron Commander's reply, nonetheless, suggest that night flying exercises were being carried out in meteorological conditions for which the pilots were not adequately prepared and for which the aircraft were inadequately equipped.

Having been given both sides Liston referred the matter to the Director of Military Aviation. Commandant G. J. Carroll replied to the effect that he would discuss the matters raised with OC Air Corps on his next visit to Baldonnell and that, in the meantime, night flying should be suspended. However, apparently before Carroll could discuss the matter with Liston, the School Commandant, who was responsible for the direction of the annual training of pilots, made a number of suggestions that probably pre-empted such discussions. Captain W. P. Delamere's main recommendation was to the effect that night flying should only take place when 'there is a good moon and reasonable visibility, i.e. sufficient to allow a clear horizon in all directions' and, secondly, that if night flying was to take place in very dark conditions that appropriate blind flying instruments and instruction in instrument flying should be provided.[69] In July and August of 1933 night flying continued 'during suitable periods of the moon'.[70] Though the DMA does not appear to have given a formal direction on this particular matter it would seem that the director's main function was to compensate for the Commanding Officer's lack of aviation expertise.

THE CAPTAIN P. J. HASSETT AFFAIR

Captain P. J. (Laddie) Hassett was a former IRA member who had War of Independence service in Clare. He had been a lieutenant in the 30th Infantry Battalion prior to being attached to the Air Corps for pilot training in June 1926. When he qualified as a pilot in 1928 he achieved the highest marks of the officer group on the 1926–8 course. During the first air firing and bombing exercises, held at Kilworth in 1932, he achieved the best score of the 16 pilot officers taking part.[71] Records confirm that he carried out a keen study of aviation, civil and military, and held the second civil pilot's licence issued by the Saorstát Éireann.[72] During the early 1930s, as lieutenant and captain, he was an energetic flight commander and squadron commander in No 1 Training Squadron and later in the 1st Army Co-operation Squadron. In the latter he had a leading part in training for the army co-operation role.[73] In 1931 he contemplated retiring due to lack of promotion and intended going into commercial aviation with Iona, the first civil air carrier in the State.[74] He demanded an interview with the Minister for Defence in regard to securing a retirement gratuity, the granting of which was apparently at the Minister's discretion. The Minister refused point blank to allow him to retire suggesting that it cost the State £5,000 (a gross exaggeration) to train him as a pilot. The interview did, however, achieve a positive result for Hassett. As a former IRA officer with a proven good record he should, according to the Minister, have already been promoted to the rank of captain and was so promoted within a month.

Arising from the robust nature of his dealings with GHQ and the Minister, and the questions regarding his promotion most likely raised by the Minister within GHQ staff, Hassett's name was apparently noted by the headquarters staff. In October 1931 Major J. J. Liston gave an introductory address to the officers at Baldonnell. Accounts suggest that he had Hassett in mind when he declared that he was going to put a stop to officers writing to GHQ. Hassett observed that Liston subsequently continually sought opportunities to take disciplinary action against him.[75]

Notable confrontation was avoided until May 1935 and then arose out of Hassett's command of the Air Corps's participation in a display in the Phoenix Park. With pilots rehearsed and detailed for the display Hassett contended that Liston interfered in the flying arrangements to such an extent that he persuaded a number of pilots to render themselves unavailable on the day of the actual display. While the display was completed with a changed line-up of pilots, Hassett remained convinced that Liston had succeeded in having officers refuse to fly in accordance with the flying detail. Hassett reported the matter to GHQ and tried, but failed, to have it formally investigated on his terms.[76] Later Colonel Mulcahy reopened the matter in the context of the Air Corps investigation of 1941–2:

> Colonel Mulcahy, in his evidence . . . stated 'the younger officers of the corps refused to fly at the Phoenix Park where a public display had been arranged and advertised, and for which they had been detailed and had carried out some weeks' practice'.[77]

The Committee, having access to the original GHQ confidential file on the matter, found that Liston had been party to a discussion amongst a number of younger officers on the night prior to the display. However, it also found that the pilots (four officers and one cadet) had indicated a desire to withdraw from the display on the basis of lack of experience but had made themselves available the following morning only to be replaced by older officers on Hassett's instructions. The 1941–2 Committee found that the matter had been investigated at the time (1935) and that it had been recommended that, owing to the peculiar circumstances surrounding the whole affair, no disciplinary action should be taken.[78]

With no conclusion being made at the time relations between Hassett and Liston apparently remained very cool. However, they were both soon to leave the Air Corps. Liston was replaced by Major P. A. Mulcahy, on 3 June 1935, while Hassett was transferred to the Cavalry Corps within months. On 3 June 1935, the day that Mulcahy was appointed, Hassett ceased to be OC 1st Co-operation Squadron and was posted to the appointment of OC Technical Workshops. On 5 September 1935 he received orders to report to the Director of Cavalry and was posted to that Corps from 9 September 1935.[79] While the precise circumstances of Hassett's departure cannot be easily discerned, matters related to the proceeding of two courts of inquiry held in 1935 and his annual confidential report for the period 1 January 1935 to 9 September 1935 were cited. On 6 January 1936 Hassett was paraded by Mulcahy so that he could be given the details of the confidential report concerning that part of 1935 in which he had served in the Air Corps. Hassett took notes:

> Military conduct and general compliance with regulations; Fair.
> Suitability for present [Air Corps] appointment; Unsuitable.
> If not suitable recommendations for alternative employment; See results of courts of inquiry. This officer has been transferred to the Cavalry Corps.
> Ability, executive and/or administrative; has shown good executive and administrative ability.
> General rating; unsatisfactory.
> Special note on officer of outstanding ability; [*nil*]

Recommendations and remarks; [*nil*]
Date; 6 Jan. 1936 [signed] P. A. Mulcahy, Major. Director of artillery & A/OC
Air Corps.[80]

While the findings of two courts of inquiry were cited as being the reasons for
the unsatisfactory rating this did not afford any explanation to the subject
officer. Despite being a witness at both courts of inquiry he was not party to
either the proceedings or the findings of either and, as an officer who was
allegedly adversely commented upon, had no way of knowing what precisely
had been said. Nor did he have any recourse to an appeal mechanism. In a
vain attempt to have the rating changed, or even have the basis of a bad report
explained, he wrote to an unspecified higher authority in such terms and
demanded the withdrawal of the adverse report.[81]

The purpose of courts of inquiry was to investigate all manner of
accidents, particularly road traffic accidents involving military vehicles, but
this did not specifically include aircraft accidents. Insofar as aircraft accidents
were concerned, a court was usually required to enquire into the circumstances surrounding the accident to aircraft number XX on a stated date at a
stated location. The court of three officers, that took evidence under oath, was
also asked to report on the 'cause of the accident, the extent of the damage
and cost of repair' and to state whether the accident was 'due to negligence on
the part of any person or persons'. The attitude of higher authority, to those in
receipt of flying pay, would suggest that findings under the heading of
negligence were welcome. Witnesses were questioned individually in private
and, until the DFR was amended on 10 April 1937, those who might be
subjected to adverse comment would not be made aware of such evidence.
The court of inquiry was, at best, a quasi-judicial process, the proceeding and
finding of which were used more for disciplinary purposes rather than for
furthering safety. In certain circumstances the findings could, in theory, be
cited to state whatever higher authority wished.[82]

In the case of Captain P. J. Hassett the main court of inquiry cited was
that into a flying accident that occurred on 2 May 1935 and resulted in the
death of a young officer. Hassett was held partially responsible for the
accident. He insisted that the accident had happened after the pilot had
departed from the flying exercise for which he had been detailed, in effect
disobeying a lawful order. The second court of inquiry is understood to have
concerned an aircraft engine damaged while being serviced.

There are uncertain factors that contributed to Captain Hassett's transfer
out of the Air Corps. These reasons would appear to be broader than those
pertaining to the aircraft accident and the court of inquiry. P.J. Hassett, as a
pilot from a distinctly IRA and infantry background, was unique in that he

had embraced the aviation ethos to an extent not matched by his infantry colleagues or, indeed, by some of the Air Corps cadets with whom he had trained. As a flight and squadron commander he had demonstrated considerable enthusiasm for the Air Corps' army co-operation role, for air firing and for improving flying standards. He advocated and encouraged instrument and night flying even though the role of his squadron did not require such disciplines and the aircraft were ill equipped for the purposes.[83] In this regard he was far-sighted, enthusiastic and progressive to an extent that might not have been appreciated by his superiors or indeed by some of his flying colleagues.

The antipathy between Liston and Hassett was probably a reflection of the antipathy between GHQ and the pilot group as a whole. The receipt of flying pay by pilots was always a contentious issue with those not in receipt. Of perhaps greater significance, however, was the fact that Hassett had adopted the culture and ethos of aviation to such an extent that he no longer represented the infantry ethos that GHQ hoped to inculcate in the Air Corps. He may well have been seen as disloyal to his infantry roots. It is more likely, therefore, that he was posted out of the Air Corps because of his allegiance to the aviation culture rather than for perceived responsibility in the matter of the fatal accident of 2 May 1935. This course of action was possible because proceedings of courts of inquiry remained confidential, coupled with the fact that an officer had no redress under the 1923 Defence Act except in the case where the alleged wrong was done by his commanding officer.[84]

The question arises as to why he was not formally charged. Had Hassett's role, and degree of responsibility, been such that it warranted his being permanently removed from his chosen Corps it might have been considered appropriate to formulate formal charges. It may have been simply that, due to the fact that that the 1923 Defence Act did not legislate specifically for any aspect of military aviation, an appropriate charge could not be framed.[85] It is also possible that Hassett, as a pilot from a subculture (of ex-IRA pilot officers), was subject to the prejudices of the other subcultures within a relatively small officer body and was isolated on that basis.

While a transfer to another corps, albeit with adverse reflections on reputation and character, might not seem a very severe punishment, this was not the case here. When Hassett was posted to the Cavalry Corps he lost eight shilling (flying pay) per day for the remainder of his military career – in effect, a summary deduction of over £3,200 – which would have been considered a very substantial sum. Captain P. J. Hassett went on to serve 23 years in the Cavalry Corps and reached the rank of lieutenant colonel. Long after his death in 1959 'Laddie' Hassett was remembered by his corpsmen as having made a valuable contribution, and was considered a fine and loyal officer.[86]

It is important to note that the manner in which Hassett was treated was in sharp contrast to that of another flying officer who had a long and colourful career in the Air Corps and who eventually retired on age grounds in the same rank that he had joined. In his case he was found, amongst other things, to be most irresponsible, undisciplined and generally unfit for his Air Corps appointment. In 1942 it was recommended that he either be transferred to another corps or that his services be dispensed with entirely.[87] In the event the officer, a graduate of the Cadet School, Military College, served the remaining 17 years of his career in the Air Corps. He was from time to time transferred out of the Air Corps only to be attached back; from 7 October 1935 to 18 February 1936, for example. Though such a paper posting did not involve actually moving to another Army unit the resultant loss of flying pay, almost £54 in this instance, served as monetary punishment for whatever misdemeanour he was deemed guilty – without burdening the system with the niceties of due process.[88]

Although the summary manner in which these two officers had been treated may have been interestingly similar it is the dissimilar outcome that is considerably more pertinent. In the first case, an obviously diligent flying officer appears to have been hounded out of the Air Corps in circumstances where justice was not seen to have been done. In the second case, an individual whose competence as an officer and pilot was questionable due to his poor performance and potential and who was, in fact, considered worthy of dismissal, retained his commission in circumstances about which there was no dispute. Based on such evidence, it could be argued that P. J. Hassett was too good for the Air Corps and the second officer not good enough for the Army.

P. A. MULCAHY'S EARLY COMMAND

On 3 June 1935 Major P. A. Mulcahy, a GHQ staff officer and younger brother of General Richard Mulcahy, was appointed Acting OC Air Corps in addition to his position as director of artillery. From the records available the rationale for this General Staff decision is not obvious. Being similarly situated to his immediate predecessor, Mulcahy had no qualifications that might have made him suitable for an air appointment and, much like those who had appointed him, would have had little concept of the nuances of military aviation. Later sources would suggest that the substantive reason for Mulcahy's appointment was related to the perception that Air Corps pilot officers were undisciplined. Under Mulcahy flying accidents were to be treated as disciplinary matters rather than flight safety issues. In addition, the fact that GHQ staff and Liston, and later Mulcahy, could not relate to flying officers in aviation terms suggests a cultural divide that could have been interpreted by the GHQ as indiscipline. It is also possible that understandable friction between the disparate pilot groupings may have been similarly viewed. Whatever the reasons, a number of sources point to Mulcahy having been appointed primarily to impose order. The most obvious was the amendment to Air Corps standing orders issued on his first day in office:

> The Commanding Officer expects from all officers under his command such undeviating support as will ensure the prompt execution of all orders he may deem necessary to issue for the maintenance of discipline in the corps.[89]

That Mulcahy considered it necessary to make an order with such an emphasis on his first day in the appointment strongly suggests that his orders, from GHQ, were to regain and retain a level of discipline that might have been perceived by GHQ not to have existed during Liston's command. A similar impression comes from an unusual source. A brief Air Ministry intelligence summary (in 1940) noted that Colonel P. A. Mulcahy had been transferred to command of the Air Corps to 'tighten up discipline'.[90] The latter opinion may well have originated with Mulcahy himself. From June 1940 Mulcahy had an unusually open relationship with the British air attaché, who would have passed any and all details to London. Mulcahy's disciplinary qualities were later endorsed by a comrade artillery officer who recalled that artillerymen had been 'ever proud to recall having served under "Muller" in his martinet days'.[91]

In contrast the retired colonel C. F. Russell, who edited and published *Aviation* magazine for three years (1935–7), expressed considerable alarm at the appointment of Mulcahy and at the fact that the separate position of Director of Military Aviation, as recently held by a flying officer, had been abolished:

> Army General Headquarters never have, and do not now understand the peculiarities of an air unit in regard either to its technical requirements or the methods of air command. . . . Flying personnel cannot be commanded and will have no real respect for anyone over them who is not an active flying officer. No greater mistake could have been made, therefore, than the appointment of a non-flying officer to command the Air Corps.[92]

Russell, believing that the Minister was unaware of such considerations, suggested that he 'look into the Air Corps organisation and administration, independent of Army General Headquarters which has proved hopelessly ignorant of the problems involved'.[93] If GHQ or the Minister were aware of Russell's observations they clearly ignored the advice offered. As will be shown, however, Russell's words were to prove prophetic.

While records suggest that J. J. Liston was more concerned with admini-strative aspects and generally left flying matters to the squadron commanders, Mulcahy was more proactive and became greatly concerned with air related matters. Not long after being appointed Mulcahy wrote to the COS stating a brief case for the abolition of the cadet scheme of pilot intake that had been used intermittently since 1925 and proposed that pupil pilots should only be recruited from the ranks of newly qualified Cadet School officers or from the ranks of young officers already serving in army units. His sole line of argument was that Air Corps cadets were handicapped by having insufficient basic military training. The proposal was annotated as having been approved by the Minister on 16 October 1935.[94] Without delay Mulcahy submitted a draft defence forces regulation that purported to replace DFR 7/1927, the pilot flying course syllabus. The draft DFR was, in effect, only a brief preamble to the syllabus proper, more a prospectus than a syllabus. He also submitted for the approval of higher authority a newly drafted training syllabus for the young officers' *ab initio* flying training course, based on the earlier DFR that he proposed should be cancelled'. Mulcahy incorrectly stated that the change to an officer-only scheme of intake and flying training required the cancellation of DFR 7/1927 and its replacement by one along the lines of the draft submitted. This blatant misrepresentation of the situation should have been obvious to higher authority. Since 1926 several classes that included both officers and cadets had received *ab initio* training together and had qualified in accordance with the 'Syllabus of training – pupil officers and cadets in the Army Air Corps' that had been specifically drafted to provide for the training of both categories of pupil.[95] Thus a new DFR was not required if recruitment was confined to officers alone.

The Department, if they were aware of the misleading nature of Mulcahy's assertion, did not dispute the point, probably on the basis that the change in DFR would not affect any section other than the Air Corps, and that the commanding officer was presumed to know what was best for the corps. On 21 May 1936 Frank Aiken, Minister for Defence, signed DFR 40/1936. In accordance with paragraph three the 'Young officers' flying training course' was fundamentally a list of the ground school subjects that bore little rela-tionship to the previous, detailed and comprehensive, syllabus. The term 'Airmanship: Flying training' constituted the complete definition of the associated practical flying training.[96] It was, as stated above, a prospectus rather than a syllabus.

There is compelling evidence that the cancellation of DFR 7/1927 and the promulgation of DFR 40/1936 was a selfish contrivance by Mulcahy to facilitate him to undertake flying training and receive flying pay. He commenced flying training three weeks after the signing of DFR 40/1936. Some years later

Captain T. J. Hanley was asked whether the commanding officer's right to be graded as a pilot was governed by DFR 7/1927 or by DFR 40/1936:

> DFR 40/1936, he first received instruction from me on 10 June 1936. His instruction continued until 21/1/'37 and he flew solo on several occasions during this period. After 21/1/'37 he got no further instruction from me and discontinued solo flying. In March 1938, he received 2 hours 20 minutes instruction from Lieutenant McCullagh but was not allowed to go solo.[97]

Hanley also stated that his commanding officer did not undergo the flying and ground school tests prescribed in the young officers' flying course syllabus for the training of new pilots as signed and issued by Mulcahy himself, on 16 October 1936. He also commented that he did not know who had certified that Mulcahy had qualified as a pilot, was engaged in flying duties and entitled to draw flying pay.[98] In this respect it appears that Major Mulcahy provided his own certification to the effect that he had undergone flying training in accordance with DFR 40/1936, while ignoring the fact that the DFR was not the syllabus proper. In August 1936 he received notification that financial sanction had been received from the Department of Finance, apparently in response to the said certification and his application for flying pay. He was to be paid four shillings per day, the pupil pilot rate of flying pay, from 3 June to 6 July 1936 and at the qualified pilot rate of eight shillings a day from 7 July 1936 – despite the fact that he had less than eight hours flying experience and had not yet flown solo.[99] By granting Mulcahy flying pay, the Department of Defence had confirmed him to be a pupil pilot for five weeks and a duly qualified pilot, thereafter. Between 10 June 1936 and 31 March 1939 Mulcahy's flying amounted to a total of 135 hours and 30 minutes, 40 hours and 35 minutes of which was during the financial year 1938–9.[100] Captain Hanley's evidence indicates that all Mulcahy's flying, between March 1938 and 10 January 1941, was as a passenger or accompanied by a properly qualified pilot. It should be noted that no other officer qualified for the receipt of flying pay solely in accordance with the terms of DFR 40/1936. Not surprisingly, the matters of Mulcahy's flying training, receipt of flying pay and his wearing of pilot's wings were to have significant adverse effects on pilot morale during the Emergency.

CONCLUSIONS

While the demobilisation and reorganisation process of the post-Civil War period was somewhat complicated by the mutiny or Army crisis the Departments of Finance and Defence saw no compelling reason for the retention of military aviation. The military, as demonstrated by General Mulcahy and by O'Duffy's scheme of reorganisation, were quite ambivalent about the matter and left the decision ultimately to the Executive Council. Notwithstanding the roles projected for it the new Army Air Corps was but a simulacrum. Major T. J. Maloney's brief on the matter of reorganisation would strongly suggest that the 1924 establishment was temporary, making it little better than a care and maintenance organisation. Temporary in this instance would, in fact, be ten years, as it was to be October 1934 before an Air Corps, consisting of a headquarters and two squadrons, as originally proposed by McSweeney, got appropriate approval.

While the period 1924 to about 1930 was spent in the aviation and organisational doldrums a more enlightened element in GHQ saw the necessity to develop an operational squadron to complement the flying training school. Though the formation of a fighter squadron was initially decided upon in 1926, an army co-operation squadron was eventually informally developed from 1930, and formally established in 1934.

During the years 1922 to 1931 a succession of five flying officers held the appointment of officer commanding, generally for two years or less. It must be noted that there is little or no evidence of the individual influence of these officers in matters that could be construed as constituting policy. Policy, to the extent that it could be said to exist, was reflected in the day-to-day decisions of the Council of Defence and GHQ. The appointment of Major J. J. Liston as officer commanding in 1931, and, in particular the circumstances surrounding that of P. A. Mulcahy in 1935, strongly suggest that the General Staff were not convinced that an air officer could maintain the strict code of discipline expected of all officers. However it is not clear on what precise basis discipline was perceived to be the problem. Formal charging of officers was apparently avoided. As the proceedings and finding of courts of inquiry are retained it is not possible to gauge the extent to which these were used by GHQ as disciplinary processes during the 1930s. Although the practice does not appear to have been very common the posting of an officer to an appointment outside the Air Corps, resulting in the loss of flying pay at the rate of eight shillings per day for an arbitrary period of time, was the preferred punishment for misdemeanours.

The move into army co-operation gave the Air Corps a new focus in the early years of the 1930s while the sending of students on pertinent RAF courses was, no doubt, a very welcome and open-minded departure on the part of GHQ and the Department of Defence. The dispatch of a young second lieutenant on a RAF flying instructors' course in 1932 was a particularly enlightened move though it is not known where the credit for initiating such action should lie. The adoption of an army co-operation philosophy brought to the fore Captain P. J. Hassett, an ex-IRA, ex-infantry officer who seems to have embraced the air ethos to a greater extent than the Army leadership might have expected, or been able to comprehend. His departure from the Air Corps in somewhat obscure circumstances does not reflect well on the systems of military justice and discipline that pertained during the 1930s. His banishment was probably facilitated by tensions among the various subcultures in the pilot body of the Corps. Accounts suggest that the appointment of P. A. Mulcahy as OC Air Corps after Liston was done solely with a view to imposing military discipline. His subsequent qualification as a pilot suggests a deliberate manipulation of regulations to his own financial advantage. As the Investigation Committee would later find, Mulcahy, by virtue of his unorthodox qualification as a pilot and his receipt of flying pay, caused great discontent among duly qualified pilots – discontent that would ultimately help to undermine his authority.

SIX

PILOT INTAKE
1922–45

—

Collins had been greatly reluctant to recruit the additional pilots so obviously needed by the Air Service, notwithstanding the fact that it had only two pilots during the early stages of the Civil War and that one of these was primarily involved in various aspects of setting up the service, the purchase of aircraft and administrative matters generally. This reluctance can be explained by the sensitivity about the recruitment of the only Irish pilots available who, by definition, were ex-British servicemen generally without pre-Truce IRA service. Though the number of pilots, on Michael Collins's authority, had been brought up to ten by 4 December 1922 this was done against the background of anti-British sentiment that pervaded the National Army. From the earliest weeks of the Civil War there was evidence of differences in culture and ethos between the pre-Truce IRA that involved the command echelons of the Army, and the new and hastily established Military Air Service. This mutual antipathy, that would in time significantly influence GHQ's perception of military aviation, was even more evident at a local level. The first manifestations of this appeared at an early stage in Baldonnell with the ideological and physical separation of infantry and air personnel literally into separate camps. This separation was highlighted by the duplication of military functions and of the standard institutions of a military post. Cultural differences were probably frequently highlighted but seldom recorded. A notable exception to this was a substantial subtext to the mutiny of 1924 as recorded in the inquiry of the time and in subsequent studies.[1] J. C. Fitzmaurice, one of the ex-RAF group of officers engaged by General McSweeney, provides a slight flavour of the atmosphere that existed in Baldonnell in the early months:

> . . . and we Irishmen who had held His Majesty's commissions were treated with great distrust by the politicians and the majority of the old IRA officers . . . The then director of military intelligence debased his office by arranging the appointments of subordinate officers on non-technical ground duties at our headquarters to carry out a campaign of snoopery and witch-hunting of a most loathsome kind.[2]

Fitzmaurice was singularly outspoken in respect of the relationship between the ex-RAF officers and the old IRA and was particularly scathing in his assessment of the character of the infantry officers at Baldonnell, individually and collectively, in the early autumn of 1922. He suggested that 'an air of hedonism prevailed the atmosphere' amongst a group of officers who 'apparently had distinguished careers as guerrilla fighters against the Black and Tans' and who 'bore exalted ranks that they had apparently conferred upon themselves', and regarded their appointments in the National Army 'as a form of life pension'.[3] In being so critical of IRA officers Fitzmaurice reflected a superior attitude on his own part. Fitzmaurice, who was born in Dublin and raised in Portlaoise had, on joining the Air Service after serving in the British Army and the RAF, retained few signs of his Irish upbringing. His account of a forced landing in Cork in late 1922 is telling. While awaiting assistance from his base he had difficulty in assuming his 'best Irish accent' in trying to placate the 'more valiant yokels' who had surrounded his aircraft. His accent had become 'impaired' by his service in the British forces.[4] It seems likely that in addition to his accent his outspoken and condescending attitude only served to intensify the opposition and ire of those of solid nationalist background. The official attitude to ex-British personnel was reflected in a report of August 1922 that tended to minimise the problem:

> The commander in chief reported that only a limited number of ex-officers of the British Army had been taken into the Army, that they were Irishmen and that they were employed mainly in instructing capacities and in some cases in an assisting capacity.[5]

Against this background the logical early expansion of the air operation during the Civil War was delayed. Pilot appointments were not advertised in the national press but were made known by word of mouth to attract some Irish ex-RAF pilots. Eventually a total of 13 such pilots had been commissioned into the Air Service, though a maximum of 11 served at any one time.[6] The fact that all were Irish by birth or descent appears to have made little if any difference to the attitude of the more republican echelons of the infantry Army, despite the fact that many former members of the British forces were still serving elsewhere as officers in the Army in the early years. No doubt a concentration of officers of a different military culture and background, engaged in a military discipline with which a former guerrilla army could not identify, was easily identified as a suspect group.

OFFICERS AND OTHER RANKS, 1922–3

An examination of the process that followed the initial intake of qualified pilots – that is, the *ad hoc* recruitment and training process of 1922–3 and subsequent courses – indicates that GHQ used its authority to promote a policy that extolled the merits of infantry culture, while endeavouring to subjugate what was perceived as an alien one. On 20 December 1922, while the Civil War was still raging and many months prior to the formal confirmation of existing appointments in an informally organised Air Service, GHQ issued its first written edict on aviation matters. This advertised the fact that there was a limited number of vacancies for pupils in the Aviation Department of the Army and invited applications from officers, between the ages of 18 and 23, indicating that candidates would undergo an exacting medical examination. Applicants were reminded that no rank above lieutenant would be granted. Applications, with the recommendation of the command GOC, were required to reach the Adjutant General not later than 31 December 1922.[7]

As already mentioned, it should be noted that the Air Service of mid-December 1922 had no defined status in the military scheme of things. The ten pilots then in service were in effect employed on a contract basis having been 'admitted on approval and if not satisfactory' would have been 'dispensed with at once'.[8] The informal organisation had the No 1 Squadron divided into two flights. The operational aircraft, the Bristol Fighters and the Martinsyde F.4 Buzzards, were being operated by 'B' Flight whose main focus was the air operations, with five pilots and five aircraft, from bases at Fermoy and Tralee. That left only five pilots, including McSweeney at Baldonnel, where 'A' Flight, using the Avro 504K training aircraft, had begun pilot training pupil pilots on an informal basis in October 1922.[9]

No records survive to illustrate the nature of the order directing McSweeney and the Air Service to undertake pilot training, or the parameters within which such a flying course was to be constructed. There is little doubt that the 'A' Flight of the single squadron Air Service of late 1922 lacked the basic prerequisites for a training task. The hastily established and rudimentary air element clearly lacked any tradition in flying training. It had no structures or adequate staff to undertake instructional duties. Without question, however, the most basic deficiency was that of a flying school with a syllabus appropriate to an *ab initio* flying course. In this regard the most that could have been available to Commandant J. J. Flynn, OC 'A' Flight, was a schedule of exercises or flights gleaned from his own experience that he considered should be completed by each student.

An examination of a manuscript record of pupil pilots (compiled about 1961) suggests that the Adjutant General was not inundated with applications from officers anxious to become pilots. While no contemporary listing of this intake has been found it appears that the students reported, and commenced training, at various dates between October 1922 and the end of July 1923.[10] No ground school subjects, to support the flying programme, were taught. Of the 14 pupils who apparently commenced training under the scheme at least five were non-commissioned personnel, who were designated as 'flight cadets' during training, while six others, holding commissioned rank, came from various Army units. The balance was made up of three Air Service lieutenants, Tom Nolan, McSweeney's first observer, and two former Air Service NCOs who had supposedly been granted commissions by General McSweeney in order to qualify to undergo the course.[11]

The flying training of Lieutenant Timothy J. Nevin was probably a typical case in point. He commenced training on 18 June 1923 and flew some 20 instructional flights with various instructors before completing his first solo flight on 12 August 1923 after eight hours and 55 minutes of flying time.[12] Flight Cadet Daniel J. McKeown commenced flying in mid-July 1923 but did not fly solo until 16 December 1923, after almost 13 hours of dual instruction.[13] If a formal schedule of flying exercises was followed no such pattern is evident in pilot or aircraft log books. The above two pupils graduated to the service aircraft, the Bristol Fighter, by about 40 hours flying.

It is not obvious at what stage a pilot was considered qualified. In Nevin's case it appears that he succeeded in getting his pilot's certificate and wings in December 1923 by which time he had between 30 and 40-hours flying experience.[14] However, he died on 24 January 1924, as a result of injuries received in an accident in Bristol Fighter BF II the previous day, shortly after commencing training on the aircraft type.[15] In McKeown's case no end to the course of instruction is indicated in his pilot's log book. In effect, it is not possible to chart the progress or qualification of those who trained in 1922–4. A total of six appear to have qualified as pilots while three others were subsequently employed as observers or navigators.[16] Only four of the 14 remained in service after the demobilisation and reorganisation processes of 1924. The aggregate Air Corps service of the six successful pilots was about 60 years.

Although the successful pilots and observers of the first intake apparently did no ground school subjects this deficiency was eventually put right. Some of these officers, along with some of the original (untrained) observers, who remained in service after the reorganisation of 1924, completed ground school examinations in 1925.[17] Another small number took ground school subjects and examinations in conjunction with the cadet and officer intake of 1926–8.[18] No pilot from this intake achieved a rank above that of captain in the Air Corps while three of the pupils were killed in flying accidents during training or subsequently. Given the small number who qualified, the seemingly low standard achieved and the brevity of subsequent service this intake must be considered a failure. The results serve to show that the decision of the General Staff to have pilots trained in such adverse circumstances during a civil war was perhaps misguided. During the reorganisation process the first intake of pupil pilots was the subject of adverse comment:

> I am informed that although every effort was made during the past 12 months to secure the right type of prospective pilot no satisfactory results were obtained, and a considerable amount of time and instruction were expended without any corresponding return.[19]

This was somewhat disingenuous. The original advertisement implied that officer volunteers would be interviewed and selected by the Director of Aviation. In the event, it appears that no educational standard was specified and there was no selection process as such. Due to a lack of interest among officers applications appear to have been accepted from personnel of all ranks. As a result, the first pupil pilots seem to have been volunteers. The recruitment process and training regime that followed indicates a certain naivety on the part of GHQ in regard to the prerequisites of pilot training. It will be noted that C. F. Russell, who would have had a constructive influence on the concept of undertaking pilot training at such an inopportune time, had been posted to the Railway Protection and Maintenance Corps in mid-September 1922. The precipitous action of GHQ in ordering the training of new pilots so early in the formation of the air arm was clearly aimed at having the ex-RAF pilots train their own replacements. Had this first flying course borne the fruit expected by GHQ there is little doubt that most, if not all, ex-RAF pilots would have been demobilised as quickly as possible after the Civil War.

OFFICERS AND CADETS, 1926–8

With the formal establishment of the Army being put into effect in October 1924 the leadership had to come to terms with an Army Air Corps that owed its existence to Collins's Civil War intelligence requirements rather than to any ideologically based concept. In view of the failure to obtain a satisfactory return from the first pupil intake, General O'Duffy, in his scheme of reorganisation of 1924, proposed a cadet scheme for the Air Corps. This scheme would evolve to become the cadetship intake system for the Army generally:

> While suitable candidates from the Army should get preference, it is deemed advisable to secure as far as possible candidates direct from school between 18 and

21, and possessing the following qualities – physical fitness, courage, keenness, decision, self reliance and intelligence.[20]

In a somewhat contradictory manner the same explanatory note stated that there was no alternative but to train a small number of infantry as pilots.[21] Towards the end of 1924 the Department of Finance was made aware of a proposal to enrol cadets for training as pilots in the Air Corps:

> The necessity for the employment of additional personnel in the Army Air Corps has been established and the proposal that suitable candidates for the Air Service [*sic*] might be obtained by the engagement of a limited number of suitable unmarried young men for training has received the assent of the Minister for Defence.[22]

It was proposed that a number of young men would undergo an exhaustive course of training and that commissions would only be granted to those cadets who qualified in every respect as flying officers and showed other required qualities during training. The proposal forwarded to Finance for sanction indicated that, while the Air Corps only had vacancies for four officers at that time, authority was being requested for an establishment of ten cadets at a rate of pay of six shillings and six pence per day. Although Finance gave approval in principle for the cadet scheme in December 1924, details of selection procedures and conditions of service were requested. In reply the Department of Defence explained several conditions that would be attached to the proposed competition. The fact that the Defence Forces (Temporary Provisions) Act 1923 made no provision for the rank of cadet was circumvented by the decision to consider cadets as Class III privates with pay of two shillings and six pence per week, plus the four shillings per day flying pay due to NCOs or soldiers of the Air Service undergoing instruction. It was also suggested that the men selected would be issued with officers' uniforms, without rank markings but 'have the privileges of an Officers' Mess'. To ensure a proper educational standard it was recommended that the men chosen should undergo a special examination by the Civil Service Commissioners.[23] On 18 May 1925 the Department of Finance approved the revised scheme 'regarding the employment of a certain number of cadets in the Army Air Service [*sic*]'.[24] In keeping with O'Duffy's original recommendation that suitable candidates from the Army should get preference in cadetship competitions, the upper age limit was extended by one year for candidates who had at least six months service in *Óglaigh na hÉireann*, the Irish Volunteers, *Fianna Éireann* or in the National Forces and by two years in the case of any candidate who had given 12 months such service and was still serving in the Defence Forces on 1 January 1926.[25]

The notice advertising 'Cadetships in the Army Air Corps' was carried in the country's main newspapers in September/October 1925, with 24 October 1925 as the closing date for applications. In addition to carrying the DOD advertisement, some papers included a news item drawing attention to the new career opportunity for the young men of the country. The *Limerick Leader*, under the headline 'Saorstát Army Air Corps – cadetships' reprinted the Department's substantial information sheet that accompanied the application form. This document presented a detailed description of the career 'in the new art of aviation' on offer and also gave the major subject headings extracted from the syllabus of training that was being drafted by Russell. Considerable detail of the course of military, ground and flying training that awaited the successful applicants was included with the ground school subjects being given particular mention. This, and other newspapers, portrayed a career that must have appeared very attractive to the youth of the country then ravaged by unemployment.[26]

Another newspaper, under the heading 'encouraging civil aviation', put a different slant on the decision to recruit cadets for training as flying officers:

> In pursuance of its policy of encouraging the development of civil aviation in the Irish Free State the Ministry for Defence gives notice of a number of vacancies in the Army Air Service [*sic*]. Although the cadets are to be trained as flying officers with the Army Air Corps it is understood that the civil aspect of their training will take precedence of the purely military side of aviation.[27]

This connotation does not appear to have reflected the intentions of the General Staff and the Department of Defence. While the new State may still have been anxious to encourage civil aviation, the public records of the time do not suggest a civil aviation aspect to this first formal intake of pupil pilots.

Following the advertising of the cadetships some 773 copies of the application form and the supporting documentation were distributed. Only 140 completed application forms were received by Defence by the closing date.[28] As early as March 1925 the COS had indicated that the 'officer commanding, Army Air Service [*sic*] would only act as a technical advisor' to the selection committee.[29] The selection board that consisted of four colonels (including C. F. Russell) reported to the COS on 26 January 1926. The board was apparently less than impressed with the quality of many of the 140 applicants. In particular they found fault with the more menial backgrounds of some of the applicants. They also observed that 'candidates graded entirely unsuitable were those whose utter incapacity was quite apparent such as half simpletons, out-of-work, and those whose character was obviously of the lowest'. Some 45 applicants were declared ineligible or had withdrawn their applications before

23 December 1925 while 48 of the remaining 95 subsequently withdrew or failed to turn up for interview. With a total of 29 candidates being rejected by the selection and medical boards only 18 were summoned to take the Civil Service Commission examination. This included some serving personnel. In addition, many officers who were well above the age limit, had made known, to GHQ, their desire to become pilots.[30] Only nine of the 16 who took the Civil Service examination were deemed to have achieved a pass mark and, thus, suitable for the award of a cadetship. The selection board attributed the poor quality of the applicants to the effects of unemployment and the relatively good pay for a cadet, as well as the fact that there was no pension scheme to attract the type of man the Army needed.[31]

Though the cadet selection had been completed before 31 January 1926 it was to be June before the flying course got under way. This delay was apparently caused by the Machiavellian actions of the Army leadership in promoting the eligibility of serving officers who did not meet the age requirements for the cadetship competition. Delaying the enrolment of the cadets, Defence made further representations to Finance in late February:

> In connection with the recruitment of suitable persons as pilots and observers in the Army Air Corps, I am informed by the Chief of Staff that the result of the recent examinations for cadets for *this service has not quite provided the most suitable type of man.* He mentioned that even in addition to the cadets to be selected for the Air Corps, applications have been received from young infantry officers who were desirous of training as pilots and observers. Generally speaking this type of officer would be under twenty-five years of age, and would only be accepted for training after very careful scrutiny into the *bona fides* of the application, and after a thorough medical examination. The advantages of having officers within the ranks of the Army trained in pilots' and observers' duties are obvious . . .[32]

In considering the latest proposal, particularly the italicised passage, Finance consulted with the Civil Service Commission and observed that two or three candidates who were selected for the Commission's examination could be identified as already serving. The most telling observation was to the effect that the reduction in the number of eligible candidates was as a result of 'thorough sifting', by the Army, of some 90 applicants. The inference was that the Army had an ulterior motive in reducing the number of suitable cadetship candidates.[33]

In supporting the case for over age officers neither the COS nor the AFO mentioned the fact that, the interest of a significant number of infantry officers in becoming pilots, would have been as a direct result of the attractive pay conditions. The flying pay of a pupil pilot represented a 50 per cent increase in pay, while a successful pupil would be rewarded a pay increase of almost 90 per cent. The more attractive pay considerations (flying pay at four shillings and eight shillings per day) had not applied back in 1922 when pupil pilots had first been sought.[34] It might, therefore, be construed that the cadet-ship applicants had been subjected to a cull in order to denigrate the standard and to facilitate and promote the candidature of mature officers with a view to having them declared eligible to undergo a flying course. In response to the Finance query as to whether or not infantry officers would be exempted the educational test that applied to cadets, the AFO replied:

> It may be stated that it is proposed that the transfer of such infantry officers to the Air Corps is to be of a temporary nature, while permanent transfer, not subject to educational test, will be conditional on the necessary technical qualifications being gained during the training course, as confirmed by a practical test.[35]

This somewhat obscure reply confirmed that officers would not have to meet the cadetship educational standard and would only be attached to the Air Corps during training. Those who passed the prescribed flying and ground school tests would be, on qualification, posted into officer vacancies in the Air Corps. GHQ made certain not to allude to the fact that all successful officers would have Army seniority over all successful cadets, thus ensuring that the future leadership of the Air Corps would be in the hands of officers with acceptable IRA service and of an infantry ethos. Even more significant to note was the fact that the officer group, while in training would not be subject to the cultural influence of the ex-RAF officers, as would be the case with young and impressionable cadets.

It was 12 November 1926 before the Department of Finance gave approval for the flying training of ten officers at any one time but stated, however, that the original Defence proposal of 24 February 1926 did not clarify 'whether it was intended to retain such trainees when qualified for the purpose of filling any vacancies on its establishment' and requested further information on the point.[36] In replying the Department of Defence prevaricated somewhat:

> I have to inform you . . . that it is hoped some of the Officers now under training will prove sufficiently suitable to warrant their retention as [Air] Corps Officers. It is, however, premature to undertake a final selection, but it is expected that the preliminary tests will be completed in about two months' time, when you will be informed of the result.[37]

In the meantime the nine cadets reported to the Army School of Instruction on 12 April 1926. After the basic military training aspect of their course they reported to Baldonnell on 27 June for flying. No less than 17 officers, apparently volunteers, were already attached to the Air Corps for instructional purposes, having reported to Baldonnell in late May.[38] These officers varied in age from about 23 to at least 27.[39] They held the ranks of lieutenant, captain and commandant.[40] In effect, while the Air Corps originally had vacancies for four flying officers in late 1924 when the cadet scheme was first proposed, the flying course began in June 1926 with 26 students. This was at a time when the Air Corps officer establishment provided for a total of 21 appointments, only six of which were vacant in June 1926.[41]

The 17 officers had been attached to the Air Corps despite the fact that the Department of Finance was not to grant the appropriate sanction until November 1926. On 1 June 1926 Colonel Russell approved the results of an assessment test, in Algebra, Geometry and English, undertaken by 16 of the 17 officers. While a pass mark of 35 per cent had been laid down, 12 officers achieved marks between 40 and 85 per cent. The other four were granted a 'Pass' mark though numerical values were not recorded. The latter four officers plus another who had achieved 75 per cent in the assessment test are recorded as having been returned to their original units within a few weeks of the start of the course. Twelve officers, including one who had not taken the assessment test at all, commenced the course proper.[42]

Mindful of the fact that the first flying course had been a failure, both GHQ and the Air Corps were anxious that the flying and ground school syllabus to follow was appropriate to producing successfully qualified military pilots. GHQ, whose staff had no qualifications or expertise in any aspect of aviation, had delegated the drafting of this syllabus to Russell as OC Air Corps. However, as the drafting of the syllabus for the pilot training course had not been completed on time, the course commenced and proceeded on the basis of a current draft. Russell completed his work on the document in October 1926. He wrote to GHQ enclosing a syllabus of training for pupil officers and cadets to be completed in two years. He indicated a certain degree of frustration arising out of his dealings with GHQ in relation to the drafting of the syllabus. He suggested that his qualifications and expertise as a pilot were being questioned by those with no knowledge of aviation matters.[43]

On 18 March 1927 DFR 7 (of 1927) providing for the 'Syllabus of training, pupil officers and cadets in the Army Air Corps' was duly signed by the Minister. This regulation, a detailed and comprehensive syllabus, was the first of only two regulations specific to the Air Corps issued by the Department of Defence. It laid down a two-year course divided into elementary and advanced stages. The 16 ground subjects were subdivided into appropriate areas of instruction and study. The marking scheme for ground school examinations was specified. A full range of tests applying to the flying of both training and service aircraft was laid down setting the standards that were to be met before a pupil officer or cadet was deemed qualified to wear the flying badge or wings.[44]

In the meanwhile the course progressed. Subsequently just seven of the 12 officers achieved the qualifying standard in the written examinations. One of these did not pass the requisite practical tests in flying and was returned to his Army unit. Only six, or about 35 per cent, of the officers duly qualified. Of the cadet group of nine, six qualified as pilots. A seventh cadet qualified as an observer in accordance with an amended syllabus having been 'found too small of stature to carry out effectively the duties of a pilot'– this was due to the fact that he could not reach the rudder pedals! One cadet had been discharged on medical grounds earlier in the course. The ninth cadet had been discharged after he was found guilty of an offence in the Dublin District Court. It was recorded that he was being 'dispensed with as a result of a court prosecution for "cheat and fraud in obtaining admission to the Army Air Corps"'.[45] Apocryphal comment suggests that another man had taken the Civil Service Commission examination on his behalf.

Comparisons between the two groups of successful students reveal interesting results. In the ground school examinations, with a pass mark of 50 per cent, the officers averaged 52.4 per cent while the cadets achieved an average of almost 64 per cent. The best mark achieved by an officer was only marginally better than the worst cadet (1,005 against 1,004, out of a possible 1,640) indicating a significant difference between the two groups, at least in terms of ground school. While the tests in flying were on a pass or fail basis, historically those who did better in ground school subjects proved the better pilots, suggesting that the cadets of 1926–8, individually and collectively, graduated as the superior pilots.[46]

Notwithstanding their better performance the cadet were, by definition, junior to their officer colleagues and would remain so for the remainder of their careers. One of the successful officers of the 1926–8 flying course subsequently commented on the course:

> In accordance with a policy of changing the atmosphere at Baldonnel it was decided to transfer in young officers of IRA service. In 1926, under this scheme 17 officers, of which I was one, were transferred to the Air Corps. The ex-RAF personnel made it difficult for us but despite this, after the two years prescribed course 12–14 qualified as pilots.[47]

Although the recollection of the number of officers who qualified is wide of the mark, the comment seems to confirm that the fundamental reason for training serving officers was to neutralise the influence of the ex-RAF element in the Corps.

The successful officers, previously only attached to the Air Corps, were posted into appropriate appointments on reaching the pilot-qualifying standard in June 1928. In September 1928 Defence made the Department of Finance aware of the outcome of the cadet course that had also finished in June. They pointed to the fact that the seven cadets were due to become commissioned officers but that this could not happen as only two vacancies were then available. The Secretary, on behalf of the Minister, indicated that in a recently proposed revision of the Air Corps establishment there would be vacancies for an additional eight second lieutenants and recommended that 'financial sanction should be accorded for the appointment of these cadets to commissioned rank in anticipation of sanction of the revised scheme of organisation'.[48] In this respect we find the GHQ/Department of Defence being less than honest about his motives with the Department of Finance. The necessity to commission seven cadets as officers, something they were obliged to do, was used to press the case for an increase in the establishment. This increase was made necessary by the fact that six Army officers, now qualified as pilots, had already been absorbed into the organisation, in effect, filling the appointments for which the cadets had originally been recruited and trained. Finally, an increase of a total of 13 pilot officers was negotiated, approved by Finance and put into effect on 1 December 1928.[49] The cadets were eventually commissioned on 5 November 1928.[50] From an Army point of view the campaign had been a success. GHQ had succeeded in pulling the wool over the eyes of Finance and had trained six infantry officers, of an acceptable IRA background, and posted them into the Air Corps with the appropriate seniority to further the infantry ethos.

CAPTAIN M. J. O'BRIEN

With the commissioning of the successful cadets of the 1926–8 recruitment for training ceased for a number of years, although a small number of pilot officer appointments remained unfilled. From about 1933 proposals for the establishment of an army co-operation squadron created a specific requirement for pilots though the training of a single pupil had been initiated earlier. Captain M. J. O'Brien had transferred to the Air Corps and had functioned as an observer from 12 March 1929. On 23 February 1931, presumably on the authority of Major J. J. Liston, he commenced training as a pupil pilot. On 28 April 1933 he was certified as having successfully passed the pilot's flying tests specified by DFR 7/1927.[51] In May 1933 it was reported to Finance that, while he had completed the flying tests, as the only pupil officer under instruction, difficulties had been encountered in making systematic progress with his ground instruction. The Department of Defence requested sanction to extend the course beyond the two-year programme specified by the regulation and to continue to pay him at the rate of flying pay appropriate to a pupil pilot. Finance approved the extension and the continuation of flying pay at the pupil pilot rate of four shillings per day, from 24 February until 27 June 1933 when O'Brien was due to complete his ground school subjects.[52]

OFFICER AND CADET CLASS, 1934–5

The first formally organised *ab initio* flying course since that of 1926–8 commenced in January 1934, with eight young officers and a single cadet as the pupil pilots. This occurred in the context of proposals being made for the establishment of an Army Co-operation Squadron of cadre strength and status within the approved strength of the Army:

> The existing establishment provides for the Corps Headquarters, the office of the Director of Military Aviation and the Air Corps Schools but no provision is made for a tactical unit capable of co-operation in the field with other arms of the forces.[53]

The Department of Defence stated that an increase in the number of flying officers in the Air Corps could not be affected except by recruiting cadets for training as pilots. Sanction was requested for the recruitment, with the assistance of the Civil Service Commission, of ten Air Corps cadets.[54] In view of the manner in which the Army had manipulated the cadet and officer intake of 1926 it is of note that the Department of Defence again cited the cadetship method as the only viable one for pilot intake and training. In considering the matter the Department of Finance noted that the recruitment of ten Army cadets in the financial year had already been approved and that, while the 'Provisional war establishment' then in course of compilation, included provision for an Army Co-operation Squadron, no authority existed in the current Peace Establishment for such a flying unit. The additional cost of £964 was also seen as a difficulty.[55] Another Finance opinion suggested that 'the appointment of Air Corps cadets would, I think, be more useful to the Army than the piling up of additional numbers of infantry cadets'. It was also suggested that the additional funding, required for flying personnel, would not be available in the current financial years and that in any event a sufficiently strong case for such a tactical unit had not been stated by the Department of Defence. Notwithstanding these adverse comments the Department of Finance apparently did approve an intake of ten cadets for the Air Corps in addition to at least 12 allowed for the infantry.[56] In August 1933 a cadetship competition was held and 39 candidates sat the Civil Service Commission Examination, however, only nine reached the required standard. Of these, seven failed the medical examination and one failed to impress an interview board. The single successful candidate, Cadet Malachi Higgins, was to complete his Cadet School training during 1934 and commence flying training in 1935, with the succeeding class.

At this juncture Defence appears to have adopted two schemes of pilot intake at the same time. As a result of only getting one cadet the Army was authorised to try to make up a class of six pupil pilots from whatever source. They started by offering Air Corps cadetships to infantry cadets already in training in the Military College. Six cadets were found suitable but five failed the medical examination. Cadet Lorcan J. Byrne was awarded an Air Corps cadetship on 5 November 1933. However he did not subsequently train as a pilot and was commissioned with the 6th Army Cadet Class (1932–4).[57] Cadet D. K. Johnston, who had been an infantry cadet since 14 November 1932, had his application to transfer to the Air Corps approved in November 1933 and was the single cadet with the Officer Class of 1934–5.[58]

Also selected for this course was Lieutenant Andrew C. Woods, whose commissioning 'in pursuance of the provisions of Sections 10 and 19 of the Defence Forces (Temporary Provisions) Acts 1923–1933' was first proposed by the Minister for Defence, in March 1933. This appointment was pursued by the Department of Defence despite the objections of the Department of Finance whose concurrence, on financial grounds, was required before such an appointment could be made. The Department of Finance pointed out that while the government had authority to appoint officers it had become the standard practice to hold open competitions in accordance with Civil Service Commission regulations. They also raised objections to the effect that both educational and medical standards were possibly being ignored. They also stated that the Minister's proposal seemed 'to be against general public policy' and was 'really a personal exercise of patronage by the Minister for Defence' Frank Aiken.[59] Further strongly worded objections by the Department of Finance did not prevent the matter being placed on the agenda for the Executive Council meeting of 22 May 1933, where the commissioning was approved. It was published in *Iris Oifigiúil*, on 26 May.[60] Second Lieutenant A. C. Woods was posted to the Air Corps on 2 October 1933.[61] Joseph Kearney was granted a commission in similar circumstances though without objections from Finance. The Minister for Defence recommended him to the President of the Executive Council as being suitable on the grounds that he had rendered 'good national service', was a 'native Irish speaker' and a 'qualified air pilot'.[62] The Executive Council discussed and approved the appointment at the meeting of 16 November 1934. He was to be a student on the 1935–6 wings' course.

About December 1933 the Department of Defence had informed the Department of Finance that the Minister for Defence had 'under consideration the question of a scheme for the training of officers as pilots for co-operation squadrons in the Air Force [*sic*] on similar lines to that recently adopted in the British Air Service as the only way to create a reserve of co-operation pilots':

> Owing to the fact that the training of this type of pilot is particularly difficult and that pupils must possess an intimate knowledge of military matters, it is not considered feasible to create a reserve of co-operation pilots from volunteer or civilian sources. It is accordingly proposed to second the Air Corps certain officers specially selected from other units, who would undergo training in flying duties for 12 months, after which period they would, if successful in their training, be graded as pilot officers and serve a further year with the Air Corps.[63]

The proposal indicated that Army Co-operation pilots would revert to their parent units after the second year and return to the Air Corps for refresher training for one month each year for six years, thus creating an 'efficient reserve of co-operation pilots'. The Department of Defence requested financial sanction for flying pay for an initial four officers. The pupil rate of four shillings a day during training, and eight shillings per day for the years as qualified pilots, would be paid subject to the appropriate approval. This scheme was not interpreted by the Department of Finance as a substitute for a cadet intake scheme but rather for one that proposed 'the training of members of the Officer Training Corps and Volunteer Reserve as personnel for an Air Force Reserve' – a case that had supposedly been mooted earlier.[64] Arising from this initiative three second lieutenants, K. T. Curran, F. F. Reade and M. E. McCullagh, from various Army units, were selected for flying training that commenced in January 1934.[65] Subsequently five more young officers, apparently in place of cadets not recruited, were attached to this class and training started on 18 January 1934. Of the total of nine pupils on the 1934–5 course seven, including Cadet D. K. Johnston, qualified.[66] In time, with officers not being returned to infantry units, the two schemes appear to have merged though the full circumstances cannot be gleaned from the files. In March 1936 the Department of Defence stated that it was not proposed to persevere with the scheme for 'the formation of a reserve of Air Corps Co-operation pilots', and that they proposed to affect the permanent transfers of the three pilots who qualified under the reserve scheme.[67]

OFFICERS AND CADETS, 1935–6

With the approval of the Department of Finance and the availability of ten vacancies created by the formal establishment of the 1st Co-Operation Squadron (Cadre) with effect from 22 October 1934, in addition to those vacancies created by retirements since 1928, a further six pupils, four officers from the OTC and Volunteer Reserve and two cadets commenced training on 1 April 1935, and qualified in March 1936.[68]

SYLLABUS CHANGES

From June 1935 new influences were brought to bear on the recruitment and training of military pilots. While the Army leadership had paid lip service to the concept of cadet entry in the period from 1924 to 1935 thereafter no pretence would be made in efforts made to minimise the possible influence of professional aviators on those entering the profession. This change was due to the actions of Major P. A. Mulcahy who had been appointed acting OC Air Corps and DMA on 3 June 1935. As Director of Artillery he had no qualifications, experience or expertise in aviation matters. He had come to the Air Corps following an unsettled period during which Liston had been in command.

Two of the more significant matters in which Mulcahy took an interest in the early stages of his command were the matter of pilot recruitment and training and the closely related matter of the syllabus of flying and ground instruction. In September 1935 he wrote to the COS stating that 'it was not in the best interest of the Air Corps that commissioned officers or cadets should be appointed to it without having sufficient training in military duties' and that 'cadets should not come to the Air Corps at all'. He cited no reasons for such a view. He stated categorically that only officers qualified and commissioned in the Cadet School of the Military College could or should be trained as pilots and recommended accordingly.[69] This approach can only be seen as a measure to ensure that pilots were primarily of an infantry disposition and that aviation motivation was a secondary consideration. By having officers complete their formative training in a strictly infantry atmosphere impressionable young men would be spared the suspect influence of the more air-oriented groups – the remaining ex-RAF pilots and cadet element of the 1926–8 Class. Having received approval in principle, Mulcahy informed the COS that the decision required that the flying training syllabus as laid down (DFR 7/1927) be cancelled and that it be replaced by a DFR that he would draft. In this regard, Mulcahy appears to have been deliberately misleading. The syllabus drafted by Russell that was in use since 1926, by its very title, applied equally to the training of commissioned officers and to cadets. A decision therefore, to train only commissioned officers, would not necessitate a change of syllabus.

After nine years in use it is reasonably certain that DFR 7/1927 would require amending. This, however, needed to take the form of revision and expansion to reflect the developments in aviation technology and the theory and practice of flying as well as advances in such areas as navigation and instrument flying. While such progressive changes were not reflected in DFR 40/1936 they were to be incorporated in the actual syllabus that was to be used for subsequent 'wings' courses.[70]

During the latter months of 1935, while he was negotiating with GHQ on the matter of a new DFR, Mulcahy instructed the School Commandant to draft a new syllabus. This task was completed by 4 November 1935 and its receipt in GHQ was noted. A minute sheet, still attached to the original draft syllabus, indicates that the document had been forwarded to GHQ for the approval of Army training staff, and it was suggested that it be designated as a training instruction (TI). A staff officer retained the document until April 1936 before returning it, without significant changes, to Mulcahy. Pencilled annotations indicated minor changes in wording and layout. The staff officer indicated that he 'had been pecking at it when its issue as a TI was contemplated' but that his superior had indicated that the syllabus that would give effect to the proposed 'Schools DFR [40/1936]' was not suitable for issue as a Training Instruction. It was suggested that 'it could be issued by the Corps simply as "Notes on the young officers' course" or some such title'. He further suggested that his annotations might be 'of some help to Captain Delamere in his further attention to the matter'.[71] This ruling, in the words used, was actually quite contradictory. On the one hand it was implied that DFR 40/1936 constituted a syllabus and that the draft syllabus was only supporting notes. On the other hand it was suggested that the draft syllabus had the effect of implementing the DFR 40/1936 and was, in effect, an essential part of the DFR. Examination of the two documents suggests that DFR 40/1936 was little more than a brief prospectus, not a syllabus.

Very shortly after the return of the draft syllabus Mulcahy forwarded it to Captain Delamere with instructions to examine it carefully. Delamere made a number of minor changes and the final draft was typed. On 25 September 1936 Mulcahy forwarded, 'for approval, the syllabus for the young officers' flying training course', to GHQ.[72] In the absence of a response it can only be presumed that some form of approval was granted. As the DFR and the new syllabus were complementary they should, more correctly, have been issued as a single document – but this would not have suited Mulcahy's purpose.

From inspection of the final draft it can be stated that the syllabus, that was to be first used for the young officers' course of 1937, represented a significant improvement on that of 1927. Though much shorter than the previous the syllabus was more precise in defining the scope and content of both flying and ground school. It brought ground school subjects and flying disciplines in line with advances in technology and flying techniques while specifying the (UK) Air Publications that would be required texts for flying and for ground school subjects. The qualifying standard laid down for pilots' flying technique placed greater emphasis on the ability of the pilot where previously the flying tests were closely related to the performance characteristics of the aircraft.

3RD YOUNG OFFICERS CLASS, 1937

When Mulcahy came to the Air Corps it was, like the Army generally, in the early stages of preparation for the anticipated Emergency. Such preparation should have included a significant increase in pilot numbers. However, during Mulcahy's first 18 months as OC no pupil pilot intake occurred. Mulcahy appears to have been preoccupied with organising the changes in intake policy, DFR 7/1927 and in the new flying course syllabus. He also took the opportunity to undergo an abbreviated 'wings' course and, in five weeks, qualified for the receipt of flying pay at the rate appropriate to pilots completing the year long flying course. It was January 1937 before the next group of pupil pilots, the 3rd Young Officers' Class, commenced training. Early that month some 14 second lieutenants, drawn from the graduates of the 5th, 6th, 7th, and 8th Cadet School classes, reported to Baldonnell for flying training. Prior to the commencement of the course four were rejected on the basis of a medical examination or interview and were returned to their original units on 18 January 1937.[73]

While the class was in training the rate of flying pay for qualified pilots, which the successful individuals had every reason to expect on qualification, was reduced from eight shillings per day to five. DFR 7/1937 of 8 February 1937, which authorised this reduction, was made retroactive to 31 October 1936. As a consequence the eight successful pupils of the 3rd Young Officers' Flying Course, who suffered this reduction in pay, perceived themselves to have been wronged. The course report on the 1937 Class recorded that the reduction in flying pay was an inappropriate decision that had caused 'dissension and distraction' amongst the student pilots during their course.[74] There is no suggestion that Mulcahy had an active part in this reduction in pay. However, there is no evidence that he concurred with the opinion of the School Commandant or that he took any action to have the decision reversed. This reduction in flying pay was to have repercussions in the context of the Air

Corps investigation of 1941–2. The decision however, may have suited the Minister for Defence. It should be recorded that the Minister and Mulcahy agreed that Air Corps pilots were not at all inclined to retire in order to fly with the newly established Aer Lingus – the inference being that they were too well paid. These facts suggest that higher authority believed a reduction in flying pay might make pilots consider more deeply a career in civil aviation.[75]

NAVIGATION

The manner in which the matter of air navigation was dealt with by P. A. Mulcahy is indicative of the many inadequacies of his command. The purchase of two Avro Anson medium range reconnaissance aircraft in 1937 should have presaged a new approach to navigation. There is no evidence to suggest that P. A. Mulcahy appreciated the necessity of advancing the standard of air navigation theory and practice to that appropriate to such aircraft. Whereas army co-operation called for the fundamental disciplines of map reading and dead reckoning navigation, longer range reconnaissance with the Anson put a greater emphasis on the third basic element of navigation, the fixing of position by means of plotting position lines on appropriate navigation charts.

In 1937–8 a young pilot, Lieutenant Jim Devoy, was nominated to undergo two courses of training in navigation with the RAF. As with previous courses with the RAF the records do not show how this came about. The context suggests that the Air Ministry had offered a place on each course gratis. Devoy summarised his participation:

> I completed two navigation courses in England in 1937 and 1938 at the RAF School of Navigation, Manston. The first was known as the short navigation course and the second as the specialist navigation course. I qualified in both courses. The short course is approximately equivalent to the civilian 2nd class navigator's certificate and the standard of the specialist's course is approximately equivalent to the standard civilian 1st class navigator's certificate.[76]

Having completed one course of three months duration and a second of six and a half months, Devoy was designated as the navigation instructor in Air Corps Schools. On his return from the advanced course, in July 1938, he recommended the running of a course of navigation for as many officers as possible – without response. He continued to press his superiors on the matter and was eventually asked to make a written submission for the attention of his CO. In April 1939 Devoy detailed the unsatisfactory level of navigation equipment available to pilots, while stating that navigation, as part of all pilots' training, was in effect grossly neglected. In particular, he was very dissatisfied with the general standard of practical navigation. He respectfully suggested that he be given permission to arrange a short course in theoretical and practical navigation for the R & MB Squadron, and another for the Army Co-operation Squadron. Emphasising the importance of meteorological information to the safe navigation of aircraft he recommended the appointment of a meteorological officer, the receipt of the short wave coded reports and forecasts from Rugby and the purchase of a range of navigation equipment.[77] Shortly afterwards he was instructed to run a very short navigation course. When he requested a greater length of time to cover a greater amount of an intended syllabus he was told that additional time could not be spared. As a result he drew up an abbreviated syllabus based on the time available.[78]

By the beginning of July 1939 the School Commandant was able to report on the results of the short navigation course conducted in the period 5 June 1939 to 3 July 1939. The course had been conducted for nine pilots of the 32 pilots then in service. Despite Devoy's earlier recommendation supplies of navigation equipment, particularly mathematical tables and instruments were still very limited. For reasons outside the control of the school the flying programme had been considerably reduced, mainly due to an unspecified number of 'special flying missions' and difficulty with aircraft tyres. While the programme called for nine flights for each of the pupils to practise practical navigation only four each were actually completed. Although good progress was reported in terms of more advanced instruction in interception problems and elementary instruction in continuous navigation out of sight of land, it was observed that the officers could not be considered to be qualified navigators.[79] In effect, an abbreviated and basic course was even further shortened and had been run for about one quarter of the qualified pilots in the Corps. To judge from the brevity of the course, lack of equipment and insufficient practical navigation, a small proportion of the flying officers of the Air Corps achieved a very modest level of proficiency, where, in reality, a significantly higher standard for all was what was required.

The question arises as to why it had taken until June 1939 to initiate navigation training. Had the two RAF courses become available because of an initiative on the part of Mulcahy, or on the part of pilots who might conceivably have influenced him, it is probable that the newly qualified officer would, on his return, have been immediately tasked to instruct in navigation for the maximum possible number of pilots. In the circumstances, however, it is possible that the places on the course were made available by the UK authorities and that a pilot was nominated and sent with no particular thought as to how he might subsequently be employed. Bearing in mind that Devoy had to prompt Mulcahy into authorising a navigation course, it seems probable that the commanding officer had little or no appreciation of current navigation practice or of its application to the operation of reconnaissance aircraft such as the Avro Anson. With an instructor duly qualified in navigation to the specialist level applying in the RAF, it is not obvious why Mulcahy did not immediately proceed to have all pilots trained to at least the basic standard for which Devoy himself would have settled.[80]

When asked by the Investigation Committee if the fullest use had been made of a qualified navigation instructor Mulcahy proceeded to mislead the Committee:

> He has been engaged as an instructor in the school both flying and navigational and I considered that it was more important that he should be available to the Schools than that he be employed elsewhere.[81]

Stating, in effect, that it was not possible for Devoy to carry out courses in navigation for service pilots due to other instructional duties, Mulcahy seems to have satisfied the Committee. The members were clearly not aware of the actual situation. Between January 1938 and August 1939 no pupil pilots were in training in the flying school. Devoy was not involved in flying instruction or navigation training with pupil pilots as implied by Mulcahy and would have had ample time to train and qualify many pilots to an acceptable standard. It is somewhat ironic that the 'wings' course syllabus, drafted by W. P. Delamere, and authorised by Mulcahy in September 1936, specified the Air Ministry *Manual of Air Navigation* of 1935 as the reference text for instruction in air navigation. This manual was a more than adequate guide as how to proceed in navigation training for the expected Emergency.[82] It is doubtful that Mulcahy, who had received no ground school training of any description, was familiar with this essential text. In his ignorance of air navigation, and of its application to long range reconnaissance, he apparently saw no need for more advanced navigation techniques than the map reading skills applying to army co-operation.

SHORT SERVICE PILOTS

In the context of pilot training it might be considered that the initiation of the short service scheme in August 1939, represented an expression of greater concern, on the part of the Minister and Mulcahy, for the future of civil aviation than it did for the provision of pilots for the expected Emergency. In the four-year period between 3 June 1935 and the outbreak of war on 3 September 1939 only eight pupil pilots, who had not already been in training on the day Mulcahy took over, qualified as military pilots. This number represents the successful students of the class that trained in 1937. They had been recruited in anticipation of the 50 per cent increase in the officer establishment, from the 30 provided for in the 1934 establishment to the 45 provided for in the establishment of 1 April 1937. The latter establishment introduced the '1st Reconnaissance and Medium Bomber Squadron (Cadre)' that consisted of a headquarters and a single flight totalling 30 (all ranks) personnel.[83] The maximum number of officers permitted under the 1937 Peace Establishment did not specify which appointment should be filled by pilots, observers or line officers. Following the qualification of the class of 1937 there was a total of 32 pilots in service in January 1938.[84]

In the meanwhile, early in 1937, P. A. Mulcahy had attempted to initiate action to substantially increase pilot numbers in preparation for the future expansion of the Air Corps. Noting that it had been difficult to get ten students for the 1937 course, he stressed the Air Corps' dependence on Cadet School graduates – a dependence that he had helped to bring about. He recommended that, in order to ensure a proper supply of pilots to the Army, 25 cadets should be appointed specifically for posting to the Air Corps after they had successfully completed their Cadet School course. He emphasised that it would be at least four years before such cadets would be of real value.[85] In response Mulcahy received only a verbal reply:

> Memo. C/S Staff Officer rang at 16.00 hrs 12/2/37. 15 cadets will be reserved for Air Corps on next lot of 30 to be appointed at once. If suitable material amongst those in training at College now, Air Corps will get. PAM 12/2/37.[86]

As this verbal reply was not subsequently supported by written confirmation Mulcahy might have surmised that the matter of Air Corps pilots was not high on the list of priorities of the COS or GHQ at the time. Indeed, there is no sign that Mulcahy himself was greatly concerned. He next communicated with GHQ on the matter some 11 months later. In January 1938 he reminded the COS that he had previously requested the recruitment of 25 cadets specifically for the Air Corps and had been promised 15 but was not aware that any had been appointed. He indicated that it would be appropriate to start another wings course in late 1938. This, however, would require a change of Mulcahy's ideology:

> I am satisfied that the younger we get prospective pilots for training the better will be the results. I am also of the opinion that if we are to ensure that a requisite number of pilots are to be available for service in the event of war, we must modify our present military training of cadets and concentrate on the flying training.[87]

Mulcahy proposed that Air Corps cadets be recruited for five years service – six months military training, 12 months elementary training and three-and-a-half years advanced and tactical flying training – followed by a permanent commission or transfer to a reserve force. This suggestion, in the context of a previous memo to the COS on 28 September 1937 that proposed a reserve of 200 pilots, was the basis of the Short Service Scheme that would eventually commence in August 1939.[88] At this stage Mulcahy must have recognised that GHQ had no intention of assigning Cadet School graduates to the Air Corps, for flying training or otherwise, as it was undoubtedly considered that the Army's need for such officers was paramount. Not surprisingly, therefore, between 1937 and 1945 the Cadet School qualified some 275 officers, none of whom was posted to the Air Corps until the Emergency was past and Army demobilisation was beginning.[89]

Three months after his previous letter, not having received written replies to his communications of 10 February 1937 and 11 January 1938, Mulcahy wrote to the COS suggesting that it would be appropriate to start a flying training course in September 1938 and requested an early decision on the matter so that appropriate new regulations could be drafted. The COS's response was to request a copy of the letter of 11 January 1938. At best he had forgotten about the matter and at worst had ignored it. Apparently arising from verbal exchanges with the ACS, Mulcahy subsequently submitted a detailed draft regulation for 'Short service commissions – Air Corps'.[90] While the final details of the scheme, and the conditions of service of the cadets were still to be worked out, the Department of Defence outlined the proposal to the Department of Finance in October 1938:

> I am directed by the acting Minister for Defence to state that he has had under consideration the question of augmenting the officer personnel of the Air Corps for the twofold purpose of providing sufficient pilots for the extra aircraft now on order . . . and creating an adequate reserve of this class of officer.[91]

The correspondence went on to state that a scheme had been prepared to provide for the appointment of officers to 'short service commissions in the Air Corps'. The service conditions provided for six months as cadets followed by 30 months as officers, with a possible extension of a further two years followed by seven years on the reserve. The payment of gratuities to provide for adaptation to civilian life (possibly not in aviation), were also provided for in the scheme. Permanent commissions would be granted in a limited number of cases. While the scheme was proposed in the context of unspecified immediate needs the requirement to build up a substantial reserve of pilots received greater emphasis:

It is estimated that a reserve of 300 pilots will be required and it is hoped to eventually reach this figure under the proposed scheme. For the moment, however, it would not be possible to cater for more than 20 cadets every eighteen months and it is desired to commence the scheme on this basis at an early date.[92]

In practice, while 300 pilots might eventually have been trained, Finance calculated that Defence's target of a reserve of 300 pilots could not be achieved. This was due to the fact that successive intakes of officers would go off the reserve after ten years at which stage numbers would level out at 100. In fact, with the traditional failure rate of about one third, for which the Department of Finance did not allow, an active reserve of even 100 was unlikely ever to have been achieved.[93] The Department of Finance, in studying the financial and other implications, compared the proposed scheme very favourably with that operated by the RAF but considered the intended gratuity structure to be too generous. Nonetheless it was also observed that 'one result of the proposal would be the regular infiltration of the new blood of youth which is particularly desirable in an air force'. Finance perceived the civil aviation aspect of the scheme as being very important:

> It is a further advantage that after their three or five-year period of initial service there will be a steady turn out of competent flying men to take their places in any civil commercial flying enterprises that may be expected to develop in this country.[94]

Noting that during the recent international crisis, authority had been granted for the purchase of 35 new aircraft that had committed the state to an immediate and considerable expansion of the Air Corps, the Department of Finance considered that a Short Service Scheme was preferable to permanent expansion.[95] In conveying the Minister's approval in principle to the proposed scheme the Department of Defence suggested that proposed gratuities be £200 when retiring after two years commissioned service and £300 after five. It was further stated that 'the minister would also like to receive an assurance that the Army authorities have available a ground force of mechanics, fitters etc. adequate to maintain sufficient aeroplanes for the training scheme contemplated'.[96] After a delay of five months, at a meeting on 14 April 1939, the Minister for Defence, Frank Aiken, mentioned to the Minister for Finance, Seán MacEntee that 'the matter of a scheme for the appointment of officers to short term commissions in the Air Corps was still under consideration by the Department of Finance' – intimating that a reply was outstanding. MacEntee subsequently reminded Aiken that approval in principle had been granted back in November 1938 and that amendments to regulations, and clarification on certain other points, that had been requested were outstanding.[97] Ignoring the contacts between the two ministers, but adopting a degree of urgency not previously obvious, Defence immediately wrote to the Department of Finance enquiring about, and expanded on, their submission of 23 November 1938:

> I am directed . . . to refer to the proposed scheme . . . for short service commissions in the Air Corps, and to state that as the need for additional officers for the Air Corps is now one of the utmost urgency the Minister proposes that this department will arrange for the recruitment of the cadets . . .[98]

To expedite the selection process it was proposed to dispense with the Civil Service examination for cadet entry and to accept, for interview and consideration by a military selection board, candidates between the ages of 17 and 19 years with Leaving Certificate (Pass) or Matriculation and those, up to the age of 23 years, with a university degree. The Secretary, Department of Defence also indicated that the necessary personnel and equipment, including machines, would be available to cope with the training of 20 cadets while the erection of a new 'cadets' quarters building costing £17,000' was being arranged by the Commissioners of Public Works.[99]

Between May 1938 and the initiation of the scheme in August 1939, and with no obvious sense of urgency, much correspondence had been directed by the Air Corps to GHQ apparently responding to verbal queries on the matter of the conditions of service to apply to the short service commission scheme. The main points of concern were the length of commissioned service, the age limits and the level of gratuity to be paid on transfer to civilian life. While the Air Corps recommended four-and-a-half years of commissioned service to ensure an adequate level of flying training and experience before transfer to the Reserve, the Department of Defence insisted on 30 months with a possible extension of two years. They also insisted on an age limit of 17 to 19 on entry. This factor, as the Air Corps predicted, was to eventually restrict the number of cadets in the first class recruited to much less than the 20 required. Defence, apparently taking note of the conditions pertaining to the RAF scheme recommended a gratuity of £200 after two and a half years and £300 after four-and-a-half years commissioned service. Eventually the influence of Finance would decide the more contentious issues.[100]

Although the proposed Short Service Scheme required that the flying school cater for overlapping classes of 20, the Schools establishment that was included in the reorganisation agreed in late 1938 and approved in March 1939, only provided for a single class. The 1940 War Establishment, however, would double the size of Schools and allow for a second intake of 20 pupils once the previous class had completed the elementary stage of training.[101]

With conditions of service for cadets still the subject of correspondence OC Air Corps informed the COS in March 1939 that 'we have taken delivery of our new training aircraft and are in a position to start the training of the first class'.[102] During the early summer of 1939 the conditions of service and necessary amendments to regulations were finally agreed between the two departments.

IST SHORT SERVICE CLASS

The first short service cadetships were eventually advertised in the *Irish Press* and *Irish Independent* on 1 June 1939. Due to the age restriction and the fact that the closing date, 24 June 1939, was prior to the promulgation of the Leaving Certificate results of that year only 42 applications were received. Twenty-five of these were deemed ineligible on grounds of education or age, or both. Two candidates failed to turn up for medical and interview while three more of the last 17 failed the medical examination leaving 12 at the interview stage.[103]

The interview board was made up of four Air Corps officers and Major G. J. Carroll who was at this time seconded to Aer Lingus as General Manager.[104] His presence on the interview board emphasised the fact that the longer-term aim of the Short Service Scheme was to provide pilots for civil aviation, in effect, Aer Lingus. Eleven successful candidates were attested on 16 August 1939 and commenced flying training on 21 August 1939 – only days before the outbreak of war.[105] Early in 1940 the first short service class completed the first term of military flying training and were deemed suitable for commissioning. Financial approval was granted on 22 February 1940.[106] The 11 cadets were commissioned at Baldonnell on 12 April, and nine successfully completed flying training by 1 August 1940.[107]

1940 SHORT SERVICE CLASS

While seeking financial sanction for a second intake the Department of Defence advised that, with the new accommodations being built it was appropriate to commence arrangements for an intake of 20, and that to ensure a 'bigger field of choice' the age bracket should be expanded to 17 to 21 years and to 23 for university graduates.[108] In due course a second class, of 20 cadets, was recruited and commenced training on 7 May 1940. The recruitment of an over age candidate with 70-hours military flying training with the Italian Air Force was also approved.[109] Fourteen of this class subsequently qualified as flying officers.

SERGEANT PILOT COURSE, 1943–5

In 1943, based mainly on a recommendation of the Investigation Committee and still under the aegis of the Short Service Scheme, a class of 31 other ranks commenced training to become sergeant pilots. These were selected, by interview, from 'seventy-five candidates with the requisite qualifications'.[110] Class A was comprised of 15 pupils, 12 newly recruited and three with previous Army service. Class B was made up of four Air Corps privates and 12 privates from Army units. The course ran from November 1943 to 22 December 1945 with 20 pupils qualifying as sergeant pilots – twice the number provided for in the 1943 establishment.[111]

4TH YOUNG OFFICERS' COURSE

In 1945, with the perceived Emergency now in the past, GHQ reverted to the officer intake system last used in 1937. The 4th Young Officers' Class, consisting of eight young officers, was selected and had commenced training by July 1945. While it marked a return to the officer-only intake policy it did not, however, mark the end of the Short Service (Cadet) Scheme. This was subsequently resurrected, at Aer Lingus's request. Nine further classes, with 93 cadets, were recruited between 1953 and 1961.[112]

A comparison between the numbers of pupil pilots recruited and training in the period October 1922 to December 1937 and the numbers recruited and trained in the period 1939 to 1945 is most revealing. During the earlier period, by means of seven generally poorly organised intakes, 64 pupil pilots commenced flying training with about two thirds being successful. During the Emergency, by way of a carefully organised Short Service Scheme with three intakes, a further 63 pupil pilots commenced training and again about two thirds were successful. No more than in the case of the training initiated during the Civil War it is debatable whether it was wise to undertake training on such a scale during the Emergency. However, unlike the situation that pertained in 1922–3, the training of pilots during the Emergency was facilitated by the availability of appropriate syllabus, structures, instructors and other essential resources.

CONCLUSIONS

While the Army leadership may have had sensitivities about the recruitment of ex-RAF pilots as authorised by Michael Collins in the 1922–3 period, they had little option but to complete the recruitment of more ex-RAF pilots in the latter part of 1922. Thereafter GHQ would endeavour to ensure that, initially at least officers of suitable IRA background would fly the state's military aircraft. As the first intake, of a nondescript collection of young officers and other rank volunteers, was almost a total failure, O'Duffy's scheme of reorganisation introduced the concept of cadet entry. The scheme also allowed the Army scope to skew the intake procedure in favour of mature officers with pre-Truce IRA service and infantry values. The intake of 17 officers in addition to the nine cadets, at a time when only four to six vacancies existed in the Corps, was a cynical use of authority on the part of the Army leadership. It aimed to foster the primacy of the infantry ethos and to ensure that ex-IRA pilots, and not those (cadets) influenced by the training of ex-RAF pilots, would be a dominating influence in the future leadership of the Air Corps.

During the 1930s, while no cohesive intake policy was ever expounded GHQ's preference in the matter of pilot intake for the Air Corps was to have newly qualified Army officers trained as military pilots. The underlying philosophy was based on the assumption that such pilots would not be required to operate other than in a battlefield reconnaissance role of the type that had evolved during the First World War (and which would be out of fashion by the next major war) and that only officers trained in the Military College would be able to understand the nuances of infantry tactics and operate army co-operation aircraft in the manner required. However, this intake method was abandoned for the period 1938 to 1945. GHQ apparently gave little consideration to the Air Corps' pilot requirements and considered that Cadet School graduates were much too valuable to waste on the flying of aircraft in time of war. While P. A. Mulcahy first proposed the Short Service Scheme in February 1937 his own ambivalence and bureaucratic inertia delayed the first intake until August 1939. The initial course, which was intended to have 20 students, provided only nine trained pilots. The manner in which the scheme evolved confirms that the main aim of the Short Service Scheme was, purely and simply, to have sufficient civil pilots available after the war.

Mulcahy's failure to pursue an adequate level of navigation training was but one example of his inability to appreciate the nuances of pilot training and aircraft operation. This failing, and the many other shortcomings of his command, would be major contributors to the demoralisation of the Air Corps during the early Emergency.

SEVEN

AVIATION POLICY AND PLANNING
1935–40

—

By the time P. A. Mulcahy had become acting officer commanding in June 1935 the Air Corps had survived the trauma and machinations of the demobilisation, mutiny and reorganisation processes of 1924. Thereafter a care and maintenance corps had stagnated somewhat due to lack of policy and purpose. The purchase of six new Bristol Fighters in 1925 did not result in the establishment of a fighter squadron as had been the Minister's original intention, while the period between the reorganisation of the Army in 1924 and the Emergency was characterised by a number of unsuccessful efforts on the part of the Army's General Staff to have the Government declare a policy in relation to external defence.[1] An air policy of sorts came about only by default. In the early 1930s reconnaissance was adopted as the Air Corps' main army support role. This role evolved through the training regime around the Vickers Vespa, a machine that had been developed (in the UK) from 1924, specifically for the army co-operation role.[2] Eight Vespas had been purchased between 1930 and 1931. Confirmation of this Army support role eventually came with the formal establishment of the Army Co-operation Squadron (Cadre) in 1934.

While air policy was not a major consideration it did at least have one strong advocate. Colonel M. J. Costello summarised his concerns of the early 1930s:

> Some time in 1930 this matter of policy as to the future development was under discussion and I endeavoured at the time to obtain a decision as to the amount of money which would be available annually for the development of the Air Corps and to have the policy which would govern its development settled. The net result of these discussions was that it could not be said in advance what sum of money would be available from time to time for the Air Corps, nor could anything definite be obtained on the question of policy [other] than a general decision that there would be an Air Corps.[3]

This general position, in effect, reflected the original decision, taken during the reorganisation process of 1924, to retain a token air corps. As the Army was not in receipt of a definitive policy, the Air Corps, as a minor corps, was unlikely to have its functions, in peace or war, defined. Although the consideration of air defence matters was to remain a minor concern for the Army leadership, in general, Costello would continue to demonstrate his belief in the necessity to develop a properly equipped and trained Air Corps as part of a substantial conventional defensive force. Costello's proposals for a greatly expanded Air Corps have to be viewed in the context of the 'suicidal defence planning' of the period as identified and defined by Theo Farrell.[4]

FUNDAMENTAL FACTORS AFFECTING THE STATE'S DEFENCE

A more conservative view of the Army's defensive responsibilities and capabilities was detailed by Colonel Dan Bryan (GHQ Intelligence Staff) in his 1936 study of the question of defence policy in future hostilities in Europe. This intelligence assessment warned of the ramifications of such hostilities for the internal and external defence considerations for the country.[5] The study was directed primarily against the 'utter insanity' of a group of senior officers who were 'talking extensively about a military war against the British and the successful manner in which such a war could be waged'.[6] Bryan suggested that Saorstát Éireann, relying solely on its own resources, could not defend itself against a reasonably strong enemy, except for a very limited period. Munitions and all manner of military equipment and supplies would soon become exhausted. In contrasting the strategic position of Ireland during First World War with its future position it was suggested that the development of new weapons such as aircraft and submarines had made the protection of naval bases more difficult and, in effect, greatly enhanced the strategic value of Irish harbours on the north, west and south coasts. In the context of military aviation, Bryan indicated ignorance of the functions of the considerable range of aerodromes and airfields that had been developed by the War Office during First World War and used during the War of Independence. He indicated that he was unsure whether these had been used for training or local defence. He did, however, correctly deduce that several coastal air stations had been developed for the US Naval Air Service and were used in the anti-submarine defences of the British Isles during World War I. Regarding possible future hostilities Bryan considered what air force measures would be required:

> Under the circumstances generally assumed . . . it is quite certain that even more extensive air forces than during the 1914–1918 period would have to be located on or near the Irish coasts. Reconnaissance, patrol, anti submarine duties would have to be undertaken . . . Because of the more serious threat it would also probably become necessary to employ aircraft on reconnaissance for possible raiding forces,

and to provide fighter aircraft to deal with hostile air attacks on shipping off the Irish coast and other centres situated on the coast.[7]

These measures were identified by Bryan in relation to Britain's defensive interests and needs and the likely aviation roles that might need to be performed in the maritime areas to the north, west and south of Ireland by fighter and reconnaissance aircraft operating from 'main bases' 'located in Scotland or Ireland'. He also suggested that the British would expect the military aviation of Saorstát Éireann to undertake some defensive measures:

> Great Britain would also expect that the Saorstát should undertake the aerial activities necessary for purely local Irish coastal control and defence. In areas used by the British fleets or on the main trade routes, her attitude to Irish activities would probably depend on the general relations and degree and nature of Saorstát co-operation and in particular on the capacity of Irish air forces to undertake such functions.[8]

Bryan's predictions regarding the possible roles for the Air Corps in a future emergency were to prove prophetic, as shall be seen.

ARMY PLANNING FOR WAR

In the absence of a government approved policy, the period from 1936 to the start of hostilities was characterised in GHQ by a considerable level of planning for the raising, equipping and training of a large conventional field Army.[9] In September 1936, due to the realisation of the country's military inadequacies, an expansion programme was put forward notwithstanding the caveats stated in 'fundamental factors' regarding the defence of the country. The plan was to complete existing units of the Defence Forces in the shortest possible period as a basis for the development of a long-term defence policy. The main proposal provided for the equipping of the six brigades of a notional war establishment. The expansion proposed for the Air Corps envisaged a major increase in personnel and aircraft numbers. While considering that the existing Air Corps was minuscule and really only an adjunct to the ground forces it was put forth that Air Corps numbers would be increased by 200 officers and 1,200 other ranks in an expanded 'Air service or force', a proposition that would in effect, more than quadrupling the current strength. The scheme included provision for approximately 100 aircraft organised in nine squadrons at a capital cost of £883,000. The capacity of Baldonnell, in terms of technical buildings and accommodation, thus, would have to be doubled. With the construction of three additional aerodromes the cost of 'aerodromes and other buildings' was put at £665,000. The increased cost of aircraft maintenance and spares was estimated at £77,500, which, with the addition of pay and allowances would bring the day-to-day cost of the Air Corps up to a total of £431,100.[10] Such expenditure should be viewed in the context of total defence spending, in the financial year 1936–7 of £1,373,257, and of £73,426 spent on the Air Corps for the same year.[11]

Whether the Army leadership appreciated it or not their proposal was nigh on impossible even if the Government had immediately authorised the necessary expenditure and the expansion in personnel. Such an expansion would have called for upwards of 170 new pilots – an impossible target in the context of the minimal training capacity and the actual output of the training squadron since 1922. It is doubtful if the planners had considered at any length the practical aspects of creating an air force starting virtually from scratch. Notwithstanding the urgency that the Army endeavoured to generate, and their warnings regarding complacency about the international situation, the submission to Government in late 1936 made no impression against the background of the caveats in 'fundamental factors'. Finance argued that there would be no war and that if there were, and if Ireland were invaded, defence against the superior force of a major power would be futile. The Government viewed national defence in terms of Anglo-Irish relations and, therefore, saw no need for a major expansion to form a conventional defence force.[12]

A CHANGE OF COURSE

As early as 1935, whether under the direction of GHQ or simply with its acquiescence, there was the first indication that the Air Corps was examining an air role other than one falling within the remit of army aviation. On 1 July 1935 two flying officers, Commandant G. J. Carroll and Captain W. P. Delamere, accompanied by Mr R. W. O'Sullivan, the Air Corps' civilian Assistant Aeronautical Engineer, attended the Society of British Aircraft Constructors' exhibition and display at Hendon. The main purpose of the visit was to familiarise the Air Corps with the state of development of military aircraft and equipment. Among the matters subsequently reported on were the rapid improvements in aircraft development, including the movement towards cantilever monoplane aircraft, closed cockpits, retracting under-carriage and supercharging as an aid to increased engine power which would give significantly enhanced performance in terms of altitude and speed.

The Air Corps personnel particularly noted that 'in the twin engine class the type which had most interest for us was the Avro 652A coastal reconnaissance and bombing aircraft'.[13] The reasons for interest in this particular aircraft, basically a civil passenger aircraft still in the process of development for military applications, are not obvious. For the Air Corps such an aircraft was a major departure from the biplane trainer and the army co-operation types usually purchased following successful introduction by the RAF. The Army's emphasis – first espoused in O'Duffy's reorganisation scheme and subsequently reiterated in the context of the establishment of the 1st Army Co-operation Squadron – was on the necessity to perfect the associated skills and techniques of traditional close reconnaissance and co-operation with ground troops. It may be the case that the General Staff recognised that future hostilities, whether the country was involved or not, would necessitate the prior development of some capacity for coastal reconnaissance and, as a result, had requested assessment of the most appropriate aircraft available. It is noted that this aircraft assessment took place nearly a year before the circulation of Bryan's study known as 'Fundamental factors'.

Despite the traditional role of the Air Corps the Army estimates for 1936–7 included provision for £15,000 for the purchase of two twin-engine long-range reconnaissance and bombing aircraft.[14] The supporting case stated that such an aircraft had been the subject of evaluation for a number of years and that, with the exception of target towing for artillery, the Avro 652A Anson met all the Air Corps' requirements. Mulcahy recommended the purchase of two aircraft that were required for training in aerial navigation, long distance and coastal reconnaissance and for 'wireless, bombing and gunnery'. They were to cost £7,800 each with an additional £500 for unspecified additional equipment. He also suggested that the balance of £1,100, not provided for in the annual estimates, could be met by foregoing the purchase of two elementary trainers on the basis that he had no immediate plans to undertake the training of additional pilots.[15]

The aircraft were duly purchased and taken on charge on 20 March 1937. While they had been purchased for navigation training and long-range patrol, strangely the aircraft were not fitted with any form of direction finding equipment when delivered, or subsequently. The modification to have a loop aerial installed was issued by A. V. Roe on 23 November 1938 but the purchase and fitting of the equipment was not pursued by Mulcahy despite the urgings of his subordinates.[16] Similarly, no ground direction finding facilities were available to the State's military aircraft. As a result, these navigational deficiencies were to severely limit the effectiveness of reconnaissance operations during the Emergency. Two more Ansons were taken on charge on 19 January 1938, followed by a further five on 2 February 1939.[17]

In the meantime the 1934 Establishment had been amended by the addition of a second service squadron, the '1st Reconnaissance and Medium

Bombing Squadron (Cadre)' with effect from 1 April 1937. This consisted of a headquarters and a single flight providing for only 30 personnel – six flying officers, eight NCOs and 16 privates – and did not provide for navigators, wireless operators or gunners. As a result of these skills deficiencies the squadron could not develop its roles effectively. The combination of poorly prepared aircraft and a small and inappropriate establishment meant that the squadron was ill equipped to train and prepare for the coastal patrol role it would subsequently undertake. The addition of this new training cadre brought the total Air Corps Establishment up to a total of 399 all ranks – 45 officers, 91 non-commissioned officers and 263 privates.[18]

THE MINISTER'S PRIORITIES

On Saturday 17 July 1937 Major P. A. Mulcahy was summoned to the office of the Minister, Frank Aiken. Initially Mulcahy gave a verbal report on progress in the Air Corps under his command. The meeting, however, did little to advance the case for military aviation policy or strategy and did not suggest where the Air Corps might stand in the anticipated Emergency. Mulcahy noted the principal points discussed:

> He agreed with me that a definite policy of [army] expansion, to take place over a period of years, must be laid down before satisfactory Air Corps expansion could make any headway. He stated that he hoped to get such a policy agreed to before long, but that as such a policy entailed very heavy financial commitments, it was a matter that could not be decided upon in a hurry.[19]

As OC Air Corps Mulcahy might have been expected to emphasise the necessity of expanding the Air Corps sooner rather than later but appears to have been happy to accept the subordination of the Air Corps' future to an Army policy that had yet to be formulated. Sensing his priorities, Mulcahy gave tacit agreement to the Minister's belief that military aviation had distinctly lower priority than land forces in preparation for the expected Emergency. Civil aviation seemed to be the Minister's main concern:

> He realised the difficulty of procuring and training pilots for civil air companies and favours training some of our apprentice-mechanics as NCO pilots with a view to supplying the companies' demands. He realises that this will take quite a long time and as the present officer pilots are unwilling to resign their commissions and accept jobs with the civil companies, he is inclined to consider detailing serving pilots for short periods of duty with civil companies.[20]

The discussion had also introduced the possibility of the secondment of military pilots to Aer Lingus and of a short service pilot training scheme for the benefit of civil aviation. While no firm decisions were taken, the tone of the discussions makes it abundantly clear that, with war looming, the Minister for Defence and the OC Air Corps considered that the aircrew requirements of Aer Lingus took precedence over the plans and preparation for the employment of military air resources in time of war. The leisurely fashion in which the short service scheme was subsequently established underlines the civil aviation emphasis of the scheme. The discussion also indicated that finance would be the biggest consideration influencing future decisions on air defence matters.[21]

Notwithstanding the lowly priority of military aviation the Minister and Mulcahy gave some consideration to the matter of expanding the number of aerodromes. They agreed that 'about four more military aerodromes should be established' but that due to the heavy expenditure involved 'it would have to be considered when the general Air policy was being considered'. Mulcahy explained his concept:

> In this connection I emphasised the necessity of being able to state that we would send a squadron to Aerodrome X in say 1939 and a squadron to Aerodrome Y in say 1941, so that arrangements could be made to make these places suitable for occupation before these dates.[22]

As will be discussed in chapter 9, when an Air Corps detachment was sent to Shannon in late August 1939 the requirement to have aerodromes prepared well in advance will be seen to have been totally neglected. The last matter agreed was fundamental:

> He asked me had we considered the question of sea planes versus land planes for our purposes. I stated that we had not considered this matter in any detail, but that it was my opinion that we were committed to the continued use [only] of land planes. He stated that that was his opinion too.[23]

Despite agreement on this basic principle, Mulcahy was subsequently to be ordered by the Minister to buy amphibious aircraft and to operate them out of Shannon. It is not easy to reconcile the Minutes of the Conference of 17 July 1937 with the Army plans for an expanded and better-equipped Air Corps as proposed in September 1936. On the one hand, Mulcahy had agreed with the Minister that Air Corps expansion could be postponed pending the expansion and equipping of the ground forces and on the other hand he was co-operating with General Staff in their planning for the expansion of military aviation required for conventional defence against possible invasion. As intimated by

O'Halpin and Farrell, it would appear that Army planning was based on the policy the Army hoped the Government would endorse, while the Government, in keeping with its rapprochement with Britain, saw little need to prepare, in this instance, for conventional air defence.

CID'S DEFENCE ADVICE

Accounts suggest that a later proposal (see below) for an establishment consisting of two-and-a-half first line units, including a flying boat unit, appears to have been greatly influenced by advice given by the UK Committee for Imperial Defence in January/February 1938. In response to a request from the Irish Government for guidance in regard to defence strategy and expenditure, CID forwarded a paper suggesting that Éire should spend only £1.4 million in capital expenditure on defence, plus a recurring total annual expenditure of about £2 million to build up adequate air and land defences. In regard to the air forces the CID advised:

> It is understood that the air force of Éire is in a fluid state but that the intention is to organise it into three squadrons on a volunteer basis . . . it is suggested that as the forms of attack are limited to seaborne raids, and possibly long range air attacks from shore bases on the continent, two of these squadrons should be equipped with a type of aircraft suitable for general reconnaissance and bombing and should be stationed in the south and west of Éire . . . the third should be equipped with fighter aircraft and should be stationed in the vicinity of Dublin.[24]

These proposals, made on the assumption that the United Kingdom and Éire would be allied in resistance to a common enemy, would be reflected in the composition of the Peace Establishment of 1939 and the War Establishment of May 1940, and seem to have influenced the eventual numerical strength and disposition of the somewhat token level of the air resources eventually raised for the Emergency.

The CID recognised that the organisation of such squadrons and their equipping with modern aircraft would involve a considerable increase in the appropriation for the maintenance of air forces. Although suggesting that economies were available if the country exercised her rights as a member of the British Commonwealth of Nations by engaging the Air Ministry as a purchasing agency it was indicated that the capital costs of a general reconnaissance squadron, with 19 Blenheim aircraft and ground equipment, general stores, spares and mechanical transport costing £15,000, would come to a total of about £290,000, with an annual expenditure of £77,000 for maintenance and personnel. While indicating that the newly developed Spitfire or Hurricane were the appropriate fighter aircraft to acquire, the CID calculated the cost of a fighter squadron on the basis of the Gloster Gladiator, a much less potent biplane. Nineteen Gladiators and appropriate stores were estimated to cost £119,000 plus £63,000 per annum for personnel and maintenance. The capital cost of equipping three squadrons was cited as £714,000 with annual maintenance cost, including that of personnel, of £217,000. In addition to land-based aircraft the CID recommended that £500 be spent on undefined facilities for mooring 12 flying boats at Bantry, 'with a very small provision for maintenance', and that the flying boats would be operated for trade protection purposes. No estimate was made of capital or maintenance costs associated with flying boats at other locations.[25]

It must be noted that, of the £1.4 million capital expenditure that the CID had recommended for equipping the whole Army, in excess of half, was aimed at developing air defence resources. In any event the CID recommendation, which was perceived by the Army leadership as a relatively modest level of spending on land and air defence, was used by the administration to undermine the more ambitious and expensive Army plans for a well equipped conventional force. Despite the relatively modest level of spending suggested by the CID and, while aspects of the British advice regarding air defence will be seen to be reflected in later establishments, the commensurate level of expenditure was unlikely to be approved by Finance at this time. From this evolving government defence strategy it can be concluded that defence preparations remained more symbolic than practical.[26]

COSTELLO'S AIR CORPS PLAN

In January 1941, while answering questions relating to military aviation policy, Mulcahy recalled that he had met the Minister on 3 December 1936 and on 17 July 1937 and on both occasions had pressed unsuccessfully for a statement on aviation policy. On 29 November 1937 Mulcahy proposed unspecified changes to Air Corps organisation aimed at incorporating the 'Apprentice Mechanic' and 'Short Service Commission' schemes. His (unseen) manuscript submission included estimates for 'aerodromes, buildings and ground equipment and aircraft'.[27]

In early March 1938 Mulcahy had discussions with Colonels Liam Archer and M. J. Costello regarding Air Corps expansion. Later that month Mulcahy received a secret memorandum from Costello who was acting on behalf of the General Staff.[28] The memorandum put considerable detail on a previously agreed outline plan for the expansion of the Air Corps and requested the 'estimates of the capital and maintenance costs' for its implementation. The ACS outlined the rationale for the expansion:

> We have neither the financial nor industrial resources to create a large Air Force and the demands on the available resources which will be made by Land Forces including Anti-Aircraft Units are such as to require the modification of your proposals for the expansion of the Air Corps as presented by you to Colonel Archer. . . . At the same time it is possible that a situation may arise in which it would be necessary for us to expand rapidly and in which the necessary machines and other equipment would be available. It is, therefore, proposed to organise and maintain the framework for such expansion to strength approximate to that outlined by you.[29]

Costello's proposal was for two-and-a-half 'First Line Units' – 'a pursuit squadron for the defence by air of Dublin and the Shannon Scheme' as a deterrent to bomber attacks; a 'Coastal Patrol Squadron, Shannon Airport' for 'spotting' for coastal defence artillery; 'attack of enemy vessels' and 'co-operation with a marine coastal patrol' and one '1/2 Reconnaissance and Medium Bomber Squadron, Dublin' with no specified role. It was proposed that these squadrons would be maintained 'at full strength' by duplicating the aircraft numbers bringing it up to a total of 84. Baldonnell was to be the only 'permanent station' with Fermoy, Oranmore and Gormanston as 'temporary camps'. It was proposed that 'Seaplanes or Flying Boats' were to be based at Shannon and 'to have a string of landing places around the coast', as well as on inland lakes and at harbours. A capability to supply 'some aircraft' for land forces anti-aircraft training was also proposed.[30]

It was suggested that the short service pilot scheme would be put into effect in addition to the training of officers of the Volunteer Reserve. In the event of an emergency it was proposed that a considerable proportion of the fitters employed in the motor trade would be made available for technical duties. A puzzling aspect of the plan was the proposal to eventually organise and train, including the first line units, a total of four pursuit, four coastal patrol, two reconnaissance and medium bomber squadrons and to provide the necessary infrastructural and ground organisations required. As had been the case with the scheme recommended to Government in September 1936 this expansion plan did not specify the status of the additional units or how, when or in what circumstances they might be raised.[31]

Mulcahy was apparently encouraged by what, as he later stated, he had interpreted as constituting a 'statement of policy'.[32] A month later, in response to Costello's request, he submitted figures of estimated expenditure under four main headings. Under the heading of transport, he listed a total of 55 air support vehicles, in addition to 11 already on charge or ordered, at an estimated additional cost of just over £39,000. Under general stores, listing aircraft among a wide range of aviation equipment, he suggested provision be made for 26 pursuit aircraft at a total cost of £182,000. The submission indicates that the Gloster Gladiator was the type proposed. The Air Corps had taken delivery of four Gladiators six weeks earlier, on 9 March 1938. Four more, ordered for delivery in the financial year 1938–9, were never delivered.

Twenty 'seaplanes' of an unspecified type, were to be provided for at a cost of £300,000. Mulcahy allowed £100,000 for an additional ten reconnaissance and medium bombers – Avro Ansons, costing up to £10,000 each. Four transport aircraft and four target-towing aircraft (for anti-aircraft artillery practice) were to cost an additional £68,000, and 20 training machines another £40,000. No less than 150 parachutes, costing £10,000, were also required. The total cost of a long list of aircraft and associated ground equipment, excluding the estimated £50,000 worth already in stock or on order, was put at £822,300. A 'reserve of fuel and oil', estimated on the basis of a very ambitious 200 hours per aircraft per year, was predicted to come to a total of 500,000 gallons, though no estimate of cost was made. A separate and unrelated 'estimate of annual consumption of Petrol, Oil and Ammo' suggested a provision of £65,000 for 750,000 gallons of aviation fuel, £3,750 for oil and £650 for ammunition.[33]

On receipt in GHQ Costello amended some of the figures. He specified 20 fighters, 20 seaplanes, six reconnaissance aircraft and 11 trainers at a total cost of £462,000. Spares, tools and equipment, including signal and photographic equipment added £152,000 to the estimate. Transport (£45,000), ordnance and ammunition (£165,000), buildings (£50,000) and reserve fuel and oil (£170,000) added a further £430,000. The capital cost of Costello's plan came to £1,044,000. On its receipt in Finance, in keeping with the Government policy of limiting defence expenditure, the (undated) document 'Air Corps – Capital Expenditure' appears to have received little consideration.[34]

O'HIGGINS'S PLAN, 21 MAY 1938

Notwithstanding the fact that his plan for Air Corps expansion had been accepted by Aiken[35] and forwarded to Finance, Costello and his plan had already fallen into disrepute in GHQ. During the investigation of 1941–2 he explained the circumstances. He stated that, at the second of a series of defence co-ordination meetings chaired by de Valera in September 1938, he was informed by An Taoiseach himself that his plan had been abandoned. The COS had added that the scheme 'did not have the approval of any responsible officer' and that a new plan was being prepared by Colonel O'Higgins and Colonel Archer, with Mulcahy's guidance. Costello claimed that he had not known that such decisions had been taken in GHQ – possibly as much as six months previously.[36]

Costello also explained to the Committee of Investigation the rationale for the scale and scope of his original proposal. He put forth that he had carried out his planning on the understanding that £10 million was being made available, on the authority of the Minister, for capital expenditure on the Army over a number of years, and that ten per cent of that would be made available for the expansion of the Air Corps. He proposed to buy the aircraft to equip two-and-a-half squadrons – a pursuit squadron to consist of 13 aircraft, a reconnaissance squadron and a coastal patrol squadron with ten aircraft each while reserves of the same magnitude would be in place to ensure that all squadrons could be maintained at maximum aircraft strength at all times. He assumed all the aircraft required to equip one of each type of squadron would be purchased immediately to ensure against rising costs and to ensure homogeneity of equipment during a period when aircraft were undergoing rapid change.[37] This latter policy would, in fact, result in condemning the Air Corps to enduring the Emergency with obsolete aircraft.

Costello proposed that the Air Corps, which would include only 40 permanent pilot officers, would be expanded by the raising and training of three quarters of all personnel on a reserve basis. This was to include 104 short service flying officers who would be recruited directly into the Air Corps and trained as pilots before returning to civilian life after a maximum of three-and-a-half years in service. One hundred officers of the Volunteer Reserve were also to be recruited directly into the corps to undergo a one-year flying course. The plan suggested that technical personnel would come from the existing boy apprentice scheme supplemented by a number of graduates of the technical schools. While the absence of a civil aircraft industry was recognised, Costello considered that retiring reserve pilots need not necessarily be absorbed into flying positions but that unidentified industrial concerns could absorb a considerable proportion. A central aspect of Costello's scheme for expansion of military aviation was the assumption that the Air Corps would benefit from the development of infrastructure for civil aviation by the Department of Industry and Commerce. As was later to transpire, he wrongly assumed that this development would provide the Air Corps with aerodrome facilities at Limerick and Cork, and possibly at Galway and provide a flying boat base at Rineanna/Shannon.[38]

Subsequently, denying knowledge of the decision to abandon the Costello plan, Mulcahy stated that as far as he was aware Costello's proposals constituted the basis of the organisation and establishment of the Air Corps in 1939.[39] The main point of difference between the two schemes was that Costello's plan was for a small permanent force with a reserve of pilots and technicians who could be called up in an emergency while, in contrast, the O'Higgins plan was for a much larger permanent force.

The O'Higgins plan was formulated in time to be considered, as part of an Army plan for the reorganisation of the Defence Forces, at a meeting of the general staff on 18 May 1938. The main proposal was for an Army of 49,805 all ranks centred on a field army of 25,605. O'Higgins proposed that the Air Corps element would include provision for no less than 1,500 personnel and ten squadrons:

Air Corps
1 Fighter Squadron)
1 Flying Boat Squadron) 1st Line.
1/2 Reconnaissance Squadron, organised without machines
3 Fighter Squadrons
3 Flying Boat Squadrons
1 1/2 Reconnaissance Squadrons
Total: 1,500 all ranks.[40]

While Costello's plan included provision of two-and-a-half permanent front line squadrons O'Higgins included only two such squadrons. In addition, the roles of the proposed '1st line squadrons', as implied by the nomenclature, did not fit in with the two existing squadrons, one nominally an army co-operation squadron and the other a reconnaissance squadron equipped with Anson aircraft. It was not indicated how a force of 1,500 could be recruited, trained and equipped.[41] The plan took into account the UK advice regarding flying boats. Mulcahy subsequently recalled that 'we had no seaplane base at which to base seaplanes and the minister insisted we should have seaplanes'.[42]

The Army's plan of 21 May 1938, which included the un-priced O'Higgins proposals, was associated with a supplementary vote of £600,000 for defence equipment. Included in the Air Corps allocation of £150,000 (apparently to be spent over two years) was provision for £30,000 for four fighter aircraft, £40,000 for two reconnaissance aircraft, and £60,000 for six advanced trainers. The proposals to spend £600 on 50 sets of flying clothing and £1,800 on 30 parachutes suggest a more realistic reassessment as to what level of expenditure would be considered appropriate by Finance. A significant provision was that of £1,500 for 'temporary hutments for 40 cadets', suggesting that proposals for a short service pilot scheme were still in preparation despite the rejection of the Costello plan.[43] Thus, this latest projected level of capital expenditure on the Air Corps was more realistic and apparently influenced the estimate provision for 1939–40, while the concept of raising 1,500 Air Corps personnel was becoming less likely. At this juncture in 1938, while GHQ continued to plan for Army and Air Corps expansion, there was, as yet, no national defence policy and therefore no Government approved policy, strategy or plan for military aviation.[44]

AIR CORPS CAPITAL EXPENDITURE: AIRCRAFT PURCHASES

The next Defence estimates, those approved for 1939–40, made a provision for the Air Corps that better reflected Government and Finance policy on defence generally and reduced the cost of Costello's plan by almost 90 per cent. However, the capital expenditure proposed went well beyond the £75,000 approved by supplementary vote.

Air Corps – Capital Expenditure	£
6 single engined training aircraft (Lysander) @ £7,900	47,400
4 single engined aircraft (Gloster) @ £6,700	26,800
1 Link Trainer	1,450
Special spares aircraft	18,610
Special purchases for Air Corps	20,050[45]
[Total	£114,310]

£1,890 for educational equipment, maintenance and engine and aircraft spares, plus £3,100 for general stores, workshop equipment, tools and parachutes brought the Air Corps estimate for 1939–40, which did not include pay and allowances, to £119,300.[46] The actual total cost of running the Air Corps for 1939–40 came to £159,520. Only £61,981 of this constituted capital expenditure – less than 55 per cent of the approved provision.[47]

The £150,000 earmarked for military aviation translated into proposed capital expenditure of £75,000 for each of the financial years 1938–9 and 1939–40. Against an original estimate of £55,850 for capital expenditure in 1937–8, only £46,635 was actually expended.[48] During that financial year six new aircraft were purchased and delivered – two Avro Ansons and four Gloster Gladiators. Two Ansons had been purchased in 1936–7 and a further 12 Ansons were ordered in 1938–9 to complete a reconnaissance squadron of 16 aircraft. Only five were subsequently delivered and the remaining seven were retained by the British. In 1937–8 eight Gloster Gladiators had been ordered but only four supplied.

The 1938–9 Defence estimates provided £49,440 for Air Corps capital expenditure though the supplementary vote raised this to at least £75,000.[49] Despite the fact that the Department of Defence figures from 1942 suggest that the total amount spent on Air Corps aircraft and equipment (in 1938–9) was £80,250,[50] in reality the material dealing with the supplementary vote of 1938 indicates that at least £86,700 was spent on aircraft alone. At least £14,500 was spent on the purchase of ten Miles Magisters required for the newly approved short service pilot scheme. The five Ansons cost about £42,500 while three Walrus aircraft delivered in March 1939 to equip the Coastal Patrol Squadron cost £29,700 – a total of £86,700.[51]

In 1939–40 the approved estimate for capital expenditure was £114,310, however, only £61,980 was subsequently spent. Of this figure £47,400 was spent on six Lysanders while another four Gladiators were withheld by the British. With £1,450 being spent on a Link trainer and, in all probability, £3,100 on various stores and equipment it is not clear how the balance of £10,050 was spent – possibly on 'special spares' or 'special purchases', for which no less than £38,660 was provided for in the 1939–40 estimates.[52]

WALRUS COASTAL PATROL AIRCRAFT

The purchase of the three Vickers Supermarine Walrus amphibious aircraft in March 1939 appears to represent a policy decision made by the Minister influenced by the CID advice of early 1938. On 14 March 1938 Mulcahy wrote to the COS informing him that he had just been made aware of proposals 'for the formation of a seaplane unit for Coastal Patrol' and had been instructed to consider the matter. Mindful of the fact that the Minister had agreed, on 17 July 1937, that the operation of flying boats was not a consideration for the Air Corps and that his proposal for Air Corps expansion made on 28 September 1937 had made no provision for such a unit Mulcahy asked if the proposed unit was to be considered in addition to the organisation he had planned. He also sought clarification as to the advantages accruing from such a decision and, in particular, the specific duties of such a unit.[53] The Costello plan that Mulcahy received one week later specified that a coastal patrol squadron based at Shannon would do 'spotting' for coastal artillery, attack enemy vessels and co-operate with marine coastal patrols.[54] Mindful of the fact that the Airport Construction Committee, of which he was a member, was due to meet on 29 March 1938, he outlined some of the infrastructural implications of basing a coastal patrol squadron at Shannon. Mulcahy suggested that the Airport Committee dealing with the question should be made aware that, in making decisions on the locations of airport buildings, adequate provision should be made for military exigencies – '22 officers, 43 NCOs, 144 men, squadron offices, stores, workshops, hospital, photographic section, etc'.[55] Soon afterwards Costello rang Mulcahy informing him that he was authorised 'to inform the Chairman of the Airport Committee that it was intended to station a Squadron at the Shannon Airport and to request that the necessary accommodation' should be provided for. He also indicated that the Minister for Defence had undertaken to speak to the Minister for Industry and Commerce on the matter.[56] As it was Government policy to minimise defence expenditure no building works, specifically for the Air Corps, were incorporated in the Department of Industry and Commerce's plans for the development of Shannon Airport.

It is not obvious that any great thought was put into the selection process that identified the single engine Walrus amphibious aircraft. In all probability cost was the main factor in selecting a very small aircraft that had been developed 'to operate from a ship fitted with a catapult, an aircraft carrier or from a land or marine base'.[57] Provision having been made in the 1938–9 estimates three Walrus amphibious aircraft were purchased, and delivered on 4 March 1939.[58] This was just prior to formal approval of the establishment of the Coastal Patrol Squadron (Cadre) of 51 personnel, which included six officers, 17 NCOs and 28 privates, with effect from 6 March 1939.[59]

The suitability of the Walrus as a coastal patrol aircraft can be gauged from the outcome of a number of flights carried out in May 1939. Apparently as part of the process of introducing the aircraft type to service, surface and air reconnaissance, 'with a view to alighting in the vicinity' the fort at Castletownbere and Lough Swilly were the subject of ground and air reconnaissance. At Castletownbere ground reconnaissance was carried out on 12 April while test flights took place on 12 and 22 May. Though the co-operation of the local artillery unit and use of 'the surface craft' were found to be satisfactory no assessment of sea conditions appears to have been made. On 22 May the Minister for Defence visited Bantry Bay, however, the purpose of his visit was not stated. Operating into and out of Castletownbere appears to have been reasonably satisfactory in the weather of May 1939; the same cannot be said for Lough Swilly. The general area around the forts was found to be quite unsuitable for alighting and anchoring. Fort Dunree was very exposed while the depth of the water precluded safe anchorage. The western aspect at Fort Lenan resulted in an exposed anchorage at practically all times and thus unsuitable. With the test flights only concentrating on two major inlets in the month of May 1939 the report did not give an overall assessment of year-round flying boat operations in Atlantic waters. Nor was the Walrus tested on rivers and lakes. However it seems probable that a small aircraft like the Walrus could not cope with the Atlantic swell when anchored or coming ashore in the waters of the south-west, west and north-west of Ireland even in the more benign weather conditions of summer. It is even more probable that their use, even in the most sheltered waters, would have been totally out of the question under winter conditions. It would appear, from Mulcahy's evidence to the investigation committee that three Walrus aircraft had been bought to be used as training machines while simply paying lip service to the Minister's wishes regarding their potential to operate off coastal waters.[60]

In retrospect Mulcahy's position on seaplanes appears contradictory. His own recall of the period, given in writing or verbally to the Investigation Committee, shows that he was not in favour of seaplanes on 17 July 1937. In March 1938 he had, in effect, been ordered to develop a seaplane unit and capability. Subsequently, in January 1941, he stated that 'in 1937 my idea was to have seaplane reconnaissance at the Shannon'.[61] The Committee did not comment on the fact that Mulcahy had originally been against the use of seaplanes when conversing with the Minister on 17 July 1937 and, having been directed to purchase such aircraft in March 1938, claimed that it was his idea all along. As the operation of the Walrus aircraft had been so ineffective it is actually not clear why Mulcahy would have wanted to claim responsibility for their introduction to service.

ORGANISATION

Parallel with the consideration of the Costello and O'Higgins plans for Air Corps expansion, the reorganisation of the Corps was being addressed by Mulcahy and GHQ. On 21 April 1938 Mulcahy submitted a draft establishment table which included provision for the permanent organisation indicated by the Costello plan. He recommended an organisation consisting of a total of 877 all ranks – Air Corps HQ (28), Corps Depot, (224) Corps School (111), a pursuit Squadron (167), Coastal Patrol Squadron (141), Reconnaissance and Medium Bomber Cadre (138). The addition of 48 personnel to constitute station staff at Shannon completed the proposed establishment.[62] Later in 1938, referring to the same plan, he submitted a slightly changed draft table:

> I have the honour to refer to the ACS's Secret communication of 21 March 1938 and to forward herewith for approval a new Establishment for the Air Corps. . . . the previous . . . is totally inadequate and unsuitable for our new commitments.[63]

He proposed an increase of only 19 personnel (896 all ranks) for the same organisation. He used the opportunity to emphasise that, 'with the advent of

Fighter aircraft and radio [telephony] in aircraft, we require our own Artificers and Radio personnel', and indicated that those in new trades, when established, would require training.[64]

Two weeks later Mulcahy's establishment table was examined and amended by Costello, O'Higgins and Mulcahy in conference. A modified table, reflecting a reduction of 37 per cent (to 567 all ranks), was finally agreed. With the concurrence of the others, Costello signed a brief submission which emphasised the necessity to obtain approval for the new establishment and the need to start recruitment as soon as possible, and forwarded it to the COS. The COS endorsed the case and forwarded it to the Secretary, Department of Defence to seek the sanction of the Department of Finance.[65]

On 8 November 1938, following discussions on the matter with the COS, Mulcahy submitted another (unseen) revised establishment table, for 'a Cadre organisation for the Air Corps', on the basis that all squadrons would remain in training mode for 'the next 12 to 18 months'. The context suggests that the cadre status had been directed by the COS in keeping with a Government strategy that would not require squadrons with operational capabilities. It was to be March 1939 before a substantial case for revised peace establishments for all Army headquarters, formations, corps and units was forwarded to Finance. The supporting case for the Air Corps (of 564 all ranks) was brief:

Table 7.1 Summary Air Corps PE, 1939

Unit/Ranks	Officers	NCOs	Privates	Total	1937 PE All ranks
Air Corps HQ	9	5	2	16	26
Depot & Workshops	10	63	178	251	224
Fighter Sqn (Cadre)	11	16	47	74	62 (Army Co-op. Sqn)
R & MB Sqn (Cadre)	17	32	61	110	30
CP Sqn (Cadre)	6	17	28	51	–
Air Corps School	10	17	35	62	57
Total	63	150	351	564[66]	399

A number of aircraft have been delivered recently and additional numbers are expected in the very near future, with a very large amount of stores. This increase in equipment involves an increase in personnel and the Establishments now proposed will only barely cover the equipment which is expected to be delivered within the next twelve months.[67]

In due course the 1939 Peace Establishment for the Army (8,014 all ranks, including 564 Air Corps) was approved by the Minister, which came into effect from 6 March 1939. In terms of personnel the 1939 Establishment represented a major departure from the substantial organisations recommended by both Costello and O'Higgins. It confirmed the training cadre status of the three (actually two and a quarter) service squadrons. As had been the practice from earlier times numbers and types of aircraft were not specified. Traditionally the number and type of aircraft in each squadron was dictated by the capital expenditure, usually miserly, allowed by Finance in annual Defence votes. As the Emergency approached the Air Corps Peace Establishment of 1939 bore many characteristics of the token establishment of 1924. In particular, the Corps had little or no operational capability while the care and maintenance aspect of the earlier establishment was reflected in the cadre status of the three service squadrons.

The formulation of the 1939 Peace Establishment was apparently only an interim measure leading to a much larger war establishment. In June 1939 GHQ outlined a war establishment that included three operational squadrons, based on the existing training cadres, operating no less than 54 aircraft. The Fighter Squadron was to have 22 aircraft and was to 'be employed in the defence of Dublin'. The Reconnaissance and Medium Bomber Squadron was intended to operate 16 aircraft for 'coastal patrol duties and special duties as necessary' while the Coastal Patrol Squadron would operate another 16 aircraft and be 'required for the patrol of the coast'.[68] It should be noted that only 22 aircraft in total had been specifically purchased for three first line units in preparation for the Emergency. A total of nine Avro Ansons had been purchased for the Reconnaissance & Medium Bomber Squadron. Three Walrus aircraft had been purchased for the Coastal Patrol Squadron. Four Gloster Gladiators were delivered in March 1938 to the Army Co-operation Squadron, which was re-designated Fighter Squadron in March 1939. Six Westland Lysanders, erroneously purchased as advanced training aircraft in July 1939, were also assigned to Fighter Squadron. In due course the Air Corps War Establishment of June 1940 would provide for the three operational squadrons and a total of 1,286 personnel, though 366 were 'not to be raised'.[69]

GOVERNMENT STRATEGY

During the immediate pre-War years it indicated that while the Army was planning for the expansion, training and equipping of a substantial field army to defend the country against invasion from whatever quarter, the Government, while maintaining de Valera's neutral stance, was working on a co-operation strategy with Britain that would concentrate on intelligence and counter-intelligence. During August and September 1938 the Irish administration had many personal contacts and written communications with defence planners in Britain and had sought and received much advice and guidance on defence and defence planning. Through the Department of External Affairs, de Valera received advice and co-ordinated his policies on strategic matters with those of Britain. In addition to agreeing policy on intelligence and counter intelligence matters, policies on food planning, censorship, petrol rationing and eventually military co-operation, were agreed between the two governments. Similarly de Valera received, mainly in the form of manuals, all manner of information on preparation for war and on the conduct of war. In effect, he subordinated a largely passive defence strategy to that of the UK.[70]

Arising out of a series of meetings of the committee on emergency measures, held between 7 September and 14 October 1938, the Army's plans for the defence of the State by a large conventional force were superseded by a passive defensive strategy put forward by de Valera. These meetings, all chaired by de Valera and attended by representatives of the Departments of Agriculture, Defence, External Affairs, Finance, Industry and Commerce and the Taoiseach, were arranged to discuss measures required in the event of a European war, in the context of the defensive priorities set out in de Valera's memorandum for the Government dated 6 September 1938. This document put great emphasis on the matters discussed and agreed with Britain:

1. Supplies of food . . . essential commodities . . . regulation of external trade
2. Censorship, counter-espionage, control of communications and publicity
3. Coast watching
4. Financial and monetary control
5. Control of transport
6. Military measures
7. [etc][71]

This memorandum marked the first occasion on which the Government had formulated a broad defensive policy or strategy for the expected Emergency. The strategy that included 'military measures' as a minor aspect of the range of measures predicated on co-operation with British defensive aims had the effect of making the Costello and O'Higgins plans redundant.

AERODROMES

In the pre-War planning for the possible expansion of the Air Corps the study and consideration of the occupation of aerodromes other than Baldonnell seem to have been uncoordinated and, ultimately, fruitless. On 17 July 1937, in his discussion with the Minister, Mulcahy had suggested that about four more aerodromes should be established for a future emergency situation.[72] Mulcahy had requested that a programme for the occupation of four aerodromes should be agreed well in advance of their eventual use. Costello's expansion plan of 21 March 1938 proposed that one and a half squadrons should remain at Baldonnell while a coastal patrol squadron be stationed at a permanent base at Shannon Airport. He erroneously assumed 'that the Air Corps would have the use of Government and Municipal Airports', and that one would soon be provided at Cork.[73]

In April 1938 Mulcahy supplied the General Staff with an outline of the accommodations required at Baldonnell, Shannon Airport, Midleton, and Oranmore, indicating that Gormanston should be maintained but for no specific purpose. He made no estimate of the necessary financial provision for aeronautical facilities at Shannon but suggested that provision should be made for 'living accommodation, including married quarters, for 22 officers and 186 Other Ranks' as well as workshops, hangars for 20 aircraft, slipways and administrative buildings.[74] In December 1938 some £300,000 was apparently earmarked by the Department of Finance for the development of two aerodromes other than Baldonnell.[75] The Department of Defence made a case that reflected indecision and lack of coordination on their part and on the part of the General Staff. The department sought the provision of funds to provide for additional accommodation for a reconnaissance and medium bombing squadron stating that while it was proposed to detach it from Baldonnell it had not been decided where the squadron would be located. The Department of Defence detailed the particular requirements at such a station:

> The necessary accommodation will include the provision of four hangars with runways; a new building to house 175 officers and men; storage and office accommodation; wireless, photographic and medical huts; quartermasters and barrack services stores; petrol tanks; dining, cookhouse, recreational and gymnasium facilities; married quarters for 4 officers, 10 non-commissioned officers and 20 men; a transport shed for vehicles; and light, water and sewage facilities.[76]

The sanction of the Minister for Finance for the spending of an estimated £135,000 was sought 'for inclusion in the 1939–40 estimates for public works and buildings' with the actual works to be carried out under the direction of the Commissioners of Public Works at a location to be notified later.[77] In April 1939 the Department of Defence renewed their request in respect of provision for additional accommodation for an Air Corps reconnaissance and bombing squadron stating that it was proposed to develop the suggested accommodation at Gormanston Camp. It was noted that it was the Minister's desire that the provision of the necessary accommodation should be regarded as a matter of extreme urgency and that sketch plans and a revised estimate of the total cost of the project would be forwarded as soon as possible.[78] The revision of the works and the estimates was made necessary due to the fact that Gormanston, like Baldonnell a former RFC/RAF training depot station, was showing the effects of 20 years of neglect.[79] Consideration, by the Department of Finance, of a revised estimate of £165,000 for the reinstatement of Gormanston was delayed, however, due to difficulties with security of tenure. Due to the fact that much of the lands of the aerodrome were held on a yearly tenancy it was suggested that it would be necessary to obtain a more secure tenure before incurring any expenditure on the proposed new works; the Minister's sanction was sought for entering negotiations with the owner.[80] In July 1940 it was recorded that the Department of Defence had directed OPW, in September 1939, to defer plans for building works at Gormanston.[81] This rather belated decision was probably related to an operational decision, most likely arrived at in the latter days of August 1939, to send an Air Corps reconnaissance detachment to Shannon post haste.[82] As a result there was to be no full-time use of Gormanston by the Air Corps during the Emergency.

The net effect of all planning, discussions and consideration, involving the Minister, Mulcahy, the General Staff and the Defence, was that no facilities were developed for the specific use of military aviation in the coming Emergency. This situation was in stark contrast to that pertaining to the development of the civil aerodromes. In October 1941 the OPW, quoting from a statement of expenditure and commitments to 30 September 1941, reported that the State had invested some £607,248 on the development of Collinstown (Dublin Airport) and a further £495,585 in the development of Shannon.[83] This level of expenditure, prior to the construction of concrete runways at either location, should be compared with the £1,119,296 that Defence calculated was the total cost of running the Air Corps for the 16 years from 1 April 1926 to 31 March 1941 – an average of just £70,000 per annum.[84] This emphasis on the development of civil aviation facilities, despite the Emergency situation, strongly suggests that the concept of air defence was irrelevant in the Government's overall strategy – something that was further emphasised by the extreme parsimony of Finance when dealing with the provision of facilities for the Air Corps at Rineanna/Shannon during the early years of the Emergency.

THE POSSIBLE SUPPLY OF AMERICAN AIRCRAFT

The alleged refusal of the Air Corps to purchase and operate American aircraft, before or during the Emergency, has been a matter for adverse comment by Aiden Quigley and Eunan O'Halpin; such commentators suggest that the defence of the country had been totally undermined as a result. The circumstances that pertained to a potential purchase of this nature certainly require explanation. The fundamental fact is that the Air Corps, through the aegis of Colonel P. A. Mulcahy spent in access of the £150,000 allocated (by the Department of Finance) for capital expenditure in the years 1938–9 and 1939–40. In accordance with the policy of the COS and Mulcahy these monies were spent on British equipment, mainly aircraft. The purchase of an array of second or third-class aircraft must be viewed in the context of de Valera's passive defence strategy enunciated in September/October 1938. This strategy did not require an air defence role from the Air Corps or, indeed, the posting of any aircraft to aerodromes other than Baldonnell.

Arising out of an arms-purchasing mission to the US in 1939 Colonel Costello was able to tell the General Staff and the Department of Defence that supplies of suitable fighter aircraft, Sivorsky P35 or Curtis P36C with spares guaranteed, were available off the production line in the US. Subsequently Mulcahy and O'Sullivan attended a meeting in GHQ where this proposal was considered. Costello quotes them as stating 'emphatically that they would not have American machines at all no matter what might be the price, performance or delivery'.[85] While the Minister, Frank Aiken, accepted this judgement, which was supposedly based on O'Sullivan's advice, he instructed Costello to seek further information on aircraft types, prices, inspection facilities and delivery dates. One might ask why US aircraft would be so readily rejected. Costello recalled that the fundamental objection to US aircraft was that it was Mulcahy's, and the COS's, policy to buy British. Mulcahy also stated that the American aircraft were inferior. When this was refuted by Costello on the basis of comparison of the performance figures of British and American machines Mulcahy changed tack. He then stated that Air Corps technicians were only trained in the maintenance of Armstrong Siddeley engines, also the ground organisation could not cope with American engines and 'that there would be no guarantee that the American machines would be airworthy'.[86] While these objections appear spurious it is a fact that the Air Corps' technical training in the maintenance of aircraft was completely oriented towards British technology. With the apprentice scheme still at a very early stage, (it started only in 1936), the retraining of technicians to adapt to American specifications in hardware and tools would not have been possible. Similarly adapting to aircraft of semi-monocoque construction would have been a most difficult and lengthy task.

In addition, the apparent availability of US aircraft must be viewed against the background of diplomatic contacts about the same time. When Sean T. O'Kelly was in the USA in 1938 he was appalled to be told that he could only acquire armaments 'if he first got the order approved by the British Ambassador in Washington'. Even though the President authorised a meeting with the War Department O'Kelly was told that American needs were too pressing so nothing could be done.[87]

At the height of the invasion alert in 1940 the Irish Ambassador in Washington was reported as attempting to place an order for a large range of armaments and ammunition as well as '25 fighter aircraft, fully equipped'. Despite its restrictive policy on the supply of arms the UK advised the US that the order should be facilitated if at all possible. A large consignment of rifles was delivered from the US on 21 September 1940.[88]

It does not seem that the purchase of 25 aircraft, which would cost upwards of £200,000, had Department of Finance approval. In view of the miserly amounts spent on the development of Rineanna in 1940 it is highly unlikely that that Department would approve such expenditure. Even if such expenditure had been approved, the concept of the Air Corps taking delivery of and operating 25 modern fighter aircraft must be seen as spurious. The Fighter Squadron of 1939–40 had 11 or 12 inadequately trained pilots and could not, in any reasonable time-frame, have raised its flying skills and pilot numbers to levels commensurate with the standards required by the air defence systems of the time. It can be concluded, therefore, that the question of the Air Corps operating American fighter aircraft during the Emergency was, and still is, a red herring in the context of the ill-prepared defences of the State at that time.

THE EARLY EMERGENCY

At the start of the Emergency the Air Corps was notionally functioning under the 1939 Peace Establishment of 564 all ranks. Recruitment of privates was apparently permitted within the strength provided for by a war establishment, however, it would not get approval from Finance until June 1940. On 20 September 1939, about three weeks into the Emergency, Colonel Mulcahy replied to a verbal query from the COS stating that, 'in accordance with your telephone instructions of today, I give herewith a general report on the Corps.' The report suggests that, in terms of personnel and training, and notwithstanding the considerable notice of the outbreak of hostilities, the Corps was unprepared for the most basic wartime task. The position with regard to pilots was particularly stark. Only 32 of the 40 officers on strength were qualified pilots and the 27 of these assigned to flying units (schools and three squadrons) comprised less than 23 per cent of the number provided for in the War Establishment. While 11 pupil pilots had commenced flying training on 21 August 1939, no new pilots had qualified since January 1938. The training of 11 rear gunners and observers had only commenced in March 1939. Eleven wireless operator mechanics had commenced training for duty in Anson aircraft (which had first entered service in March 1937) in June 1939. Four of the best of the class, still only partially trained, were already flying on patrols out of Rineanna. The delay in initiating the training of gunners and wireless operators had resulted from the fact that the 1937 Peace Establishment did not provide for such trades. Mulcahy reported that, in addition to 57 mechanics being trained under the boy apprentice scheme, 'approximately 60 recruits are being trained in trades in the Depot' but that 'the material is not good and not more than 50 per cent are expected to be satisfactory'.[89]

Mulcahy cited establishment and strength figures to illustrate the extent to which the Air Corps was under strength vis-à-vis the War Establishment. He highlighted that the Corps had only 53 per cent of the total numbers permitted by the War Establishment (920) that would become effective in June 1940. The position in regard to officers, specifically pilots, was particularly stark. Total pilot numbers came to 28 per cent of the June 1940 provision. The position relating to the pilots in the combined operational squadrons was even more alarming. While the War Establishment was to allow for 101 pilots only 18 were serving with the three squadrons on 30 September 1939. This, however, was offset somewhat by the fact that the aircraft strength was about 41 per cent of that allowed. The overall (all-ranks) positions of the individual squadrons were little better than that of the officers. The Fighter Squadron had 93 personnel compared with an establishment figure of 233 – less than 40 per cent. The Reconnaissance and Medium Bomber Squadron, a detachment of which was already patrolling the west coast out of Rineanna/Shannon had 116 all ranks (or about 44 per cent) of a War Establishment figure of 265. Coastal Patrol Squadron, with a strength of 22 against an establishment of 273, could hardly have been termed even a token unit. On this latter point it should be remembered that it had been the Minister's original intention to have a seaplane squadron based in Shannon. The unit that existed (in Baldonnell) in September 1939 fell far short of a viable operational squadron.

CONCLUSIONS

In preparation and planning for the raising of a conventional three-service force for the defence of the country the General Staff foresaw the necessity of preparing plans for the expansion of the Air Corps. The agent of this process was to be Colonel M. J. Costello who rightly recognised that, if the Army was to have an air element commensurate with a realistic conventional defence of the country, such an element should be appropriate to the air mission in terms of organisation, personnel, equipment and training. Costello's plan of March 1938 was bold in its scale and concept, being predicated on capital expenditure in the order of £1 million. He had a vision of an air corps having a nucleus of three permanent squadrons and a capacity to expand to ten operational squadrons and 1,500 personnel in time of war. In reality, however, this expensive option was never going to get Department of Finance or Government approval. In fact it did not even get the approval of his peers.

Instead the General Staff placed their faith in the professional judgement of Major P. A. Mulcahy who had achieved dubious aviation qualifications following his appointment as OC Air Corps and DMA in 1935. The first significant development under Mulcahy was the assessment and purchase of Avro Anson aircraft. It is not at all clear what considerations influenced the decision to evaluate medium range reconnaissance aircraft in 1935. It is possible that at this early stage the General Staff foresaw, based on observation of the reconnaissance carried out during the Great War and the concepts indicated in the 'fundamental factors' document, the necessity of developing a marine reconnaissance capability. The purchase of Ansons indicated the beginnings of a significant ideological shift, in terms of Air Corps roles and functions, from those of an army air corps to those of air force status. However the aircraft types and numbers, and thus the air power capabilities of the three rudimentary squadrons eventually raised, were not destined to assume air force proportions.

With the Costello and O'Higgins plans made redundant by Government strategy, the Air Corps entered the Emergency, under the 1939 Peace Establishment, with three under-strength squadrons of training cadre status – in effect the training element of the scheme proposed by Costello. However, due to the apparent failure to set training goals and the actual failure to train adequate aircrew, including pilots, observers, gunners and wireless operators, the Air Corps of the early Emergency lacked adequate numbers of skilled personnel in practically all key areas. These inadequacies were exacerbated by an unstructured aircraft selection and purchase programme that equipped the Corps with, at the very best, second-rate aircraft for potentially front line operations – and in inadequate numbers. There is no evidence that the 22 aircraft acquired for service squadrons in the 1937–9 period were purchased because of their suitability for their intended roles, but rather because they were the aircraft available at the time and because only token amounts of monies had been provided by Government direction.

One of the more significant shortcomings entering the Emergency was in the number of pilots – only 32 compared with a notional establishment of 60 (in the 1939 Peace Establishment), and no less than 140 under the 1940 War Establishment. The small number of pilots did not concern Mulcahy or the General Staff prior to the Emergency. There is little doubt that the short service scheme was instigated, and continued during the Emergency, with the assumed future needs of civil aviation in mind.

The failure to develop military aerodromes was indicative of the fundamental differences in overall defence policy between the Government and the Army. The Government's emphasis on the development of the civil airports at Shannon and Dublin, while minimising expenditure on military aviation and co-operating closely with the UK, confirms the priority of civil over military aviation – even in this time of national emergency. Clearly, therefore, the short-sightedness of the Government was what led to the great inadequacy of military aviation during this crucial point in the history of the State.

EIGHT

SUPPORT SERVICES

—

As the Military Air Service (1922–4) and the Air Corps, in their actual and assumed roles, constituted army aviation, the squadrons and flights were never dispersed throughout the country as part of ground forces formations. The centrally organised Air Corps of the 1920s and 1930 functioned at a very modest level of airmanship. The *ad hoc* nature of pilot training resulted in flexible objectives and made only moderate demands on staff, aircraft and other training resources. As operational flying was practically non-existent the absence of such service supports as meteorology and communications (signals) did not constitute an identifiable problem. However, with the advent of the Emergency and a planning process that presumed air force rather than army aviation roles the failure to develop essential service support structures becomes more obvious. While not identified at the time, inadequate support services were to have a significant impact of the efficacy of the Air Corps during the Emergency.

METEOROLOGY

For reasons that are not sufficiently clear, but will be explored, it was to be 1964 before a properly organised, staffed and equipped meteorological station was established at Baldonnell to provide hourly meteorological observations on a 24-hour basis. And it was even later before a meteorological forecaster was part of the staff there. Weather observations, during daylight hours only, had been supplied by service personnel from 1941 until 1958. The Meteorological Service began taking observations at Baldonnell in June 1958.[1] From 1964 a 24-hour service has been provided by the civilian staff of the Meteorological Office of the Department of Transport and Power and its successors. This development was made necessary by the purchase of search and rescue helicopters in late 1963, the start of search and rescue operations (SAR) operations and the formal establishment of the Search and Rescue Flight (Flying Wing, Air Corps) the following year.[2] This advance in meteorological services had taken an inordinate length of time to evolve.

A fully equipped observation station, not the forecasting station suggested in the O'Duffy scheme of reorganisation, had existed at Baldonnell up to May 1922 as a standard facility on an RAF aerodrome. Four times a day it had provided the standard meteorological observations as provided by the long-established observation stations, at Valentia, Birr, Malin Head and Roche's Point which were to remain under British management until 1936.[3] The Air Service of 1922 had no capacity to make meteorological observations and had no access to a forecasting service. Though the observations taken in the Phoenix Park under the supervision of the survey blacksmith would have been of limited use to the new air organisation, it was recommended that the staff of the Meteorological Office should be put at the disposal of the Air Service and that a trained observer should be appointed and put in charge with immediate affect.[4] In August 1922 it was reported that the Department of Agriculture, which had taken over responsibility for meteorology, had arranged for the observations of the Ordnance Survey station to be passed to Baldonnell by telephone by 10.30 hours each day. It was recognised that the records of one station were not of great value. It was also noted that the Meteorological Office in London had been requested to send copies of its observations, special reports and maps to Baldonnell each day.[5]

As a result of this request the Air Service was soon in receipt of the 07.00 hour and 13.00 hour telegraphic forecasts addressed daily to the officer commanding, Baldonnell Aerodrome. However, in May 1923, the Air Ministry informed the AFO that the ministry was then incurring charges for the telegraphic and telephonic services relating to the transmission of meteorological information to Baldonnell and that the annual transmission cost to the new State would amount to about £205.[6] An account was sent to Defence:

> An amount of £44. 7s. 4d. has been expended on this service in respect of the period 1 April 1923 to 18 June inclusive, and I am to enquire whether you are prepared to accept this amount as a charge against the vote of your department.[7]

The AFO, under the mistaken impression that the meteorological stations in Ireland were funded by the State, suggested that the cost of transmitting meteorological information could be more than offset against the financial value of the meteorological reports from Irish stations and, thus, requested that the Air Ministry agree to waive the charge.[8] The Air Ministry responded stating that the meteorological information supplied was free but that the transmission costs were payable and pointed out 'that the cost of the services rendered by the meteorological observers at the various Irish stations was borne, not by Saorstát Éireann but by the Air Ministry'.[9] Subsequently, with no apparent reference to the Army or the Air Service, the AFO sought financial sanction for the initial account, paid the account on receipt of authorisation and terminated receipt of the service on the basis of cost.[10]

In 1924–5 an interdepartmental committee considered the arrangements for the collection and distribution of meteorological observations made in the Saorstát and the possible establishment of a meteorological service. Acting Major T. J. Maloney, then OC Air Corps was nominated as a Department of Defence representative with Liam Archer (OC Signals) as a 'joint representative'. Having sought direction from the COS Maloney was informed that the Army had, in effect, no interest in meteorology.[11] Notwithstanding, in his contribution to the committee Maloney emphasised the strategic importance of meteorology to ground forces that had been demonstrated during the recent war. He noted that the modern tendency was for states to place meteorological services under the defence or war departments and to assign its management to the State's military aviation service. He emphasised the increasing importance of weather forecasts in the context of aircraft flights of the order of 200 miles or more. Though not placing great stress on the necessity for synoptic meteorology and forecasting for the Air Corps he did suggest that the headquarters of such a service should be at Baldonnell.[12] The committee's report recommended that a meteorological service should be established in the country:

> That so far as synoptic meteorology is concerned the existing [British] machinery of forecasting should not be duplicated, but efforts. . . should be directed towards establishing a system of local forecasts based upon a study of the general forecast in relation to local and geographic and meteorological conditions.[13]

Despite the fact that Baldonnell was the sole aerodrome in the State for both military and civil aviation no specific recommendation was made to locate there the meteorological facilities required by international agreement. With no development forthcoming the Air Corps' relationship with meteorology was to remain largely theoretical for some time. Though the 1924 Establishment provided for a meteorological officer no officer was so qualified or appointed. An Edward Cannon made an unsolicited application to the Air Corps for such a post in August 1925. By return of post he received a copy of R. G. K. Lampfert's *Meteorology* which he was requested to review.[14] On the basis of the handwritten review supplied it was considered that Mr Cannon's knowledge of synoptic meteorology was insufficient to justify his employment. It was suggested that he might find employment in the State meteorological service, the formation of which was expected.[15]

At about this time the Air Corps' appreciation and observation of weather conditions was rather basic. On a daily basis the Duty Officer was required 'to test the air, and render a short report to the squadron adjutant as to weather conditions, and suitability or otherwise for flying'.[16] The report of Lieutenant T. J. Prendeville on 30 June 1925 was probably typical. He reported that 'the air at 500 feet' was 'gusty and bumpy' and at 1,500 feet it was just 'bumpy' while, at ground level, the wind was 'SW 15–20mph'.[17]

In June 1928 the International Commission for Air Navigation requested the General Staff to supply details of the meteorological facilities and services available at Free State aerodromes. The specific questions as to what observations were made at Baldonnell and what other sources of weather data were available were referred initially to Air Corps Headquarters. These questions were referred by Commandant G. J. Carroll to the Revd W. M. O'Riordan; MSc C. F. Father Bill (Chaplain at Baldonnell from 1925 to 1959) was the only person on the station who had an appreciation and knowledge of meteorology and was recognised as the Air Corps' authority on the subject until 1936.[18] He reported that observations of barometric pressure, wet and dry bulb thermometers (relative humidity), and maximum and minimum thermometers were taken at Baldonnell by the Air Officer on a daily basis. He suggested that, as the instruments were substandard, the observations taken were of little value. He also reported that the daily weather reports of the Meteorological Office, London were being supplied to Baldonnell but being sent by post they arrived one to three days late. As a result these reports were of use for instructional purposes only. His conclusion was blunt:

> In conclusion I would remark that the International Commission's enquiry was regarding facilities available in the aerodromes of the Irish Free State. You will see that there are no such facilities whatsoever.[19]

In the 1928–9 estimates £200 had been approved for the equipment of a first class station but apparently was not expended. In 1929 £48 was sanctioned for the purchase of meteorological instruments for the training of officers of the Air Corps.[20] One of the principal instruments purchased was a mercury barometer that was delivered to Baldonnell on 15 July 1930. The instrument began giving inaccurate readings and was returned to the manufacturers in London in June 1931. As the manufacturers could find no fault it was returned to service only to be found, in October 1931, to under-read by five millibars. After inconclusive inspection of the instrument by the Jesuits at Rathfarnham Castle, and its return to Baldonnell, the tube was found to be cracked apparently due to accidental damage in transit. After repair in Dublin the instrument was deemed to be functioning normally during the summer of 1933 but in need of calibration and certification that could only be carried out in London. In late October 1933 the barometer was taken to the National Physics Laboratory in London by Commandant Carroll. However, when it was about to be collected in November, it was reported by the high commissioner to have been found to have a problem with the vacuum.[21] Due to unexplained delays the necessary repairs were not completed until June 1936 and the account, for £3. 18s. 6d., was settled in August of that year.[22] It would appear that much bureaucratic lethargy, combined with the indifference of OC Air Corps or his subordinates, had contributed to this inordinate delay.

By 1930 the Air Corps had been receiving the 07.00 hours Air Ministry forecast by telephone at about 10.00 hours with no mention of the cost.[23] By February 1931 this arrangement was terminated as it was reported that since the previous February the wireless station at Baldonnell did not have the appropriate equipment to take the 09.10 hours Air Ministry weather report being transmitted from Rugby.[24] At the end of that year it was reported that the daily weather report from the Air Ministry, in written form, had been received throughout the year. As these reports were being forwarded through the Secretary of the department they were being received several days late, a situation that was subsequently remedied.[25] In 1932 arrangements were made to receive occasional weather reports on request, from the British station at Valentia in Kerry. While the reports were free, to obviate transmission expense to the Air Ministry, the Air Corps' requests for special reports were sent on a reply-paid basis.[26] It should be noted, however, that although difficulties with the receipt of meteorological reports and forecasts were frequent in the early 1930s, there was no evidence of any adverse effect on the conduct of flying.

The 1931–2 Peace Establishment made provision for a meteorological instructor and for a single observer of private rank. Lieutenant J. P. Twohig was made responsible for the recording of meteorological observations at Baldonnell. With Twohig's untimely death in August 1933 O'Riordan took over the task of making and recording the daily observations and, when pilot training courses were in progress, also acted as the lecturer on meteorological theory. Nonetheless, he did not consider himself competent to give practical instruction in the subject. He also advised that, while he was very willing to assist in various ways concerning meteorology, he suggested that a qualified meteorologist should be on the staff of the aerodrome to provide weather forecasts.[27]

In July 1935, when about to go on leave, O'Riordan wrote to the Adjutant, Captain D. J. Murphy, for the information of the commanding officer, to the effect that his understudy, Private James Flavin, the 'Meteorologist (private)' of the 1934 Peace Establishment, would also not be available to make the once-daily observations due to other duties. He suggested that this brought into focus the question of having a permanent meteorologist appointed and intimated that he would like to discuss the matter with the commanding officer. On the same day the newly appointed OC, Major Mulcahy, directed a pupil pilot to take over the observation duties and to consult the Chaplain with regard to making himself familiar with the work involved. Two days later O'Riordan wrote to the Adjutant reminding him that he was anxious to discuss the whole matter of meteorological observation with the commanding officer. He emphasised the shortcomings of the situation, pointing out in particular that a full range of observations, as required by pilots, needed to be made and recorded much more frequently. Though he disagreed with the concept of the observations being taken over by a pupil pilot he had referred the young officer to the *Observer's Handbook* for full and complete instructions. The Adjutant in turn studied the five files reflecting the history of meteorology since 1922 and discussed the matter with the Chaplain. He recommended to Major Mulcahy that a committee of investigation be appointed to report on the matter.[28]

Later in 1935 O'Riordan penned a study, 'Meteorological facilities for pilots in the Free State', highlighting again that there were no such facilities. He pointed out that Northern Ireland had at least one properly equipped weather station (at Aldergrove), and that if the Free State wanted to establish itself as the terminus for transatlantic flights, it was imperative that the State provide similar facilities. 'It should not be forgotten that Saorstát Éireann has definite international obligations in this matter' in accordance with the International Convention for Air Navigation. These obligations were particularly in respect of climatology, or historic weather records, current weather reports for synoptic purposes and short or long-term forecasts for specified areas:

> Quite apart from these international obligations, there is a definite obligation on the government of the Saorstát to provide meteorological facilities for its own military and civilian pilots. The need for these facilities has not been appreciated as it might up to the present.[29]

He reviewed the proceedings of the Inter-Departmental Commission which had been set up by the Minister for Education 11 years previously. He restated the terms of reference, and the four principal conclusions of the earlier study and indicated that no steps had been taken to give effect to its recommendations. O'Riordan, indicating that expert advice was available from academics in the Dublin universities and from the senior officials in Cahirciveen (Valentia Observatory), suggested the establishment of a meteorological service consisting of four or five main stations and a large number of observer stations. While he considered that the question of meteorological services was the concern of the Department of Defence alone he put forth the idea that any scheme should include locating a trained meteorologist at

Baldonnell as well as extra meteorological equipment and appropriate radio facilities operating on a 24-hour basis.

> The meteorologist would be trained in forecasting and would preferably be a civilian. He would need a staff of two or three at least. One of his duties would be the issuing of forecasts to military pilots as required.[30]

He recommended that all the existing equipment at Baldonnell could be used but that a number of additional instruments would have to be obtained at an added cost of about £500. This sum was to have included £200 for an anemometer In conclusion he stated that a broad view should be taken of the matter as not alone were lives at stake, those of citizens of the Saorstát and others, but also the honour of the country.[31]

Mulcahy forwarded the submission to the COS complete with a brief covering letter. He reported that he had discussed the matter with Revd W. M. O'Riordan and with Tom Morley of Valencia Observatory. He gave the impression that the memorandum was his idea indicating that 'as a result of these talks I asked O'Riordan to prepare a memorandum under certain headings so as to assist me in drawing up a report on the matter'. He suggested that the subject should be taken up without delay by the Department of Industry and Commerce and that the Department should be reminded of its responsibilities in the matter of meteorological services.[32]

In view of the second-hand exchanges between Mulcahy and the Chaplain and the fact that the files reflect no personal contact, one has to be somewhat sceptical about the manner in which Mulcahy, in effect, claimed credit for initiating the Chaplain's study. In his submission O'Riordan cited the loss of an aircraft and its pilot on a flight from the US to Ireland in unknown weather conditions as the main reason for highlighting the absence of meteorological services in the Free State. However, he made no reference to specific discussions with Mulcahy or Morley. While Mulcahy gave the proposal his general endorsement he did not mention, as he should have done as DMA, to the specific needs of military aviation. Mulcahy could have availed of the opportunity to emphasise that Baldonnell, as the State's only military and civil aerodrome, had an immediate and urgent need for such a service and for an appropriate meteorological station on the aerodrome for that matter. His subsequent correspondence with O'Riordan would suggest that Mulcahy did not appreciate the relevance of meteorology to military aviation and at best only saw meteorology in a civil context.

On 15 October 1935 Major J. P. Cotter, Director of Signals (DS) having been given a copy of O'Riordan's memorandum and Mulcahy's covering letter by the COS, visited Baldonnell to discuss meteorology.[33] Following discussions between Cotter, Mulcahy and O'Riordan the COS established an Army committee on meteorology under the chairmanship of Cotter with Commandant J. G. Carroll, Mr R. W. O'Sullivan (Aeronautical Engineer) and O'Riordan as members. Their brief was to co-operate with the Inter-Departmental Committee in establishing a meteorological scheme for the Saorstát and to investigate 'all aspects of meteorology as they affect military activities in this country'.[34] On 4 November 1935, for reasons that are not obvious, only Cotter and O'Riordan attended a preparatory meeting at Baldonnell in advance of an inter-departmental meeting. Some months later, apparently following a number of meetings of the Inter-Departmental Committee, O'Riordan felt obliged to write to the commanding officer:

> Since I was present at the conferences on the question of meteorological services in November last, I have been under the impression that it was intended to provide a fully equipped and staffed meteorological station at Baldonnell before the cross-channel air services started. Now, however, I hear rumours to the effect that part of the duties of Lieuts Cumiskey and Stapleton will be the issuing of weather reports to the pilots of these Services.[35]

He went on to point out that not only were the two named officers not adequately qualified for the intended duties but that no one on the station was so qualified. Lest he be held responsible, in view of the fact that he had been lecturing on meteorology to Air Corps pilots for some time, he disclaimed any responsibility for the fitness of the two officers mentioned for the duty of issuing weather reports in the absence of a meteorological station and fore-caster. He reminded Mulcahy that the instruction he had given over the years was purely of a theoretical nature aimed at giving pilots an elementary know-ledge of the principles of meteorology.[36] The inferences in O'Riordan's latest response went far beyond the matter of the competence of two pilots, then being groomed for positions as control officers. It appears that, following an unknown, but apparently small, number of meetings of the Inter-Departmental Committee held in November 1935 O'Riordan was no longer party to the discussions. It also appears that the military contributors to the main commit-tee had no concept of the application of meteorology to army or air activities and, therefore, made no particular representations on the matter of establishing a meteorological station at Baldonnell.

As it transpired, no station, for civil or military use, was to be established at Baldonnell during the four years that the newly established Aer Lingus operated to and from that location. In fact it would be several decades before Baldonnell would have a meteorological station with a forecaster on the staff.

Based on the ambivalence displayed previously in such matters, it could be inferred that the military influence at the Inter-Departmental Committee (most likely conveyed by Major Cotter) was very negative in character, and that the case for a station at Baldonnell, as proposed by O'Riordan but not supported by Mulcahy, was not projected. Mulcahy's response to O'Riordan's latest, and it seems last, communication on the matter was terse and dismissive, and contrary to the tone of his fulsome endorsement of O'Riordan's submission of October 1935:

> The fact that any responsibility might rest with you is not at all apparent. A job has to be done and I am using the materials available. I am quite aware that the officer personnel here are not expert in meteorological matters but have got to control this end of the [civil air] service and not only the two officers whom you mention but, all officers will have to take on the job of control officer.[37]

He informed the Chaplain that his efforts in the matter of meteorology were much appreciated and that he was confident of having his assistance again if and when required.[38] This response, insofar as it actually dealt with the matter of the duties of control officers, ignored the fundamental fact that Baldonnell was not to have a meteorological station despite its status in civil and military aviation. It also implied that the Chaplain's assistance was not likely to be required by Mulcahy. While O'Riordan acted as meteorological instructor to the 1937 wings' course, however, there is no record of him offering, or being asked for, further assistance or advice on meteorology during the remainder of Mulcahy's command.

The national Meteorological Service was established in December 1936 and took over the management of the existing stations from the British on 1 April 1937.[39] With no station established at Baldonnell and the station planned for Dublin Airport yet to be put in place, area forecasts for Baldonnell, taking local conditions into account, were not available. From 1937 civil and military pilots had to rely on interpretation of the general forecast available from Foynes.[40] The seemingly apathetic attitude of Mulcahy, probably reflecting a similar attitude in GHQ, can be understood in the context of an Army leadership that displayed no understanding of the strategic importance of meteorology to military ground operations. Clearly, they perceived military aviation as an army or ground forces matter that did not require knowledge or appreciation of meteorology. However, it is not easy to understand why Mulcahy was so reluctant to accept advice on a highly complex matter from one who knew. The indirect exchanges with O'Riordan strongly suggest that Mulcahy did not seek or welcome advice even in matters that were clearly outside his comprehension.

A golden opportunity, to have appropriate meteorological facilities estab-lished, was thus passed up in 1936, and little progress appears to have been made subsequently – before, during or after the Emergency. From as early as June 1937 financial sanction for the supply and installation of a remote-reading anemometer at Baldonnell, as originally recommended by O'Riordan, was in place.[41] However, as late as 24 May 1942, with an obvious lack of urgency on the part of the Air Corps and with evidence of bureaucratic lethargy in various Army offices, the equipment had still not been installed by the Corps of Engineers or even supplied by the Meteorological Service.[42]

From October 1943 the taking of observations at Baldonnell was put on a more formal basis by arrangement with the Meteorological Service. While the three military meteorologists of the 1943 Establishment also performed aerodrome control duties their primary responsibility was the taking of meteor-ological observations. These were apparently taken twice a day and made available to the meteorological service on the Service's Forms 7440 and 7441.[43] Early in 1944 R. W. O'Sullivan, the Air Corps' (civilian) Aeronautical Engineer, forwarded a brief case, supporting a proposal for the installation of a proper meteorological station at Baldonnell, for the consideration of OC Air Corps, Major W. P. Delamere:

> The present arrangement consists of telephone communication with the Dublin Airport at Collinstown by means of which information based on an analysis of the general weather situation and on observations at Collinstown is used to prepare a daily weather chart at Baldonnel . . . It takes no account of local conditions at Baldonnel . . .[44]

O'Sullivan pointed out that the local conditions at Baldonnell were very differ-ent to those at Collinstown due to the effect of the proximity of the Dublin/ Wicklow hills on the amount and height of cloud as well as the wind speed and direction. He further suggested that the Director of Meteorological Services would generally be in favour of the idea on the basis that an increase in personnel was pending. However, Major Delamere did not agree. He annotated the submission to the effect that he had discussed the matter with the Director and that it was not the opportune moment to put forward a case.[45] He was proved correct as, later that year, it was reported that additional personnel were not being recruited and that some meteorological personnel were being moved from Dublin Airport to Shannon, resulting in a reduction in the standard of service available to the Air Corps.[46] From 1 February 1945 there was no duty forecaster at Dublin Airport. In the meantime, arrangements were in place for the Air Corps to receive special forecasts from Foynes.[47]

There is an intriguing postscript to the failure to establish a meteorological station at Baldonnell until 1959. Correspondence unearthed in 1957, in
fact, reveals that a meteorological station could well have been established
much earlier. In 1957, with dwindling numbers of military meteorologists the
then OC, Colonel P. Quinn, made a detailed and well argued case for an
appropriately staffed meteorological office at Baldonnell. To emphasise the
case being made, Quinn referred to GHQ and Department of Defence files
and correspondence going back to 1945, 1942, 1939 and 1937.[48] The submission
was initially referred to the Director of Plans and Operations in GHQ:

> This matter, as far as can be seen, was first raised in 1939 (CR File S/91) and again
> in June 1945 when the Dept. of Defence on the recommendation of the then Chief
> of Staff wrote and asked the Dept. of Industry and Commerce to allot one
> meteorological officer to the Air Corps to act as instructor and forecaster.[49]

In its reply dated 27 July 1945 Industry and Commerce indicated that the
Minister for Finance, as far back as 1939, had granted approval for the recruitment of one meteorological officer and four assistant officers for Baldonnell. It
was stated that the appointments could not be filled at the time because of staff
shortages but that one officer and two assistants could, by 1945, be appointed.
While GHQ agreed to the 1945 proposal, Industry and Commerce did not
make the appointments and GHQ apparently lost interest.[50]

Nothing in the military correspondence of 1957 suggests that either the
Army or the Air Corps were aware of the financial sanction granted in 1939
while the Department of Industry and Commerce appears to have been remiss
in not pursuing the recruitment of the personnel authorised in 1939, and again
in 1945. Equally, Defence/GHQ could be faulted for not pursuing the matter
when made aware in 1945. However, in view of the correspondence between
the Meteorological Service and GHQ, it is curious that the principal Air
Corps file on meteorology contains no correspondence for the period from 9
March 1936, when O'Riordan was, in effect, dismissed by Mulcahy, until after
the appointment of W. P. Delamere, as OC Air Corps, in December 1942.

AIR TRAFFIC CONTROL

From as early as September 1924 the functions of an aerodrome control officer
were the subject of a daily roster. Initially those duties went well beyond the
basic one that specified that the duty officer would ensure that both the
landing 'T' and wind vane were correct, and that the aerodrome was clear of
all obstructions during flying:

At about 9 am he will test the air, and render a short report to the squadron adjutant as to weather conditions, and suitability or otherwise for flying. . . . He will be responsible for . . . warning of pilots, observers and pupils for flying duties, and detailing of machines in connection with flying for the day.[51]

Air Corps standing orders of 1927 defined the duties of the Air Officer in slightly different terms. In addition to testing the air he had to record his observations regarding the weather in the commanding officer's 'daily weather log' at 09.00, 14.00 and 17.00 hours.[52] The following year these duties were considerably widened to include those relating to the arrival and departure of civil aircraft transiting the Irish Sea. Apart from sending arrival and departure messages by telephone, the Air Officer had particular alerting responsibilities in the event of an aircraft being overdue. In effect the Air Officer, on a daily basis, monitored the conduct of civil aviation and performed the various related administrative functions on behalf of the Department of Industry and Commerce.[53] It should, however, be appreciated that civil arrivals and departures were rare happenings in the 1920s and 1930s. In fact they were so rare that the events were often reported in the newspapers of the time.[54]

At some time between October 1931 and June 1935 the aerodrome control officer had been relieved of the responsibility of making the decision relating to the suitability of the weather for flying and for any arrangements relating to the organising of flying. This task is recorded as being assigned to the orderly officer in 1935.[55] However, his responsibilities relating to the safe operation of the aerodrome were further clarified stating that sheep and cattle should be removed from the landing and take off area of the aerodrome when flying was imminent or in progress. In 1937 the duties relating to the control of civil air traffic remained substantially the same as those of 1928 though the alerting procedures relating to overdue aircraft were brought up to date. For reasons that are not obvious, as late as 1937, no wireless telegraphy facility existed for civil aviation communications between Baldonnell and with the corresponding UK station at Seaforth (north-west of Liverpool).[56]

FORMAL ATC

In the meantime the Department of Defence and Industry and Commerce had the matter of the training of personnel in civil air traffic control duties under consideration, in preparation for a State-sponsored civil air service based at Baldonnell. The Department decided that while the existing military personnel could also do civil control duties it was preferable that they should undergo specialist training. In pursuance of this matter the Department of

Defence requested that the High Commissioner in London be asked to make enquiries as to the conditions under which three officers might, as soon as possible, attend a course of instruction at Croydon, then London's civil airport.[57] The High Commissioner replied to the effect that the Air Ministry did not provide such courses at Croydon or elsewhere but that they would facilitate extended visits for familiarisation purposes. The Dominions Office had indicated that the Air Ministry was prepared to take two officers, one at a time for a fortnight each, starting on 1 January 1936. Return visits would be possible and no fees were payable.[58]

Subsequently, the two flying officers made a report on attending a 'course of instruction – Croydon Airport' that commenced on 13 January 1936. They confirmed that there had been in fact no formal system of instruction but were satisfied that they had come away from Croydon with a complete knowledge of the system of control in use there. Devoting most of their attention to the operation of the control tower they had observed the work of the control office, meteorological office and the communications department. The 'W/T and R/T station', with 'four powerful transmission and receiving sets' was the major component of the communications department. Wireless telegraphy, using the 'Q Code' of the era, was the means of communication with aircraft while the extent of the use of radio telephony is not clear. They noted the inter-relationship between the various departments and the records that were kept and obtained a supply of the special forms used in communications. They also visited the Air Ministry and obtained a number of relevant publications.[59]

As a result of these visits the duties of aerodrome control officer were performed by individual officers on a daily basis as detailed. Until the transfer of the Aer Lingus operation to the new Dublin Airport in January 1940, the task was predominantly one of facilitating the safe conduct of civil aircraft into and out of Baldonnell.[60] The Air Corps' 1939 Peace Establishment would then put this assistance in air traffic control on a more formal basis by providing two lieutenants for civil aviation duties.

In the meantime, with aerodrome control at Baldonnell being carried out by any and all flying officers, the emphasis in control matters changed. The development of the flying-boat base at Foynes gave rise to the necessity for air traffic control at that location which, in effect, took precedence over the requirements of Baldonnell (and later Dublin Airport) prior to and during the Emergency. From 1936, until he retired in 1948, Captain E. F. Stapleton (who had qualified as a pilot in 1928) was almost continuously attached to the Department of Industry and Commerce as a control officer at Baldonnell, Collinstown and Foynes and, and from January 1946, at Shannon. During the Emergency period the control officer roster at Foynes required six officers to man two control positions. By January 1944 no less than nine flying officers had spent extended periods of duty at Foynes performing ATC functions that apparently took precedence over the flying duties for which they had been trained and appointed.[61] It was to be as late as 1964, however, before a military air traffic control section, with officers solely trained for the specific function, was established in the Air Corps.[62]

AVIATION COMMUNICATIONS

During the Civil War the detrimental effect on the conduct of reconnaissance operations due to the absence of appropriate air-to-ground and ground-to-air communications, using wireless telegraphy and Morse code, was evident but eluded rectification at the time. While the Air Service had an aviation wireless officer his duties supposedly did not extend beyond providing standard Army inter-barracks communications. Even though the particular need for air communications had been identified the opportunity presented by the reorgan-isation scheme of 1924, to establish a Signal Corps element appropriate to the needs of military aviation, was passed up. Although the 1927 syllabus of training for pupil pilots required instruction in wireless telegraphy (W/T) and theory, no provision was made for aviation signals personnel until the Peace Establishment of 1931–2.[63] The Department of Defence signals unit then included provision for two W/T instructors, one lieutenant and one corporal, who were attached to Air Corps Schools for the instruction of pupil pilots in receiving and transmitting in Morse code.[64]

Throughout the army co-operation training regime of the early 1930s the emphasis put on communication by W/T was more notable. A mobile W/T station was used during air firing at Kilworth in 1932 (and presumably at subsequent air firing practices) for communication between pilots and the ground observers who provided information as to the accuracy or otherwise for the guidance of individual pilots. This service was observed to be an essential aspect of such exercises.[65] W/T communication, backed up by visual signals and message dropping, was also an essential aspect of successful artillery co-operation at this time and of early exercises conducted with the Cavalry Corps.[66] The emphasis on the use of wireless telegraphy in the context of army co-operation training starkly contrasts with the situation at Baldonnell where no ground station existed whatsoever for routine military aviation traffic. In effect, a distinctly biased policy in air communications matters towards the needs of the Army, was still clearly being laid down by the Director of Signals as had been the case in 1922.

During 1935–6 Baldonnell was very poorly equipped for military aviation. The only aviation communications facility available was a military W/T station with a range of 50 miles – though few military aircraft were equipped to use it. However the Department of Defence and GHQ facilitated all developments required to equip the aerodrome for the new civil air operation (Aer Lingus). In fact Colonel O'Higgins of GHQ is quoted as having informed Industry and Commerce and the Post Office authorities, that 'while the civil airport was at Baldonnell the needs of military flying and wireless would be subordinated to those of the civil air service'.[67] The State's first civil radio station for aviation communication was installed and inaugurated at Baldonnell to coincide with the commencement of civil air services by Aer Lingus in May 1936. The start of the service was promulgated by statutory instrument:

> As from Wednesday, 20 May 1936 a new radio station, providing a radio-communication and direction-finding service available to all aircraft, will be brought into operation at Baldonnel aerodrome, Co. Dublin.[68]

While the notice suggests that the radio station was for the use of both civil and military aircraft, the station provided both W/T and radio telephony (R/T – two-way voice communication) facilities, including a direction finding (DF) service, for civil aircraft only. Military aircraft continued to proceed in accordance with the traditional visual ground signs displayed on the aerodrome. In the period 1938 to 1941, in particular, the poor standard of airborne W/T and R/T equipment, and of the corresponding ground equipment, as well as the absence of a dedicated military direction finding service, were to impact adversely on military air operations and indeed on the morale of pilots.

The investigation of 1941–2 provided a review of the Air Corps' signals matters from 1936 to the end of 1941. The most telling remark was not given in evidence but was rather the opening question put to Captain P. J. Murphy (Air Corps HQ Signals Staff Officer, 1939–43) on 30 January 1941, some 16 months into the Emergency: 'We understand the position is that we are gradually building up a signal service within the Air Corps?'[69] Murphy answered in the affirmative. The admission that the communications facilities for the Air Corps were still being gradually built up can only be seen as an admission of the Signal Corps' failure to provide adequate communication facilities. Murphy added that when he had arrived into the Signal Company at Baldonnell in 1936 there were practically no aircraft equipped with (W/T) radio. Equipment was gradually acquired and the Air Corps radio service was, thus, eventually built up. Except for the five most recently acquired Ansons it had been quite difficult to equip the aircraft with radio as most of the radio equipment was only acquired from the UK in dribs and drabs. By the end of 1940 some 16 aircraft (six Ansons, three Walrus, one Hawker Hind and six

Lysanders) had been fitted with wireless telegraphy equipment which operated on medium and short-wave frequencies and required a wireless operator as part of the crew. While the Gladiators were fitted with R/T radios other aircraft designated by Mulcahy as fighters, such as Hinds and Lysanders, were fitted with W/T equipment. In the context of aerial combat and the fighter operations of the time, the use of W/T sets in fighter aircraft was as impractical as it was antiquated. It is apparent that the Hinds and Lysanders of 1940, designated as fighter aircraft, should have had the same R/T set as was fitted to the Gladiators of the same squadron.

The three Gladiators were fitted with TR 9B short-wave R/T sets when delivered in 1938 even though there was no ground station. After Gladiator 23 was written off as a result of an accident on 20 October 1938 its receiver and transmitter, were used as a ground station for the other Gladiators. The TR 9B radio, operating off an aircraft battery, had a range of ten miles or less. The same aircraft radio in RAF use afforded a range of 35 miles because of superior ground stations.[70]

Despite the fact that the first aircraft requiring a radio operator, the first two Avro Ansons, had been in service since 20 March 1937, the training of operators did not commence until 1939. This situation arose because the recruitment and paying of wireless operator mechanics (WOMs) had not been provided for until the 1939 Peace Establishment. Mulcahy had anticipated that the Signal Corps would post qualified operators into the newly created vacancies but, probably due to general demand for such skilled personnel, trained operators were not forthcoming. As a result the training of wireless operator mechanics for Air Corps aircraft did not commence until as late as June 1939.

The training of operators was not the responsibility of the Air Support Company (Cadre) Signal Corps but that of Captain P. J. Murphy, the Signals Staff Officer in Air Corps HQ. Initially training was provided for only 11 men though the 1939 Peace Establishment allowed for a total of 19 wireless operators in the service squadrons. Such was the urgency to complete a 12-month course in the fastest possible time that OC Air Corps instructed Murphy to curtail instruction to the actual operation of W/T sets while omitting theory and technical aspects. Defects in wireless operator training were highlighted by Murphy in the posting of a reconnaissance detachment to Rineanna/ Shannon, apparently with the minimum of notice, on 30 August 1939:

> Shortly afterwards, on the outbreak of war, owing to the lack of qualified operators in the Air Corps, in order to enable patrols to be carried out from Rineanna I had to go as one of the two qualified operators and between us we took four of the best pupils for a period of six months . . . and I carried them on all flights for the purpose of training them. When they were sufficiently trained to

carry out communication on patrol, I was recalled to Baldonnel to take up my normal duty as Air Corps Signal Officer.[71]

In effect these four pupil operators, who had the theory and technical aspect of their course suspended, were flying on operational missions within three months of commencing training. Subsequently, they achieved the required standard of operator skills by way of on-the-job training during operational missions patrolling the south and west coasts in wartime and North Atlantic winter conditions.

By January 1941 the Air Corps Signal Officer was able to report that 18 men had been trained as wireless operators for service as aircrew though only one could be graded as a 1st Class Operator. By this time the War Establishment of 1940 allowed for a total of 45 wireless operators. As a training objective this number would appear to have been unattainable in view of the fact that Air Corps did not have a dedicated signals training element and had made little progress since June 1939.[72]

In January 1941 it was reported that the position regarding the range of airborne and ground R/T reception had improved greatly with the acquisition of a more powerful ground transmitter. In September 1940 Thomas Murphy, an amateur radio enthusiast had contacted the Director of Signals with a view to demonstrating a transmitter of his own design and manufacture. The transmitter was built mainly of commercially available components but included a small number of parts sourced through the Signal Corps. The Director agreed to have the equipment demonstrated for test purposes with the possibility of it being purchased or rented. On 18 and 19 September 1940 extensive tests were carried out on the transmitter installed at Baldonnell using aircraft fitted with the TR 9B R/T set. A number of flights, by day and by night, as far north as Dundalk and as far south as Carnsore Point, and at altitudes between 1,000 and 12,000 feet, were carried out with very satisfactory results. The test results provided 'reliable communication both ways from the plane to the ground and from the ground to the plane up to about 40 miles' dependent on the altitude of the aircraft. The satisfactory results achieved with Murphy's transmitter were only possible, however, because a good receiver was made available by Lieutenant Andy Woods, an Air Corps flying officer and radio enthusiast.[73] One of the pilots involved in the tests, Captain T. J. Hanley, suggested that ground-to-air communication was possible out to 60 or 70 miles and recommended that this could be achieved using a transmitter with an output of 2,000 Watts or more and a good type of receiver.[74] When it is considered that the radio sets used for the tests were those produced by amateurs, the question surely must arise as to the commitment of the Signal Corps to air communications. It could be concluded that the Signal Corps perceived its primary function to be the provision of W/T communication services as appropriate to the Army co-operation function and, thus, concentrated on this to the detriment of those communications commensurate with air force roles, particularly that of fighter aircraft that required R/T communication. With R/T equipped aircraft in service since March 1938 it should not have been beyond the technical competence of the Signal Corps to design and build an appropriate R/T ground station (as in effect had been done by Mr Murphy and Lieutenant Woods).

The obvious deficiency in the preparedness of the Air Corps Company, Signal Corps, whose main responsibility was the maintenance and operation of ground stations, for both air and army purposes, was also highlighted when, on the outbreak of war, a 24-hour watch was introduced for ground stations. This was only possible at Baldonnell and Rineanna by using the wireless operators, who had been being trained as aircrew by and for the Air Corps, as operators of the ground station wireless sets. This continued until April or May 1940.[75] As a result, the Signal Corps, having failed to provide or train wireless operator mechanics for Air Corps aircraft, similarly failed to provide adequate wireless operators for its own ground stations and, initially at least, had to rely on partially trained Air Corps operators to carry out the most fundamental Signal Corps function required at military aerodromes.

DIRECTION FINDING

A direction finding (DF) service, as an aid to the safe navigation of civil aircraft, operated at Baldonnell from May 1936. From July 1937 similar services, on a grander scale, were made available in the Shannon area. Ballygirreen, County Clare had two direction finding stations for transatlantic traffic approaching the Shannon/Foynes area. The Foynes seaplane base had a DF station to facilitate aircraft landing in the river estuary. At Ballygirreen the Marconi DFG10 medium-wave direction finder was suitable for use by military aircraft such as the Ansons equipped with W/T wireless sets while the DFG12 short-wave station was compatible with the R/T equipment of fighter aircraft.[76] The Baldonnell medium-wave DF station could only be used by those aircraft, mainly reconnaissance types, fitted with W/T sets. As this DF facility was intended specifically for the use of civil aircraft Air Corps pilots could only avail of the service at such times as when it was not engaged with civil traffic. The Air Corps' use of the civil DF stations at Baldonnell and Ballygirreen, therefore, was mainly in the context of Anson and Walrus aircraft transiting between Baldonnell and Rineanna/Shannon.

With the operation of the Baldonnell Civil DF Station only available between 09.15 and 17.00 hours, and the with the service severely curtailed within those hours, the availability of dedicated direction finding services for military navigation purposes was to become a somewhat confused and contentious issue during the first 15 months of the Emergency. In April 1939 Lieutenant Jim Devoy had recommended that at least two direction finding W/T stations be installed in selected locations as essential aids to the safe navigation of military aircraft.[77] The matter had been the subject of correspondence from the director of signals to OC Air Corps on 24 February 1939 and 5 April 1939. Later that year a board of officers was assembled by order of the COS to investigate the proposal put forward by the Director of Signals, and agreed with Mulcahy, to purchase four (G12, short wave) Direction Finding sets. The Board, comprised of staff officers of GHQ, primarily considered the merits of short-wave DF sets for the detection of illicit radio transmitters but first addressed the Air Corps' DF requirements with Mulcahy. Mulcahy considered that three short-wave DF stations (for use with Gladiators) were urgently needed by the Air Corps even despite the fact that the effectiveness of short-wave stations was limited, particularly at night. He stated that medium-wave DF stations were not suitable for erection in the vicinity of military aerodromes as the masts constituted too great an obstruction. He did not favour homing devices in military aircraft. He objected to the concept of loop aerial in fighters or larger aircraft on the basis again that the erection of masts on aerodromes would cause obstruction.[78]

The very short-sighted position taken by Mulcahy suggests he had little appreciation of the necessity to improve direction finding services facilities for military aircraft. In particular his objection to medium-wave aerial system seems contrived. The civil radio station at Baldonnell transmitted and received on medium wave, 348 kilocycles, and provided a DF service for both W/T and R/T equipped aircraft. It was located on the eastern side of the aerodrome and its size presented no obstruction to aircraft. This being so, it is not easy to understand why Mulcahy did not insist on similar equipment being installed at Baldonnell and Rineanna, specifically for military use.

The number of ground transmitters, in Ireland, compatible with airborne homing devices such as loop aerials normally fitted to aircraft like the Anson, was limited. However Mulcahy's objection to the concept of loop aerials in Ansons could only be considered spurious and ill informed. Captain T. J. Hanley highlighted the position regarding loop aerials for Anson aircraft:

> The modification to have the loop aerial installed on Ansons was issued by A. V. Roe on 23/11/38, Anson modification No 214. On 7/1/39, [a copy of] the modification was passed to OC Workshops, to requisition the material. This was not done as OC Workshops got instructions [from higher authority] not to requisition them.[79]

This response strongly implies that Mulcahy had made a conscious decision not to authorise the purchase and fitting of equipment that, in the circumstances, should have been considered essential. Anson aircraft, had they been fitted with the loop aerial (direction finding) modification, could have made very good use of the wireless stations at Malin Head and Valentia for navigation purposes while patrolling off the west coast. Such use would have certainly increased the efficiency of patrols in marginal weather conditions and, moreover, added greatly to safety. The Board made no comment on the loop aerial decision. The Board also made no response on the matter of short-wave DF for aviation purposes while the sole recommendation was the purchase of two short-wave DFG12 sets for 'the detection of illicit transmitters'.

During the discussions the Board had indicated to Major Gantly that a serious view was being taken of the fact that a short-wave DFG12 set had been delivered to the North Wall. Apparently its purchase had not been properly sanctioned. In addition, no recommendation was made regarding its possible use.[80] In view of the two applications of such equipment it is not obvious why 'two short wave direction finding sets' that had been apparently been purchased in January 1940 were still 'lying in the stores of the Signal Corps' two years later.[81]

THE CIVIL DF AT BALDONNELL

During 1940 the position regarding DF services, at Baldonnell in particular, did not improve much and the matter of the poor service available to military aircraft drew the attention of Captain T. J. Hanley. He reminded OC Air Corps that the Aer Lingus service was shortly due to move to the newly developed Collinstown (Dublin) Airport, and that as sufficient DF facilities and staff were already available there he suggested that the DF station and staff at Baldonnell would be surplus to civil needs. He stated a substantial case to prevent the medium-wave civil DF station being closed down in accordance with the previous plan and to have it, and its Posts and Telegraphs staff, taken over as a dedicated military facility. He argued that every effort should be made to preserve the safety of aircraft:

> any personnel or equipment which exist in this country and which we think is necessary to preserve those aeroplanes should, not only be put at our disposal, but should if necessary be seized by military authority. The DF Station and staff at Baldonnel come into this category.[82]

Hanley was sceptical about the possible installation of a short-wave DF system and cited his 'considerable study of DF systems for navigation and approach purposes' over the previous three years. He also referred to poor results from tests carried out in June 1939 with the Ballygirreen short-wave DF.[83]

It is not clear how Mulcahy reacted to these recommendations. Annotations on the letter suggest that he agreed with the general thrust but would stop short of recommending the military seizure of civil aviation facilities. On 17 January 1940 the acting CO suggested that until such time as a military short-wave station was installed at Baldonnell the existing civil DF, which would continue to function as a stand by for emergency civil use after the transfer of air services to Collinstown, with the co-operation of the Department of Posts and Telegraphs, might be used to train Signal Company or Air Corps wireless operators in DF procedures.[84]

In due course the Department of Posts and Telegraphs responded. They indicated that the Baldonnell DF was still required, during the normal hours of the service of the Collinstown station, for emergency purposes only. The post office also stated that it was anticipated that the Baldonnell DF receiver would be required at Shannon Airport in a few months. Conditional permission was granted to allow Air and Signal Corps personnel become familiar with procedures and for the use of the station to give DF bearings to military aircraft. The main condition specified that, while civil aircraft were in flight on the cross-channel service in either direction, the training of Army personnel should be suspended and that the receiver at Baldonnell could only be tuned to the military frequency when there was no civil aircraft on the cross-channel service. The final condition stated that training and DF facilities could only be provided while the Post Office operator was in attendance so that, if DF services were required after normal hours of operation, prior arrangements should be made.[85] Having been asked to specify the times at which a DF service and training facilities would be required Mulcahy confirmed that the service would be required during normal Air Corps duty hours and stated that he appreciated that civil aviation must receive priority from the DF station.[86] While this arrangement might have appeared satisfactory in terms of the potential availability of DF bearings on the Air Corps medium-wave frequency, it must be noted that there were two civil cross-channel flights in each direction each day – a situation that would have obviated military use for large parts of the standard day. In regard to operator training the Signal Corps apparently failed to supply additional personnel to train in DF procedures, thus, it fell to the Air Corps to supply a small number of wireless operators who had to be withdrawn from flying duties.[87]

While the withdrawal from service and the removal of the Baldonnell DF station had been anticipated for the time being it continued to remain in service. It was intimated once more in October 1940 that, as the services at Dublin Airport were then well established, the question of the closing down of the DF station at Baldonnell for civil purposes was being considered, and the views of the Air Corps on the matter were requested.[88] Mulcahy, replying through the office of DS, stated that, as it was the only DF facility at Baldonnell it was required, in bad weather, for aircraft of the R & MB Squadron and for the training of young pilots in the use of DF during instrument flying practice. He insisted that the DF continue to be made available for military aircraft when necessary and indicated that in the event of the PO staff being withdrawn 'Army personnel would have to take over the operation of the Station'.[89] As the DS did not appear to pass the Air Corps' response to the Department Industry and Commerce, a reminder had to be sent asking for the Air Corps' view on the continued use of the station for civil stand by purposes.[90] Mulcahy reconsidered his position, and, having consulted with two of his staff officers and OC Air Corps Signals Company, recommended 'that the DF station should be retained for civilian purposes as an alternative to Collinstown' on the basis that 'the latter could be rendered useless by enemy action'.[91] The net result of this apparently perverse recommendation was that the Baldonnell DF station continued in its traditional civil aviation role until 5 September 1941.[92] It might have been considered that the DF station could have been transferred to military control much earlier on the basis that it could revert to civil use in the unlikely event of the Collinstown station being destroyed. It is significant that, despite the agitation on the part of the pilots who required a proper DF service, Mulcahy and his headquarters staff could not be convinced that such a service, under military control, was essential.

Notwithstanding the reluctance to take over the station it became more available for military use. Arising out of increased use, those primarily concerned with the quality of the service and most familiar with its limitations – Lieutenant Woods and Captain Hanley – endeavoured to improve on the cumbersome and slow method of transmitting bearings to military aircraft. With no military transmitter located at the DF station the operator was not able to transmit bearing information direct to the requesting pilot. The bearings were first sent by telephone to the Air Corps wireless station for transmission to the requesting aircraft, thus imposing an unacceptable delay. Lieutenant Woods proposed two alternative wireless solutions that would speed up the process greatly, particularly for the benefit of aircraft as they got closer to the station. Captain Hanley endorsed the observations and recommendations of his operations officer and suggested that if the system could not give four bearings per minute it should be changed without delay.[93]

The very next day the contentious position regarding the use of the station for military purposes was demonstrated by an incident, and its aftermath, involving the unauthorised use of the civil facility by Air Corps personnel. On the morning of 21 November 1940 Anson 42 was flown by Captain Hanley to Rineanna. Acting on instructions, he was conveying a medical officer on a brief visit to the R & MB Squadron Detachment there. Having checked the weather forecast prior to his departure at 11.00 hours, he indicated to the Air Corps Signal Officer that he would require DF assistance on his return, the time of which depended on the length of the medical officer's stay in Rineanna. At about 16.00 hours the aircraft left Rineanna in good flying conditions but encountered bad weather just west of Baldonnell. With cloud at 500 feet and visibility of about 400 yards and the conditions deteriorating, it was essential that DF assistance be requested.[94] In the meanwhile Lieutenant Woods established that the aircraft had left Rineanna and that it would not reach Baldonnel until about 17.30 hours and observed that the weather at Baldonnell was deteriorating to the extent that the aircraft would require DF services. Woods also established that the Post Office staff had closed down the station at the usual time, 16.15 hours. As Captain M. Egan (OC Air Corps Signal Company) had indicated that he did not know what might be done, Woods ascertained that Lieutenant Lorcan Sinnott (Signals Officer, Fighter Squadron) knew how to operate the equipment and suggested that he be allowed to operate the station with the safety of the aircraft and crew as the primary consideration. With the agreement of Captain Egan, Lieutenants Woods and Sinnott gained access to the DF station with the caretaker's co-operation. The station was opened at 16.45 hours and, in the period from 16.56 hours to 17.35 hours, Lieutenant Sinnott passed bearings to the incoming aircraft. At 17.35 hours the aircraft made a safe landing in poor weather conditions and the station was closed three minutes later.[95]

Two days later Captain P. J. Murphy submitted a report to OC AC giving a brief outline of the circumstances relating to the use of the DF station. He did so not by way of complaint but rather to highlight the urgent necessity for having satisfactory DF facilities available for military aircraft whenever necessary.[96] Without seeking written reports or explanations on the matter Major Mulcahy, wrote to the two squadron commanders concerned:

> I am informed that, on the evening of 21st instant. 2/Lieutenant Sinnott of your unit entered the Post Office DF Station and operated the station in the absence of the Post Office DF operator. You will inform this officer that his action was irregular.[97]

He also reminded OC Fighter Squadron that the DF station had closed down at 16.15 hours and that it was only with the permission of the Post Office that it could be made available thereafter. On the same day Mulcahy communicated in similar terms with OC CP Squadron reprimanding Lieutenant Woods, stating that his actions were considered to be 'most irregular'.[98] Apparently no effort was made to establish why the AC Signals Officer, a member of Mulcahy's staff, had failed to ensure that the DF station remained open after 16.15 hours.

Woods was to express his dissatisfaction with the implications of his commanding officer's reprimand when explaining the matter to his squadron commander, Captain Hanley, who had been the pilot of the aircraft. He detailed all the circumstances and, while accepting that his action was considered irregular, justified his action on the basis of the safety of the aircraft and crew:

> I take full responsibility for my actions in this case, I accept responsibility for both Capt. Egan and 2/Lt Sinnott, as these officers, knowing me to be a flying officer of experience, agreed with me. I must confess that if a similar situation were to arise again, I would still feel it my duty to do the same thing.[99]

Due to his being indisposed it was to be early January 1941 before Captain Hanley could address the issue. He first confirmed that he had, as directed, made Lieutenant Woods aware of the CO's displeasure regarding the irregular use of the DF station. As pilot of the aircraft he made a comprehensive report on all pertinent aspects of the flight and the incident. He further confirmed that the Air Corps Signals Officer had stated that he would arrange DF facilities, would fly on the aircraft to Rineanna in order to carry out an inspection there and would act as the radio operator on the flight. Having encountered the adverse weather conditions in the Baldonnell area Hanley stated that he was most thankful for the DF assistance and that he had not been aware that the regular operator was not on duty. He also stated that he would expect no less from Lieutenant Woods or any other officer left in charge. He reminded OC Air Corps that when he (Mulcahy) had been flown from Rineanna to Baldonnell on 9 December 1940 the DF station was also manned by Air Corps personnel in circumstances similar to those of 21 November. He was openly critical of his CO:

> This bears out my statement that any officer with a sense of responsibility will have no compunction to ensure the safety of an aeroplane and its crew. To me the deplorable part of the situation is, that officers who do their obvious duty in such circumstances are admonished by their superiors, and all because no proper DF facilities exist at Baldonnel for Air Corps aeroplanes. The question of providing proper facilities for the Air Corps has now been going on for years without result, and these irregularities would not occur if (a) The DF station were handed over to the Air Corps or (b) Post Office operators were stationed at Baldonnel where one of them would be at all times available.[100]

While not couched in personal terms this robust endorsement of the actions taken by Woods was, in effect, a considerable criticism of Major Mulcahy and his lack of empathy with pilots and the aviation culture. In November 1940 Captain Hanley had already taken matters a stage further when he had written to the Minister complaining about the failure of the Air Corps to purchase vacuum pumps, essential equipment on Anson aircraft.[101] The highlighting of such matters was a manifestation of the frustration of the pilot body with various aspects of Mulcahy's command.

The folk memory of the 1935–42 period indicates that the pilot body was at loggerheads with its commanding officer and that pilots were threatening to remove their pilots' wings because the CO wore wings to which he was not entitled. Hanley's critical correspondence would have been received by Colonel Mulcahy on 10 January 1941, on the same day that the Chief of Staff had convened an 'investigation into the effectiveness, organisation, equipment, training and administration of the Air Corps'. The convening of this investigation had been made necessary by 'the evidence of demoralisation, in some cases inefficiency and stagnation, and the inadequacy and unsuitability of equipment'. The Committee would subsequently conclude:

> The confidence of a large number at least of the junior officers of the Air Corps in Colonel Mulcahy has, through one cause or another, been hopelessly undermined.[102]

On 30 January 1941, as stated above, the Investigation Committee was able to comment favourably on the fact that the signal service within the Air Corps was gradually being built up, implying a considerable improvement of an earlier position. Notwithstanding these slow improvements, the situation illustrated by the evidence on signals matters suggests that the unsatisfactory state of communications and direction finding services was a large contributor to the demoralisation of the pilots and central to the distrust that existed between them and their commanding officer in the latter part of 1940.

However, to fully appreciate the poor situation regarding these services it is essential to compare the communications services available to civil aviation – the cross-channel and trans-Atlantic services – and those available to military aviation on 3 September 1939. Baldonnel civil airport had both W/T and R/T transmitter and receiver sets (medium wave) with appropriate power to serve the Irish sector of the cross-channel air route. In addition it had a medium-wave DF station to assist navigation and approach to the civil airport.[103] These facilities were to be duplicated at the new Collinstown Airport by January 1940. The Foynes/Shannon area was very well provided with the

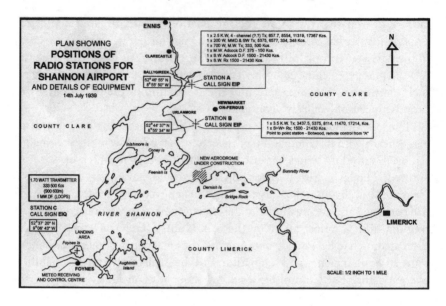

Radio Stations for Shannon Airport, 14 July 1939

communications systems appropriate to the flying range of the trans-Atlantic flying boats using them. Ballygirreen had three transmitters and receivers covering a broad spectrum of frequencies appropriate to long-range communication. It also had one short-wave and one medium-wave DF station. At Urlanmore there was a short-wave transmitter/receiver for point-to-point communication with Botwood, Newfoundland – over 2,000 miles away. This wireless was remotely controlled from Ballygirreen. Foynes had a single transmitter/receiver and a medium-wave DF for air traffic approaching and departing the seaplane base.[104]

At the same time the military aerodrome at Baldonnell had a medium-wave W/T transmitter and a receiver giving a range of 50 or 60 miles. In addition the aerodrome had a transmitter/receiver set, salvaged from a crashed aircraft, operating on short wave and limited to a range of five or ten miles, and no dedicated DF station. Rineanna had a mobile radio car that had insufficient range to maintain contact with aircraft any further than 60 miles away and had no DF facilities.[105]

The quality of military communications generally, and DF facilities in particular, which were available to the Air Corps was also in sharp contrast to those available, on Irish soil, to the RAF during the Emergency though it is not clear when these facilities became available. Towards the end of the war the UK Government noted a number of facilities it feared could be withdrawn by de Valera in certain circumstances:

There are two Post Office wireless stations (operated by Southern Irish personnel) on Southern Ireland territory, one at Valentia and the other at Malin Head; these are used as direction finding beacons by our aircraft . . . the withdrawal of the facilities would be a serious loss.[106]

Similarly when listing the 'facilities obtained from the government of Éire during the war' the Dominions Office acknowledged 'the use by United Kingdom ships and aircraft of two wireless direction finding stations at Malin Head'.[107] Although the end of the war prevented its construction, de Valera had also given the UK authorities permission to build and operate a radar station, for use in its campaign against German submarines in the North Atlantic, in the same compound on Malin Head. Had it been built the UK authorities had agreed with Colonel Archer (who functioned as the main point of contact between the Irish and British military during the Emergency) that it would have been passed off as a 'radio lighthouse' or a 'glorified marker beacon' for the guidance and safety of aircraft.[108]

CONCLUSIONS

There is compelling evidence that the main support services, meteorology, air traffic control, communications and direction finding, only developed slowly and then mainly in the context of the services essential to civil aviation. During the period under review the Air Corps was only moderately anxious to obtain appropriate weather forecasts and at no stage did pilots appear to get exercised by deficiencies in this area. While a national Meteorological Service was eventually established this was purely to coincide with the commencement of the civil air service to the UK. Though civil air services operated from Baldonnell from May 1936 to January 1940, the fact that no station was established at Baldonnell suggests not just apathy on the part of the Army leadership but outright opposition. Revd W. M. O'Riordan MSc and R. W. O'Sullivan, both expressed clear views as to the vital importance of meteorology to military aviation. Nonetheless, Major Mulcahy attached no importance to the matter and, in fact, acted in a manner detrimental to the setting up of a meteorological station at Baldonnell. The apparent apathy of the pilot body is perplexing. However, given Mulcahy's dismissive attitude towards O'Riordan, who has been shown to have had progressive opinions on the matter, it is quite possible that pilots were, in effect, browbeaten by an opinionated and disciplinarian commanding officer to such an extent that they were dissuaded from voicing conflicting opinions.

Although a rudimentary form of air traffic control existed from an early stage the air traffic control function was only formally established, for civil aviation, in 1936. Air Corps officers performed this function at Baldonnell on behalf of the Department of Industry and Commerce until early January 1940 and thereafter at Dublin Airport. Similar duties were performed at Foynes from August 1937 to January 1946. In this matter also the indifference of Colonel Mulcahy was quite evident. However, there is no evidence that the emphasis on civil Air Traffic Control (ATC) had any detrimental influence on the conduct of military aviation except to the extent that many flying officers were employed on civil ATC duties for longer or shorter periods during the Emergency. There can be little doubt, however, that the provision of services to civil aviation had greater priority than had air defence during the Emergency.

As the civil air service was being initiated in 1936, an appropriate wireless and direction finding services were established at Baldonnell for the sole use of civil aircraft. While the Signal Corps had traditionally provided W/T services for Army co-operation, from 1939 the new squadrons would have required communications technically appropriate to air force roles. The evidence suggests that the Signal Corps, the independent arbiter of what was appropriate in terms of the Air Corps' communications requirements, totally failed to identify such requirements, did not keep abreast with modern developments and, as a result, failed to develop systems and equipment appropriate to the new roles of military aviation. In view of the ease with which two amateurs, Thomas Murphy and Lieutenant Andy Woods, provided the equipment and demonstrated how R/T could be greatly improved, it is not easy to understand why the Signal Corps could not have manufactured a ground station appropriate to the needs of fighter aircraft. The failure of the Signal Corps to actually install the direction finding equipment purchased is equally perplexing.

Mulcahy's failure to insist on the equipping of aircraft and aerodromes with appropriate communications and direction finding facilities can only be understood in the context of his lack of expertise and knowledge demonstrated while dealing with these essential aviation services. His dominant personality and lack of willingness to listen to those who clearly understood the nuances of military aviation put him, literally, on a different frequency to the flying officers. The contrast, between the substantial communications resources put in place for commercial civil aviation and the rudimentary facilities that existed at Baldonnell and Rineanna on 3 September 1939, could not have been starker. The ultimate irony regarding communications, though the Air Corps pilots would not have been aware at the time, was the fact that the RAF had the use of far superior direction finding facilities on Irish soil during the Emergency.

1 William J. McSweeney and Charles F. Russell, Baldonnell, August 1922 (Air Service)

2 Postcard from the front, Fermoy, October 1922. Back row left to right included are: Sgt Bob Long, Lt C. (Tiny) Flanagan, Cpl J. Maher. Centre of front row: Lts J. C. Fitzmaurice and W. P. Delamere (author's collection)

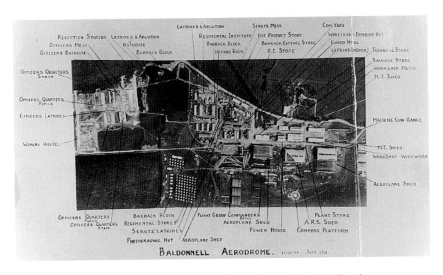

3 Baldonnell under construction, September 1918 (Royal Air Force)

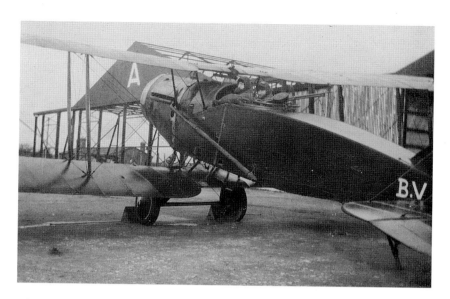

4 Lt J. C. Fitzmaurice and Lt L. C. Gogan about to depart on a mission, Fermoy, early 1923 (L. Gogan)

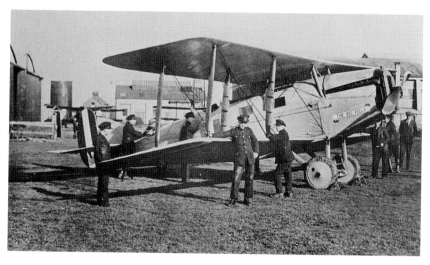

5 The Martinsyde Type A, Mk II at Baldonnell, 10 February 1923 (L. & C. Walsh)

6 Instructors, pupil pilots and technical personnel, 'A' Flight, No. 1 Squadron, Air Service at Baldonnell, mid 1923 (Air Service)

7 Two young visitors in Bristol Fighter 19, c.1926 (George Weston)

8 TGIF, Baldonnell, September 1926: Lt G. J. Carroll, Lt O. A. Heron, Mr P. A. Sheehan (*Irish Times*), unknown observer, Lt J. C. Fitzmaurice and Lt A. X. Lawlor (pupil pilot) (A. C. Woods)

9 'C' Flight, No. 1 Training Squadron, July 1933. Left to right, front row: Lt W. J. Keane, Capt. O. A. Heron, Lt P. J. Twohig. Back row: Lt T. J. Hanley, Lt J. F. Moynihan, Lt A. J. Russell, Lt F. O'Cathain (Air Corps)

10 De Havilland DH9 at Baldonnell *c.*1926 (George Weston)

11 'At home' day at Baldonnell, 18 July 1936: Major P. A. Mulcahy (replete with wings), Mrs P. Quinn, Comdt P. Quinn, Mrs M. O'Brien, Mrs A. C. Woods, Capt. M. O'Brien and O'Brien children (*Irish Times*)

12 Group of pilots at air firing at Gormanston, July 1939. Left to right, front row: M. A. O'Byrne, P. J. Murphy (Observer), N. E. Fogarty, J. Kearney, T. L. Young (Observer), D. V. Horgan, M. Sheerin, W. J. Keane, M. P. Quinlan, J. Moynihan, T. J. Hanley, W. Ryan, P. Swan. Back row: J. O'Brien, K. T. Curran, A. C. Woods, J. F. Walsh, M. E. McCullagh, O. L. O'Toole, F. F. Reade, D. K. Johnston, P. O'Sullivan, P. J. McDermott, M. Higgins, P. McCormack (A. C. Woods)

13 Lysanders at Baldonnell, 6 August 1939 (Air Corps)

14 A posed photograph of Lockheed Hudson 91 being 'armed' with 112lb bombs, c.1942 (Air Corps)

15 The building of the garrison church at Baldonnell. The completed building was dedicated on 10 December 1944 (Air Corps)

16 A line-up of eleven Hawker Hurricanes at Baldonnell, *c.*February 1944 (Air Corps)

17 Group of officers at Rineanna, 16 April 1943. Left to right, front row: Lt J. Kearney, unknown officer, Lt T. McKeown, Lt M. Carr, Lt A. C. Woods. Back row: Lt W. Ryan, Lt M. Cusack, Lt Dodd (Signals), Capt. T. J. Hanley, Lt N. E. Fogarty, Lt Sutherland (Artillery Corps), unknown pilot (A. C. Woods)

NINE

THE AIR CORPS' EMERGENCY

—

As war broke out on 3 September 1939 the Army plan for the conventional defence of the country had long since been replaced by the Government's passive defence strategy centred on co-operation with the United Kingdom. Despite expansive plans for the defence of the country prepared by the Army, Government strategy envisaged little by way of active defensive measures. As a result there was no concept of air defence and thus no identified function for military aviation. Apart from the purchase of a small number of second-rate aircraft, no preparation had been made for the deployment of air resources. Nevertheless, an ill prepared, poorly equipped and inadequately trained Air Corps had already been committed to an operational task. As will be seen this was only the first unplanned mission to be imposed on the Air Corps.

THE PILOT OFFICER SITUATION

The Air Corps entered the Emergency on the basis of the 1939 Peace Establishment that provided for 63 officers, 150 NCOs and 351 privates, a total of 564 all ranks. While the strength on 20 September 1939 (512 all ranks) represented over 90 per cent of that permitted many had only recently been recruited and were largely untrained. There were significant deficiencies in the key disciplines of a technical corps. The main personnel shortages were in pilots, wireless operators, fitters, riggers and armament artificers – in effect in most occupations essential to the operation of military aircraft. The pilot numbers, at 32, were a little over 50 per cent of the notional establishment figure and less than a quarter of the number that was soon to be provided for in the 1940 War Establishment. This number excluded the commanding officer, Major Mulcahy, whose qualification as a pilot in 1936 was highly suspect and who had not been authorised by the school commandant to fly solo after March 1938.[1] Immediately prior to the Emergency there was no perception of an overall pilot shortage or within individual squadrons. Indeed it is of particular note that Commandant G. J. Carroll and Captain T. J. Hanley spent much of 1939 in the employ of Aer Lingus. Both had returned to the service by 1 September 1939. Although Hanley served from June 1939 to March 1945, Carroll, the senior qualified pilot, returned to Aer Lingus on 23 October 1939 and, except for brief periods in 1940 and 1941, spent the greater part of the Emergency away from the Air Corps on half pay.[2]

During the first 16 months of the Emergency pilots were employed on the basis of about one third between Air Corps HQ and Schools, one third between R & MB and CP Squadrons and one third with Fighter Squadron. The training cadre status of the three squadrons indicates that they were not considered capable of performing to an operational standard.[3] During the Emergency pilot numbers increased with the qualification of a total of 23 young officers in 1940–1 (1st and 2nd Short Service Classes) but the squadrons were never to achieve anything like full strength in flying officers under the 1940 War Establishment. However, this deficiency never became a critical factor. As will be explored, the manifest deficiencies in aircraft numbers and performance, as well as inadequate support services minimised squadron operational capabilities to an extent that far exceeded any disadvantage represented by inadequate pilot numbers. After 1940 the matter of pilot numbers was no longer even of academic interest except to the extent that flying instructors had to be withdrawn from the service squadrons from time to time to ensure the progress of the short service flying courses in 'Air Corps School'.[4] Paradoxically, the training of pilots for civil aviation and the provision of ATC officers for duty at civil airports appears to have had greater priority over all other Air Corps activities during the Emergency. The records show that, during the period July 1937 to December 1944, 17 officers were detached to civil air traffic control (at Foynes, Collinstown, Rineanna/Shannon, and at Baldonnell up to January 1940), for periods ranging from one month to five years. These duties removed pilots from operational and other military duties for extended periods. Flying instructors were taken from instructional duties. Captain P. McCormack, an aeronautical engineer and pilot, carried out ATC duties for 19 months from 1943 to 1944 to the neglect of his responsibilities as the officer in charge of workshops in Maintenance Unit. As early as December 1943 six officers were seconded to the Department of Industry and Commerce for an ATC course. Of these, three were appointed to civil ATC, and had retired by November 1944. As late as January 1945 four pilots were still detached from their units on loan to civil ATC while the last one remained so detached until January 1946.[5]

RECONNAISSANCE AND MEDIUM BOMBER SQUADRON

On Tuesday 29 August 1939 Mulcahy received a verbal instruction from the COS, by telephone, in reply to which he penned an acknowledgement:

> In accordance with your instructions . . . Shannon airport will be occupied by the Reconnaissance Squadron (Cadre) tomorrow . . . The following matters are required to be arranged immediately . . . authority to use the labour camp at the airport . . . Industry and Commerce to be notified . . . Southern Command to be instructed to facilitate . . . in regard to armed guard, supply of bedding [and] rations. . . . a medical officer and medical orderly to be attached . . . arrangements to be made for mass . . .[6]

Mulcahy requested that a previous arrangement, for the refuelling of military aircraft at Shannon by Irish Shell, be approved by the QMG. He also requested that a telephone line be installed at the Squadron Commander's proposed headquarters. He requested the return of a mobile workshop which had previously been transferred to the Supply and Transport Corps. While this vehicle was considered essential to the servicing of aircraft, literally in the field, there is no record of its return. That the Air Corps got the minimum notice is confirmed by Mulcahy in his evidence to the Investigation Committee in late 1941. In response to assertions made by a number of officers that they had no training or experience in maritime reconnaissance Mulcahy explained:

> It must be borne in mind that sea reconnaissance was sprung upon us and that we moved to Rineanna to carry out coast reconnaissance at 48 hours notice.[7]

While Mulcahy's acknowledgement of the verbal order suggests that he got only 24-hours notice, in the military tradition he had probably received a warning order the previous day. The timing of the detachment, in the absence of any preparations, strongly suggests that the decision to occupy Shannon had only just been taken, not by the COS but by the Government. It is not at all clear that the decision to have the Air Corps undertake a maritime reconnaissance role, in the North Atlantic, in winter weather conditions, was assessed at any government, department or military level.

Although correspondence survives to illustrate administrative aspects of the detachment there is a great paucity of material relating to operational matters. The most obvious deficiency is that of a written order authorising and establishing the detachment and stating its mission. In an organisation that was hidebound by regulations and written orders, the absence of a written order in this instance must be taken as deliberate. It is apparent that the posting of

an Air Corps detachment to Shannon/Rineanna on 30 August 1939 had little
to do with national defence and much to do with British-Irish co-operation
and the intelligence needs of the Britain. This is only one of many instances
where there was great reluctance to record details of military co-operation
with the UK. Records relating to Air Defence Command show that the role
of the detachment had instructions 'to carry out coastal patrols of our territorial waters from Lough Swilly along the west and south coasts to Wexford
Harbour' and part of the State's intelligence gathering machinery. Other aspects
of this network included over 80 lookout posts of the Coast Watching Service
and some 759 Garda stations and a small number of military posts that observed
and recorded aircraft and shipping movements during the Emergency.[8] The
Air Corps' role is confirmed in the correspondence indicating British consideration of Irish demands for aircraft spares on the basis of the reconnaissance
patrols which the Air Corps was carrying out and the reports on German
submarine activities the country was then furnishing to the UK.[9]

The question, however, arises as to what influences were brought to
bear to bring about such a precipitous decision. Mulcahy was informed of the
situation on 28 August 1939 – almost a week prior to the signing of the
Emergency Powers Order, 1939 – and the declaration of war by the UK. The
detachment had taken place almost immediately. On or about 30 September

Table 9.1 Establishment and Strength – R & MB Squadron Detachment, Rineanna

R & MB Squadron	Officers	NCOs	Privates	Total	Aircraft
1937 Establishment (Cadre)[12]	6	8	16	30	Not specified
Proposed – 26 Mar. 1938[13]	22	43	144	209	Do
Proposed – 21 Apr. 1938[14]	22	[62]	[124]	208	Do
1939 Peace Establishment (Cadre)	17	32	61	110	Do
1940 War Establishment	37	72	156	265	
'Less [72 O/Ranks] not to raised'[15]	37	48	108	193	16
Average strength 1939–40	11	10	65	86	[9][17]
Attached personnel	1	6	6	13	
Total[16]	12	16	71	99	
Strength, R & MB Sqn 14.1.1941	10	19	74	103	6
Attached, Sigs, Cav, Arty [18]	5	8	40	53	
Baldonnell – Maintenance [19]	1	6	14	21	
Total	16	33	128	177	

the Minister for Defence stated that it had been found necessary to send Archer to London, on 25 August 1939, on business of a similar confidential nature to that first authorised by de Valera in October 1938, that is, intelligence and counter intelligence.[10] In this particular instance, however, Archer was sent to London acting as head of a joint civil/military delegation that had instructions from de Valera to discuss arrangements for the setting up of the Coast Watching Service and explain that permanent personnel could not be recruited until October 1939. The Admiralty and the Board of Trade supplied information helpful to setting up this intelligence gathering service in a context that confirms the UK's concern with obtaining intelligence relating to German activity off the west coast.[11] Nonetheless, in all probability, a delay in deploying the service did not meet UK defensive requirements.

With war looming and with de Valera withholding use of the Treaty ports, at this juncture it is probable that the resultant lack of air and naval intelligence became an urgent matter for the UK in relation to its own defence. Within days of Archer's return from London Mulcahy was instructed to send a reconnaissance detachment to Rineanna/Shannon. The circumstances strongly indicate that the Irish Government had been requested to post a reconnaissance element to the west coast.

THE RINEANNA DETACHMENT

An Air Corps detachment of three Avro Anson and two Walrus aircraft arrived at Rineanna on the evening of Wednesday 30 August 1939. The personnel consisted of 11 officers (ten pilots and a signals officer/W/T instructor) and 77 other ranks. The attachment of a medical officer and three other ranks brought the total on the first day of occupation to 92 all ranks. This total should be noted in the context of the R & MB Squadron that had a notional establishment of 110 all ranks – 17 officers, 32 non-commissioned officers and 61 privates. As such the Squadron was not equipped to function without station services such as catering, guards, transport and sundry stores. Almost immediately a pilot officer was ordered to report to Foynes for ATC duties. An officer, six NCOs and 30 men of 1st Battalion, reported for garrison duties but were not placed under Air Corps command.[20] The inherent cultural divide, relating back to the RAF roots of the Air Corps, is evident here in the Army's attitude to the Corps. During the first seven months, when the operation was at its most intense, the average strength of the detachment, including attached personnel, was less than 100. Even after the transition to the 1940 War Establishment the strength of the air element was just above 50 per cent of the permitted 193 while the number of pilots was marginally above 25 per cent. Rather than being kept at a maximum possible strength, within the limits of the approved establishment, it appeared that the Rineanna detachment was, in fact, kept to the minimum.

THE AERODROME

In August 1939 Shannon was still at a very early stage of its development as a civil airport for trans-Atlantic air services. It could best be described as a mile square of recently reclaimed marsh. Its only buildings, other than the pump houses, were some 13 timber huts – the Labour Camp. These had housed the labour force that had carried out the drainage and reclamation work during the period beginning 8 October 1936.[21] Basic aeronautical facilities, such as: telephones, aeronautical communications, direction finding station, meteorological station, hangars and runway lighting, which should exist on any military aerodrome in advance of the arrival of an operational squadron, were not present. The major deficiency was that of an aircraft hangar. In its absence, Air Corps aircraft had to be picketed in the open for the first nine months while a marquee and tents, without duckboards, were used initially to store spares and other materiel.[22] To protect aircraft from the elements some were returned to Baldonnell while makeshift covers were made for others using material salvaged from a barrage balloon that had been shot down by Fighter Squadron's Lysander 64 near Foynes, on 4 October 1939.[23] In essence, Rineanna was a primitive facility.

It was to be 22 September 1939 before the Department of Defence contacted the Department of Finance with regard to the provision of appropriate accommodation for aircraft that, of their very nature, were not intended to be parked in the open in any weather. DOD emphasised the necessity to arrange the erection of a hangar as quickly as possible, indicating that aircraft and instruments were subject to rapid deterioration.

Consideration had been given to the question of dismantling the ex-RAF hangars at Fermoy and re-erecting them at Rineanna. This idea had been dropped on the basis of dilapidation and also due to the fact that they were required for the accommodation of Southern Command troops. Consideration was also given to the erection of temporary canvas hangars but the department opted for a new and permanent hangar as the only practical solution to the problem. Having been in contact with Messrs Thomas Thompson of Carlow, and ascertaining that the company had sufficient stocks of steel to build a suitable hangar, Defence had the director of engineers draw up a specification in consultation with the firm. The sanction of the Minister for Finance was sought, 'as an emergency measure for placing an order' for 'the complete structure (including electric light)' that could 'be erected at a cost not exceeding

£10,000'.[24] About a week later Finance had considered the proposal and directed the Office of Public Works (OPW) to take charge of the project. Under the instructions of the Department of Defence OPW was to place the contract with Thompsons, without the customary tendering process, and to supervise the completion of the contract.[25]

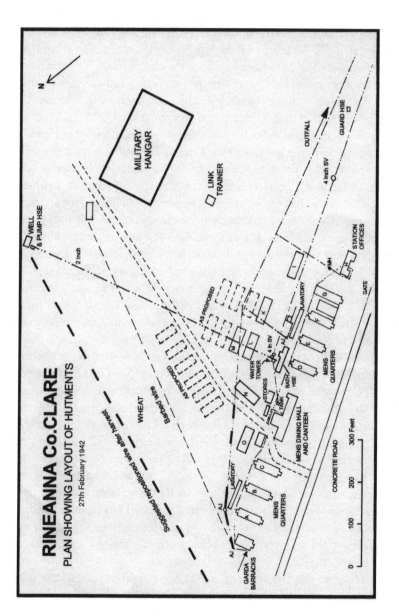

Rineanna, County Clare, 27 February 1942

Having studied the drawings and specification proposed by Thompsons, OPW identified several deficiencies in the design. They considered that the structure would need to be strengthened for erection in such an exposed location and that the provision for natural lighting was inadequate. It was also thought that the roof and walls, of galvanised corrugated iron, would permit extremes of heat and cold and cause excessive condensation. They concluded that the structure was of a type which could only be justified by the emergency situation. The OPW also noted the absence of adequate provision for site works and surmised that the Air Corps would require some sort of apron in front of the hangar that would add a further £1,000 to the cost. The OPW's preferred option was for a permanent hangar of better construction and costing as much as £22,000 but which could not be ready until July or August 1940. As the Air Corps had indicated that such a delay was unacceptable the OPW, proceeding with the lesser specification, indicated that they had arranged with Thompsons to proceed at once with the erection of the hangar. The OPW sought sanction for £11,100 to cover the cost of the hangar and up to £1,000 for an apron.[26] A Department of Finance official considered the proposal to be unattractive but, recognising that action had to be expedited to provide shelter for the aircraft, provided the required sanction.[27] After further exchanges of views on the matter of the costs, Finance sanctioned the expenditure of £500 for the apron and a total of £11,100 for the provision of a hangar, including £200 for site works.[28] The OPW estimates for 1940–1 would put the total cost of the project at £12,500 – £11,100 for the main works, £1,000 for the apron and £400 for contingencies.[29] The contract for the erection of the hangar at Shannon, costing £10,988, was placed with Thompsons in December 1939 without a competition by specific direction of Finance.[30]

The development of military accommodations must be seen in the context of the development of the civil airport at the same time. The latter project obviously had greater priority. The Airport Construction Committee, no more than any other agency, did not know what the medium to long-term policy of Defence might be and was concerned about the provision of some form of accommodation for its own administrative purposes. The Committee was endeavouring to take a decision regarding the spending of £1,800 on temporary huts in 1940, and £18,000 to £20,000 on additional temporary accommodation in 1941 and subsequent years or the construction of a first phase of a permanent building to cost about £45,000. They considered that there would be no temporary accommodation problem during the Emergency if the Air Corps could be got out of the labour hutments which they had taken over at the end of August.[31]

In February 1940 the Department of Defence sought £250 in addition to the £500 already sanctioned for the apron area. They also indicated to Finance that it would be necessary to provide a water supply and sewage system as well as drinking water and water for the washing of aircraft.[32] Financial approval for an additional £250 was received in March.[33] The matter of a water supply was further addressed by OPW with a new water main which was being laid between the well in the military camp and the new civil buildings then being built. It was suggested that a water connection be made to the hangar. However, rather than have mains sewage, it was recommended that a chemical toilet closet similar to those available in the military huts, be installed in the hangar. The water and sewage proposals, costing £100 each, were put forward as being inexpensive.[34] The OPW received sanction for the required £200 by return of post.[35] Subsequently, OPW sought and received financial sanction for a further £88 spent on the apron, £75 for a sealing coat on the apron surface and £48 for the installation of two winches to facilitate opening and closing of the heavy steel hangar doors.[36] The water connection and sanitary works, originally estimated at £200, eventually cost £282. 8*s. od.* and this was duly sanctioned.[37]

Although additional huts were built by March 1940 living conditions were still poor for officers and other ranks alike. Further modest expenditure was incurred during 1941 and 1942 as more hutments were built to provide for additional Army troops as the aerodrome assumed the status of an outpost of the Southern Command.[38] It must be noted, however, that the measured, and belated, expenditure on facilities for the Air Corps at Rineanna during the Emergency contrasted sharply with the substantial investment made in developing Dublin and Shannon at about the same time.[39]

THE AIRCRAFT

The detachment arrived in Shannon/Rineanna with three Avro Anson I aircraft out of a full complement of nine, and the two Walrus amphibian aircraft, out of three delivered in March 1939. The number of aircraft at Rineanna was not fixed; aircraft had to be returned to Baldonnell for servicing after only 20 hours flying or for anything other than minor repairs. Air Corps folklore recalls that many aircraft were rotated on Saturdays and Mondays so that married personnel could return to their families in the Dublin area at weekends. The number of aircraft was kept to a minimum due to the lack of shelter and the damage done to aircraft and instruments by the high relative humidity. The R & MB Squadron at Rineanna and the CP Squadron at Baldonnell, in effect, therefore, were run as flights of a single squadron operating from the same pool of aircraft.[40]

The Avro Anson MK I, or Avro 652A, from 1935 had been developed initially as a six seat commercial aircraft and was then further developed for service with the RAF as a general coastal reconnaissance aircraft. It was a twin-engine monoplane with a fabric covered metal fuselage and wooden wings, and had a maximum range of 790 miles and a cruising speed of 158 mph.[41] In RAF service before the war the Anson was primarily used in a variety of training roles, such as twin-engine conversion, reconnaissance, bombing and navigation training. By 1940 a small number of squadrons were operating the type in an inshore coastal reconnaissance role at various locations around the UK but only until such time as the production and supply of better aircraft – such as the Lockheed Hudson, Blenheim and Whitley – facilitated their withdrawal from front-line service. As early as the summer of 1939 due to its much superior speed, range and endurance, Hudsons had begun to replace some of the Ansons of ten Coastal Command squadrons. Thereafter, Ansons reverted to training or were used for inshore search and rescue duties, which were more suited to its capabilities.[42]

The Supermarine Walrus I was the production version of the renamed Seagull aircraft developed from about 1935. The original specification was for an aircraft that could 'operate from a ship fitted with a catapult, an aircraft carrier or from a land or marine base'.[43] Having in effect been ordered by the Minister to buy such an aircraft it appears that Mulcahy paid lip service to the concept of operating from marine bases. There is no evidence that an Air Corps specification was drawn up and it may well be that the Walrus was purchased without evaluation. The Investigation Committee of 1941–2 found the Walrus to be 'slow and underpowered', thus, unsuitable for patrolling off the west coast. The Committee also noted that the Walrus was being used only as a land-based machine due to the 'lack of sea-going spare floats' during the Emergency.[44]

OPERATIONS

Considering the secrecy surrounding the original decision and operational matters generally, it is not surprising that little is known about the actual mission and the manner in which it was undertaken. The administrative diary for the 'Air Corps Detachment, Rineanna', opened by the Squadron Commander, Captain W. J. Keane on 30 August 1939 indicates that coastal patrols began the following day. On the same day 'Capt. Hanley explained purpose and details of patrol to all officers'. The diary entry immediately before the above recorded that 'Lt. O'Byrne flew from Baldonnell with [a] dispatch'. This may well have been GHQ's written order detailing the conduct of patrols. The first patrol was forced to return due to bad weather conditions. On the same day also Captain W. J. Keane visited Foynes Meteorological

Station to make arrangements for 06.00 hours and 14.00 hours weather forecasts to be relayed via the telephone in the 'Civic Guard' (police) barracks on the airfield. Subsequently, arrangements were made to get three hour forecasts by telephone five times each day. Similarly, arrangements were made with Ballygirreen radio station, which was located six miles north of Rineanna, with particular reference to the availability of its civil direction finding service for use if and when the said station was not busy with trans-Atlantic traffic using Foynes.[45] The basic nature of the facilities at Rineanna was emphasised by an incident in early September:

> Lt. Ryan, when returning from patrol found it necessary to land in [the] dark, as [the] landing light in [the] machine (A45) had been removed. An emergency flare path was made of lighting hay, and with [the] assistance of [the] headlamps of a car, he landed successfully.[46]

After this the Squadron Commander reported that he had procured some Toledo flares for runway lighting and had erected some temporary obstruction lights in the vicinity of the aerodrome. In the absence of a proper ground station for communication with patrolling aircraft, and for the transmission of patrol reports to GHQ, these services were provided by a mobile radio car. Within days of arriving at Rineanna the Signals Officer, Lieutenant P. J. Murphy, received instructions from the Director of Signals in GHQ. The directive was based on the fact that the Department of Posts and Telegraphs had agreed that the civil aeronautical communications station at Ballygirreen would take over as the ground station for communications with the squadron aircraft. Ballygirreen was to be used, not just for aeronautical communications, but also for the normal military wireless traffic with GHQ including patrol reports. On completion of the necessary arrangements with Mr Enwright of Ballygirreen, and the installation of a direct telephone line to GHQ via Ballygirreen, the wireless van was to be returned to the Signal Corps.[47]

Many disadvantages were identified by Captain Keane and the signals officer. These points were taken up by Mulcahy after he had visited Rineanna on 9 September. He informed the ACS that the signals arrangements proposed by the director of signals were unworkable. While he was in Rineanna he had observed that an Air Corps patrolling aircraft was working with the military wireless car while Ballygirreen was working, on a different frequency, with a flying boat which was on its way to Foynes. He concluded that Ballygirreen could not be of service to military and civil aircraft at the same time. He also pointed out that confidential matters in the reconnaissance reports had to be sent by secure radio reports to Command HQ and to intelligence branch (G2) of GHQ at the end of each patrol and that such matters could not be handled by the civilian staff. He suggested that the wireless car must remain at Rineanna for aircraft duties and that Ballygirreen could handle all normal ground messages and provide a direction finding service in an emergency situation.[48] Thus, the wireless car remained at Rineanna as the only means of communication with aircraft on patrol and as a secure means of communication with Southern Command and GHQ intelligence staff.[49]

In the absence of reports on operational matters at Rineanna it is not clear to what extent the medium wave DF station at Ballygirreen was used either for navigation purposes or as an aid to aircraft returning to the airfield in bad weather. It is however known that pilots were given instruction and practice in both instrument and night flying and used the DF station in making practice approaches to Rineanna. The Squadron Commander reported that as there was no DF station at Rineanna and that the use of EIP (Ballygirreen) demanded the utmost precision as pilots familiarised themselves in approaching with the aid of QDMs and QDRs (magnetic bearings to and from the station).[50] This precision was required due to the fact that the DF station was so far removed from the airfield at Rineanna. As track error could increase as the aircraft flew away from the DF station a small error could result in the aircrew completely failing to see the aerodrome in bad visibility.

Such use might have been made of Ballygirreen DF Station on 10 October 1939; but reports suggest it was not. On that day Anson 44 was being brought back to Rineanna after servicing. Having left Baldonnell at 18.30 hours the aircraft encountered low cloud and very poor visibility en route. When the conditions got too bad to continue visually the pilot decided to carry out a forced landing in blinding rain. In doing so the aircraft bounced and struck a hedge causing some £1,245 worth of damage to the aircraft, engines, and equipment. It is apparent from the abbreviated report that the subsequent court of inquiry, with a presumably very narrow focus, did not examine the full circumstances of the accident at Ardcroney, Nenagh.[51] Had the court done so it might have inquired as to why the aircraft, returning to Rineanna at dusk and in bad weather, was not flown at a safe altitude towards the DF Station at Ballygirreen prior to making an instrument cloud-breaking procedure into Rineanna.

By 19 December 1939 a total of three Ansons had been removed from service as a result of accidents. The first, Anson 45, had been badly damaged as early as 8 September as a result of engine failure and forced landing at Ballyferriter in Kerry. The main contributory cause was a faulty fuel cock that caused fuel starvation. On 19 December 1939 Anson 43 was damaged beyond repair due to engine failure that resulted in a forced landing into Galway Bay. Ansons 44 and 45 were to remain out of service for some time due to difficulty in procuring spares.[52] As early as November 1939 Mulcahy reported that 'the coastal patrol is being maintained with difficulty' and that some aeroplanes 'are being kept serviceable by taking parts and instruments from other aeroplanes'.[53]

The question arises as to the number of patrols undertaken. Initially the Squadron carried out two per day. This frequency was soon reduced. On 5 September the Squadron Commander recorded that until further notice there would be only one patrol per day with two pilots and crews on standby. While no figures are available for the total number of patrols carried out Mulcahy gave a somewhat vague indication to the investigation:

> during last winter [1939–40] the reconnaissance squadron flew approximately 80,000 miles and covered generally the coastline from Wexford to Donegal, with particular attention to the west coast from Belmullet to the Mouth of the Shannon and the south coast from Mizen Head to Waterford.[54]

When asked what the distance represented in patrols Mulcahy suggested an average of one to two patrols per day and that the complete area had been covered once a week, and 'special areas' once daily.[55] His vague description does not appear to indicate more than a patrol per day at best. By April 1940, with three Ansons out of service *pro tem* the maintenance of the remaining six was proving difficult due to lack of spares.[56] The major difficulty was that while a 12-month supply of spares had been ordered when the aircraft were bought, spares for Ansons were only arriving spasmodically. In the meanwhile radio sets and armament were being received for other aircraft that had not been delivered. On 29 April 1940 only three of the six remaining Ansons were serviceable as the other three awaited engine spares; it was predicted that if spares did not arrive quickly the coastal patrol operation would cease.[57]

The matter of aircraft serviceability and its effect on coastal patrols was brought to a head not by Mulcahy but by the intervention of Colonel Costello who, as OC Southern Command, was Captain Keane's immediate superior. Prior to communicating with the COS, Costello consulted with Captain Keane and other pilots at Rineanna and concluded that Rineanna had 'neither the trained personnel nor the equipment suitable for the task'.[58] Acknowledging the COS's role in directing and monitoring the conduct of patrols out of Rineanna Costello indicated that it was with hesitation that he wrote on a matter that was, strictly speaking, outside his own remit. The basic point that he made was to the effect that he considered that the Ansons were almost at the end of their useful life:

> the present position is so unsatisfactory that, unless there is a reasonable prospect of maintaining a reconnaissance squadron at a reasonable [aircraft] strength the

entire position of the Air Corps will have to be reviewed. . . . In order to survive [*sic*] the limited number of flying hours left patrols are not now undertaken save in the most favourable weather conditions . . . I am sure that you fully realise the serious strain on the morale of all ranks at Rineanna which the gradual petering out of their equipment imposes.[59]

Costello also described the living conditions at Rineanna in stark terms stating that the accommodation in the camp was unsatisfactory from the point of view of the health and morale of the troops and of security. He suggested the Squadron could not be expected to survive another winter with the current accommodation conditions.[60]

The alarming aspect of the situation described by Costello was not that it had deteriorated to such an extent – in the circumstances it was inevitable – but that Mulcahy was not aware of, or had not seen fit to highlight in similar terms, the gross inadequacies relating to aircraft numbers, maintenance and supply of spares, the operation and general living conditions and resultant affect on efficiency and morale. On receipt of the letter from Costello the COS discussed the matters with Mulcahy and decided, amongst other things, to suspend coastal patrols, and withdraw the Ansons to Baldonnell where they would be used only for instruction in twin-engine aeroplane flying, navigation by radio and bomb aiming. The Walrus aircraft were to remain in Rineanna to carry out training for operation off water, instrument flying and navigation blind approach practice using Ballygirreen DF. Requests for 'special missions' were to be made to OC Air Corps who would decide whether to use aircraft from Baldonnell or Rineanna.[61]

Mulcahy later recalled that 'general coastal patrols were discontinued in May 1940' because the 'few suitable aircraft available were becoming due for complete overhaul' and that it had been necessary to conserve flying time 'so that they would be available for other missions should the situation get worse'. He made the situation out to be less futile than it actually was:

Also the necessity arose at this time for holding aircraft for special missions as ordered by the Chief of Staff and the Officers Commanding Southern and Western Commands. Such missions included the interception of belligerent aircraft and special patrols of portions of the coast line.[62]

This interpretation of the situation that pertained from about 10 May 1940 suggests that Mulcahy did not fully admit that the Squadron cadre in Rineanna, and the remainder of the two reconnaissance elements located at Baldonnell had little or no capacity for normal or special missions at the particular juncture. The Committee did not seek clarification as to what Mulcahy had meant by 'suitable aircraft'. With three aircraft out of action there were still six Avro Ansons in service – in theory sufficient to carry out the mission. The use of the term probably arises from an aspect explained by Captain Hanley:

> All Ansons have blind-flying equipment. The first four Ansons bought have only elementary blind-flying equipment which is insufficient for safe flying in bad weather. The last five Ansons have the complete blind-flying [instrument] panel but lack the vacuum pumps to operate the instruments.[63]

The points Hanley was making were to the effect that, with three of the five newer aircraft unserviceable since 1939, only two of the better equipped aircraft were available. But even these lacked the vacuum pumps that were a more reliable source of suction for gyroscopic instruments. In May 1940 these two Ansons – Nos 41 and 42 – on the basis of total flying hours were close to major inspections, the completion of which would be prolonged by the absence of spares.

About this time Captain D. V. Horgan and Mr R. W. O'Sullivan (Air Corps) and J. B. Carr of the Department of Defence spent 12 days in the UK. There, with the assistance of the High Commissioner, they made representations to the War Office and Dominions Office in respect of the supply of equipment for the Army and to the Air Ministry about, in particular, the supply of 15 advanced trainers. The delivery of six (ex-RAF) Hawker Hinds and five new Miles Magisters resulted. A major concern, however, was the supply of spares for various aircraft, including the Ansons, already in service. The Air Ministry requested lists of the spares required for a specific period and undertook to try to arrange a contract as required by the Department of Defence. A visit, by special written permission, to A. V. Roe was no more promising. Subject to 'instructions to proceed' being issued by the Air Ministry, Mr Burley promised that his company would do everything possible to assist.[64] At this time, as previously, the UK authorities were well disposed to assist in training matters but less so in regard to aircraft, equipment and spares that might be put to operational use. When viewed in the context of the intelligence value of coastal patrols this UK position appears somewhat contradictory.

No records to indicate the extent of such reconnaissance missions or, indeed, their effectiveness are available for guidance. The General Report on the Defence Forces 1940–1 made no reference to the Rineanna operation and, of course, did not note the termination of scheduled patrols. Similarly, neither Archer's summary report of March/April 1944, on relations and contacts with the British military, nor Lieutenant Colonel Childers (1947) comprehensive review of the Emergency period, make any reference to this important, though short-lived aspect of Anglo-Irish co-operation.

In due course a flight of the R & MB Squadron detachment was posted back to Baldonnell. On 22 May 1941 Mulcahy issued a secret order directing that one flight of the Rineanna detachment be attached to Air Corps Schools and that the minuscule Coastal Patrol Squadron (Cadre) replace it. The main reason for the rotation, which took place on 26 May, appears to have been the necessity to facilitate the return to Baldonnell of married personnel who had spent 21 months in Rineanna.[65]

FIGHTER SQUADRON

If the Reconnaissance Squadron in Rineanna in 1939–40 was engaged in a fool's errand – an examination of the voluminous investigation proceedings of 1941 and the report of 10 January 1942 supports no other conclusion – the 1st Fighter Squadron (Cadre) at Baldonnell will be seen to have been no better equipped for a viable wartime mission. The Squadron was established under the Peace Establishment of March 1939 by the renaming of the 1st Co-operation Squadron that had existed, informally and formally, since 1930. While the

Table 9.2 **Establishment and Strength – No 1 Fighter Squadron, Baldonnell**

Fighter Squadron	Officers	NCOs	Privates	Total	Aircraft
1937 Co-op Sqn (Cadre)	11	15	36	62	Not specified
1939 Peace Establishment Fighter Sqn (Cadre)	18	28	28	74	Not specified
Strength – 20 Sept. 1939[67]	7	12	64	83	3 Gladiators 6 Lysanders
1940 War Establishment 'Less [42 Other Ranks]	27	57	149	233	22
not to be raised'[68]	27	44	120	191	
Strength – 12 Dec. 1940	9	26	119	154	3 Gladiators
Attached	3	1	5	9	6 Lysanders
Total[69]	12	27	124	163	2 Avro 636 2 Hawker Hind 1 DH Dragon 1 Miles Magister 15 Total

Rineanna detachment had verbal orders, at least up to May 1940, Fighter Squadron appears to have had no orders, written or verbal, from OC Air Corps or higher authority. At the outbreak of war Mulcahy had reported that during the immediate pre-War period the Fighter Squadron had 'concentrated on training to fit in to the air defence scheme for Dublin'.[66] In terms of manpower the Fighter Squadron of September 1939 was nine over strength due to a surfeit of privates. At the same time pilot strength was about 40 per cent of the number permitted by the 1939 Peace Establishment. The maximum number of pilots attained [during 1940] under the 1940 War Establishment was only 12 or about 40 per cent of the approved [War Establishment] figure of 27.

The 1940 War Establishment was the first to provide for a specific number of aircraft for each squadron, in this case 22. By convention the aircraft of a fighter squadron would be of a single current fighter type. Notwithstanding, the No 1 Fighter Squadron's main equipment, on the 3 September 1939, consisted of three Gloster Gladiator I aircraft. It also had six Westland Lysander II and sundry older aircraft. The Gladiator was the last of a long line of biplane fighters to serve with the RAF and was in production from 1935 to 1938.[70] Even as it was entering squadron service it was being rendered obsolescent by the design and manufacture of high performance monoplanes such as the Hawker Hurricane and the Vickers Supermarine Spitfire. By the outbreak of War the Gladiator had been withdrawn from over 70 per cent of the RAF's UK-based front-line squadrons. The remainder were replaced by April 1940 as scores of front-line squadrons were being re-equipped with Hurricanes and Spitfires from 1938 onwards.[71]

The Westland Lysander II aircraft was delivered in July 1939. The type was originally developed in response to an Air Ministry requirement for an aircraft capable of an artillery spotting and reconnaissance role to replace the Audax and Hector types which was in service since 1934.[72] In its reconnaissance role it was well suited to the static style of warfare of an earlier era but not to the highly mobile armoured warfare of 1939–45. In the RAF context it was largely withdrawn from the Army co-operation role by 1941.[73] While the Lysander was a purpose built Army co-operation aircraft it was adapted, in RAF service, for roles such as special operations into France dropping supplies and agents that made the best use of its short take-off and landing characteristics.[74] For Air Corps it was originally purchased as an advanced trainer – though, as accounts suggest, erroneously so. When questioned on this point by the Investigation Committee Mulcahy was somewhat coy:

> To the best of my recollection the Lysander was selected as the most suitable type available at the time as an advanced trainer. The order was placed, but as far as I know the firm was unable to supply dual controls. The machines had been built

for us and we took delivery. The Lysander is a suitable machine for advanced operational training.[75]

The Committee was not satisfied with this evasive and uninformative answer and asked Mulcahy if it was normal to have Lysander aircraft fitted with dual controls. His response was brief:

It is normal to have Lysander aircraft fitted with dual controls when such aircraft are being used for flying instruction in the same way as advanced trainers are fitted with dual controls.[76]

The Committee did not detect that the second answer was much more misleading than the first and, apparently being satisfied, moved on to a totally different matter. Clearly it was so completely deceived by Mulcahy that the final report deemed 'five aircraft of the "Lysander"' type suitable for 'training purposes in the School'.[77]

Mulcahy had, thus, succeeded in concealing the true situation from the Investigation Committee, who, to judge by many of the questions put, and the answers accepted, were very naive in technical matters. Mulcahy implied that the Air Corps, when ordering the aircraft, had specified the inclusion of dual controls in a small batch built to their specification. Had this been the case the non-availability of dual controls would have been made known at the time of ordering. The Air Ministry production records indicate that the production of Lysander II commenced on or about 14 June 1939.[78] The Air Corps took delivery of the six aircraft, off the standard production run of the type, on 15 July 1939.

The truth of the matter lies in the 1939–40 Defence estimates. The capital expenditure of £47,400 for '6 single engined training aircraft (Lysander) @ £7,900' each was proposed.[79] It is not easy to understand how the Lysander II could have been purchased as an advanced trainer. It was, purely and simply, an Army co-operation aircraft. Though it was reasonably well armed for a tactical reconnaissance machine, nothing in its design, performance and handling characteristics fitted it for advanced training purposes. While the Directorate of Technical Development of the Air Ministry had specified that provision be made for a dual control conversion kit there is no record of such being produced.[80] Had such a machine been developed it could only have been used for the conversion of pilots to the type and, most decidedly, not as an advanced trainer. The rear seat even faced to the rear!

In effect, in their anxiety to expeditiously spend the monies allotted in the financial year 1939–40 the Air Corps purchased aircraft without adequate reference to detailed technical specifications. In this case Mulcahy ended up with six aircraft unsuited to any training role. The anxiety to purchase training aircraft was due to the intended intake of short service cadets which eventually took place in August 1939. Notwithstanding any misunderstanding about dual controls, it is not easy to understand how Mulcahy could have considered a classic Army co-operation aircraft as being suitable for advanced training. Advice on the selection of aircraft would normally have been available from Commandant Carroll. However, the Chief Technical Officer (and second-in-command) was on half pay while functioning as General Manager with Aer Lingus from 31 January to 1 September 1939 (and for most of the Emergency). Mulcahy's evidence to the Investigation Committee suggests, therefore, that he took his own advice on aircraft selection at this time.[81] In Air Corps service, although operated by Fighter Squadron, the Lysander was designated as an army co-operation machine. Despite it being unsuitable Mulcahy was satisfied that it could also be used as a fighter.[82] However, fundamentally the parasol wing arrangement, and stability at slow speed, that made it a very suitable aircraft for observation of the battlefield, in addition to its poor performance, rendered the type practically useless in terms of air combat. It could, if unopposed, act in a ground attack role.

OPERATIONS ORDERS, MAY 1940

Unlike the Reconnaissance and Medium Bomber Squadron (Cadre), the failings of which were evident very soon after it commenced active service on 30 August 1939, the inadequacies of the Fighter Squadron were not exposed until the summer of 1940 when the fear of a German invasion was at its height. It was at this time that the first GHQ operations orders were drafted, for all Army elements, to direct action to counter the perceived threats of IRA agitation and a German invasion. Emergency Defence Plan No 1/1940 was drafted on the basis that available information showed that the IRA was 'planning something in the nature of a major operation' and that the operation might 'involve the support of a foreign power, directly or indirectly'.[83] Additional comment would suggest that there was little by way of a firm basis for the perceived threat other than 'a study of the developments to date in the present international conflict'. The resulting Operations Order No 1/1940 of 24 May 1940 apportioned the defence of the State mainly on a geographic basis dictated by the command areas of the Eastern, Western and Southern Commands with mobile columns forming the first line of defence.[84]

The operations order assigned no mission to the Air Corps. It appears, however, that about the same time Mulcahy had been instructed to draft an Air Corps annex. Annex No III was submitted four days later for the approval and signature of the Chief Staff Officer, Operations, General Hugo McNeill. Other annexes that would concern the Air Corps were those on the 'Defence of Aerodromes' (Annex V) and 'Air Defence' (Annex VI). Assessment of the various orders is made difficult by the lack of co-ordination apparent in the drafting of the main order and the several annexes. This resulted in funda-mental responsibilities, particularly in relation to the air defence of the Dublin area, not being definitively fixed. It might be understood that the role of Fighter Squadron should have been clarified by orders relating to the air defence of the eastern region of the country and that Air Defence Command would be the appropriate agency to direct and co-ordinate the efforts of all air defence elements, including fighter aircraft. It might also be expected that the Air Corps responsibilities would be clearly set out in Annex III. But such assumptions would not be entirely valid. During the summer of 1940 the air defence of the region was mainly the responsibility of 1 Anti-Aircraft Brigade, McKee Barracks, whose orders purported to include responsibility for the co-ordination of the Air Corps aspects of air defence:

> The air defence scheme for the protection of Dublin provide for combined active defence by aircraft and anti-aircraft units and its co-ordination and development in conjunction with passive defence measures.[85]

However, neither ADC nor the AA Brigade made provision for the coordina-tion of anti-aircraft defences while the Air Corps aspect of air defence, in reality, was to be of little consequence. The air defence annex to Operations Order No 1 related mainly to the responsibilities of Air Defence Command in co-ordinating and plotting the results of the intelligence gathering functions of various agencies:

> A special scheme has been agreed upon between General Headquarters and Garda Síochana headquarters for the collection and rapid transmission of information concerning the activities of foreign aircraft seen over our territory or territorial waters. The scheme provides that look-out posts of the Marine Coast Watching Service and Garda stations will co-operate in the collection of such information.[86]

The orders required observers to supply detailed reports on all aircraft, not identified as Irish, in such spatial and temporal detail that the movements of individual aircraft, seen or heard, over land or sea, could be plotted at Air Defence Command, Dublin Castle, and any potential threat assessed. The ADC was required to keep OC Air Corps informed of all reported movements of belligerent or unknown aircraft so that the 'Air Corps Interception Service' could be called into action. To illustrate the naivety of GHQ's concept of what might constitute a defence against aircraft of an invading force it is necessary to quote a modicum of the relevant order:

> On receipt of all such information aircraft will be dispatched to intercept offending aircraft flying over Irish territory or territorial waters provided there is a reasonable chance of aircraft affecting this purpose.
>
> The pilot of [the] Irish aircraft will signal to [the] foreign pilot that he is over neutral territory and endeavour to ascertain his mission.
>
> If in communication by radio with his headquarters he will remain in [a] position of observation, report and await orders.
>
> If not in radio communication he will collect all information and proceed to [the] nearest aerodrome, where he will make an immediate report.
>
> He will not initiate offensive action but if attacked will take all necessary defensive action.[87]

This order must be seen in the context of a time when a German invasion was expected at any moment while communication, by R/T, with fighter aircraft was most uncertain. The LOPs of the Coast Watching Service had the primary function of observing and recording the movements of aircraft and ships while Air Defence Command co-ordinated and plotted the information and made an intelligence assessment. The system did not constitute an early warning system in the accepted sense and, in particular, bore no relationship to the air defence system, based on the use of radio direction finding (RDF) or radar as it would be known by 1942 (developed in the UK from 1935). In the summer and autumn of 1940 a very efficient air defence system facilitating the ground control of fighter aircraft was a deciding factor in the Battle of Britain.[88] By virtue of the visual and aural nature of the Irish observer system the identification of hostile aircraft, and the prediction of the tracks and possible targets, would be so delayed as to obviate interception by three obsolete aircraft on standby on the ground. Totally inadequate aircraft and pilot resources ruled out the possibility of standing patrols. Ultimately, Baldonnell was too close to the target area of Dublin even if appropriate and sufficient aircraft were available. For an adequate defence of Dublin several squadrons of aircraft would have been required to be based in south-east Leinster.

The implication of the above order was that Fighter Squadron aircraft were expected to respond to each and every incursion of Irish airspace by foreign or unidentified aircraft. However, in practice, the order only applied to the 'artillery zone of the AA defence of Dublin', delineated by lines joining Howth Harbour, Killiney Hill, Tallaght Aerodrome and Blanchardstown to Howth Head. With an average of over 400 belligerent aircraft being identified each month in the summer of 1940, mainly in the eastern region, such a task would not have been practical except with appropriate resources and systems.[89] As a result single aircraft, which constituted the greater bulk of sightings, were ignored. The order was subsequently formally amended to reflect the fact that the simultaneous incursion of two or more aircraft together was to be considered a hostile act.[90] It is of note that pilots were directed to take no offensive action against belligerent aircraft. This command was most likely primarily aimed at avoiding any possibility of a breach of neutrality; it was possibly also given in the knowledge that such action would have been utterly futile, anyway. There is no record of any Air Corps aircraft firing on Allied or German aircraft. The firing of live ammunition was apparently confined to the destruction of barrage balloons that were observed drifting over or near the country. Only four such incidents are noted while details of only three are known. The first balloon destroyed by the Air Corps occurred on 4 October 1939 in the vicinity of Foynes. The Browning machine guns of Lysander 64 fired about 1,800 rounds to destroy the balloon. The same aircraft and guns were used in the destruction of another balloon on 20 June 1940 off the Waterford coast when some 2,500 rounds were fired. A third balloon was destroyed near Carlow on 1 May 1941 when almost 3,500 rounds were fired.[91]

There can be no doubt that the order was totally impractical from the points of view of the number and type of aircraft available and their inadequate communications. In the unlikely event of one of three Gloster Gladiators intercepting a belligerent aircraft the Irish pilot would probably not have been able to maintain two-way communication with base due to the underpowered ground station at Baldonnell. It is not recorded to what extent interceptions were attempted. In September 1940 an Air Corps note dealing with the activity of foreign aircraft did not elucidate but indicated that adequate numbers of suitable aircraft and pilots were not available to provide an effective interception service for the Dublin area. It also indicated that if Air Corps efforts at interception were to be successful it would be essential to get more timely and more accurate reports from observer stations. The context would suggest that attempts at interceptions were very rare.[92] On at least two occasions aircraft were scrambled to intercept unidentified aircraft. Firstly, on 13 April 1940 'Air Corps fighters' were scrambled on the orders of ADC to investigate unidentified aircraft off the east coast. The aircraft turned out to be escorting a convoy and was of absolutely no threat to Irish neutrality. On the second occasion, on 29 December 1940 the guns of the Ballyfermot AA battery, which had engaged a German Ju88, had to cease firing when 'two interceptor aircraft of the Air Corps entered the zone of fire'. The aircraft, climbing slowly, were unable to make contact with the German machine.[93] This later incident, rather than demonstrating the co-ordination between the ADC and the Air Corps, shows up the lack of coordination with the artillery and highlights the impotence of the Fighter Squadron. The fact that Air Corps aircraft were fired upon on several occasions by anti-aircraft artillery suggests a deeply flawed co-ordination system. Colonel W. P. Delamere had no doubt about the reasons for the lack of adequate co-ordination of air defence efforts. In January 1941 he stated that ADC was mainly concerned with collecting and tabulating air intelligence for the intelligence branch (G2) of GHQ and that, due to the emphasis on the collection of air information for Department of External Affairs and Government purposes, operational deductions were not being made.[94]

Notwithstanding the obvious impotence of the Fighter Squadron, and the utter futility of interception as the squadron mission, the 'Air Defence Annex of Operations Order No 4/1941', a slightly revised version of the previous, was issued a year after the first. It designated to the Air Corps the task of continuing to operate the 'Air Corps Interception Service' as previously described. The Air Corps paragraph was repeated practically verbatim in spite of major obvious deficiencies.[95] To a large extent, therefore, it appears to have been a classic example of a staff officer taking out the previous order and simply changing the dates. One is reminded of the candid admission made by Mulcahy to Air Commodore T. N. Carr only a few months after this order had been renewed. Having regard to the state of the aircraft and the state of readiness of the Air Corps, Mulcahy made it clear that the Corps could be utterly ignored as a factor in the defence of the country.[96]

AIR CORPS ORDERS IN 1940

Though Air Defence Command functioned primarily as an intelligence organisation it was also the nearest the country got to having an early warning system though this emphasis was very much a secondary one. 1 Anti-Aircraft Brigade was responsible for the air defence of the Dublin region even though the air defence response was not subject to central coordination or control measures. The Air Corps annex, though drafted by Mulcahy at Baldonnell, was signed by General McNeill and had the same standing as other GHQ orders. It did not read like an operations order *per se*. The first of two paragraphs supplied 'for the information of ground troops' details of the roles and characteristics of the service and training aircraft that the order intimated would be operating in an army support role when the invasion came. The second section dealt with the 'present missions' and 'future missions':

Present Missions
Two Ansons and two Walrus . . . at Rineanna for patrols . . . to be initiated on the orders of the Chief of Staff.

Stand-to aircraft at Baldonnel and Rineanna for special duty for Chief of Staff . . . Training is proceeding with as little interruption as possible.[97]

The 'future missions', in addition to the 'present missions' then being carried out, included the dispatch of aircraft to command areas for reconnaissance, co-operation with ground troops and 'ground strafing':

In case of emergency aircraft will be dispatched to selected landing fields in various commands. Ground crews will follow with the least possible delay. Fuel supplies will be arranged by mobile tanker. Commands will be responsible for local protection.[98]

Information on hostile shipping was to be provided to 'Coast Artillery' by coastal patrols. Fighter Squadron's role in providing the Air Corps Interception Service was not mentioned except to the extent that the annex directed that fighter aircraft were to be retained for the defence of Dublin.

Although not stated as such, the obvious emphasis of Annex III was on co-operation with ground troops in a post-invasion scenario. This unwitting emphasis reflects the Army's fears, indeed expectation, of a German invasion in May/June 1940. It is abundantly clear that, in the event of invasion, a significant proportion of Air Corps resources would be dispatched to selected landing grounds in the Commands where reconnaissance would be the principal air mission in co-operation with ground troops engaged in active operations.[99]

This post-invasion emphasis is similarly reflected in supporting documents subsequently distributed by Mulcahy and GHQ. While drafting, and authenticating, Annex No III, Mulcahy had acted as a GHQ staff officer. Continuing in this role he circulated an instruction on 'Landing Fields' on 30 May 1940. He informed the Commands that he had been directed by the COS to point out that it was essential to have landing grounds near column headquarters. He suggested that column commanders should identify suitable fields convenient to their headquarters and that the locations should be made known to the Air Corps to save aircraft flying time when they were being sent to co-operate with ground troops. He indicated the minimum dimensions of the fields required by Anson, Lysander, Magister and Avro 631 Cadet aircraft. Included with the instructions was a list of 54 fields mainly located in Leinster and Munster. He indicated that these fields had been inspected at various dates between 1932 and 1937 and had originally been licensed for aerial circus work, and suggested that the list of fields 'might be of assistance when aircraft were operating with your columns'.[100] About the same time GHQ had distributed copies of a 'list of known places' which had been prepared by the Air Corps and was recommended to the Commands as being up to date. GHQ considered that any of the 139 fields, identified on One-Inch Ordnance Survey sheets, would be 'suitable in an emergency'.[101]

Notwithstanding the Army co-operation emphasis of Annex III and the number of landing grounds designated for possible use there was nothing in the orders of the time that facilitated Air Corps co-operation with Army troops. Pre-War Army co-operation had consisted, almost entirely, of target towing for anti-aircraft artillery while none of the extant Squadrons had Army co-operation as a primary role or were practised in its techniques. In particular, no planning had been done in this area and no coordinating agency or communications system was in place or contemplated. Annex III and its supporting documents represented a minimal capacity for army co-operation in a post-invasion situation and in all probability would have been hopelessly ineffective.

From an examination of Annex III and bearing in mind the other orders it could be concluded that Mulcahy was not *au fait* with all the relevant documents. In the hectic and somewhat confused circumstances of the last ten days of May 1940 this may well have been the case. Mulcahy appears to have been kept in the dark about many operational matters. He did not receive copies of operations orders as a standard practice but, if GHQ considered it necessary for him to read a particular order, he 'had to go in [to GHQ] and read and initial it'. In this regard he is certified as having seen, on 23 December 1940, Operations Order No 3 dated 17 December 1940. In relation to his familiarity with Army orders he was asked, in October 1941, 'are you informed of the plans for defence and employment of the forces, of the divisions and the commands?' It is of note that he answered in the negative.[102]

With the three Gladiators tasked to the defence of Dublin it might be assumed that Mulcahy had issued orders, written or verbal, to the Squadron or to individual pilots. However, no such orders are reflected in the EDP material or mentioned in the Air Corps report and proceedings of the Air Corps investigation. Considering the extent and detail of the instructions regarding the designation of Army co-operation landing fields the absence of orders for the conduct of interception is difficult to understand. The only instruction about the movement of fighter aircraft was issued in relation to the arrangements for the dispersal of aircraft 'in the event of the situation becoming more serious'. In the event of an attack, to prevent the destruction of the aircraft in the hangars, serviceable aircraft of Fighter Squadron were to be picketed around the perimeter of Baldonnell 'ready to take the air for defence or reconnaissance purposes'.[103]

At best pilots may have had verbal orders from Mulcahy to get airborne when ordered to do so. Mulcahy's attitude is illustrated, here, in his evidence to the committee of investigation:

Q. Why have you such a mixed collection of aircraft in the Fighter Squadron?

A. Because it was the most suitable equipment I had with which to train and keep on training the Fighter Squadron.

Q. The bulk of the equipment is training equipment?

A. Yes, it is something to progress with until something better comes along.

Q. Why should you have a Fighter Squadron?

A. Because if you do not have fighter aircraft you could never have air superiority over an area. Fighter aircraft is [*sic*] the best form of anti-aircraft defence. Except you have fighter aircraft you cannot even have local air superiority.

Q. What use would our 3 Gladiators be against a determined attack on, say, Dublin?

A. Supposing bombers came over and that our three Gladiator pilots were shot down over Dublin, it would be a certain consolation to the people and would improve their morale by letting them know that we had at least done what we could.[104]

The above suggests that Mulcahy considered that Fighter Squadron was fundamentally still the peacetime training cadre of the 1939 Peace Establishment, but expendable in the context of giving the impression of being a viable fighter deterrent. While he did not specify so in writing, he appears to have had little difficulty in committing an inadequate number of obsolete aircraft to a wartime defensive task that was practically suicidal. His attitude is further explained by the tone and content of his 'order of the day', issued on 4 July 1940, when all aircrew were required to standby on the air station awaiting the invasion:

If we fail to get into the air, if we lose our aircraft on the ground, we have failed utterly in our duty to our people. It is therefore necessary that the crews of the service squadron and detachment at Baldonnel be readily available to their aircraft at all times. . . . Let us, therefore, bear inconvenience cheerfully now so that we will be standing by to perform [–] whatever the task and whatever the hour.[105]

While Mulcahy appears to have willingly accepted an impossible task on behalf of Fighter Squadron, given the military situation in the summer of 1940 he had little choice in the matter. With fears of a German invasion running very high there was tacit agreement that British forces would come to the country's assistance. However it was de Valera's policy that before British assistance could be requested the Irish Army, of which the Air Corps was an integral part, had to take the brunt of an initial assault.[106] In such circumstances the squadron's efforts would certainly have been of little effect.

The attitude of GHQ to the effectiveness of the Air Corps in 1940 is reflected in a GHQ 'map manoeuvre' exercise, undertaken in preparation for the updating of defence plans, at which Mulcahy was Assistant Director in charge of air operations. The exercise 'German estimate' of 'the enemy forces and disposition' concluded that 'as regards opposition to our attack, the Irish air force may be regarded as non-existent'.[107] At the same time and while holding the above opinion, GHQ, through the aegis of the air defence order, purported to defend the Dublin area by means of a largely mythical Air Corps Interception Service.

DEMORALISATION

By the summer of 1940 the operational capability of the reconnaissance detachment was so severely degraded that it made little or no contribution to the intelligence gathering effort off the west coast while the morale and *esprit de corps* of the unit was recognised and reported, by Costello, as being extremely fragile. In Baldonnell, during what Mulcahy described as the 'invasion nervous' months of the summer of 1940, given the suicidal task assigned to Fighter Squadron and the feeble response to orders to scramble, the morale and confidence of this unit can have been little better. From Mulcahy's final submission to the Investigation Committee on 21 November 1941 it transpires that, during the latter part of 1940, his command was under severe strain due to alleged irregular communications from two junior pilot officers to persons, including the Minister, outside the Army. Mulcahy cited a visit by the Minister (Oscar Traynor) and COS to Baldonnell on 23 October 1940. The visit was in connection with certain allegations about Mulcahy's command of the Air Corps, made in writing to the Minister.[108] While the original complainants are not identified the subsequent investigation points to Captain T. J. Hanley and Lieutenant A. C. Woods. They apparently complained about standards of navigation and instrument flying, the standard of aircraft equipment such as instrumentation, direction finding equipment, communications generally and the failure to purchase Link trainers or to provide the vacuum pump and loop aerial modifications for Anson aircraft.[109]

After the Minister's visit the two officers were invited to write to him to detail and explain the exact nature of the complaints. In the absence of both sets of letters it is not possible to be sure about the precise matters highlighted. From evidence given to the Committee the failure to purchase and fit the loop aerial modification to the Anson was a major bone of contention. In November 1940 the unrest manifested by officers' complaints indicated a state of demoralisation, inefficiency and stagnation over the weak leadership, as well as the deteriorating situation due to the inadequacy and unsuitability of equipment.[110] This situation would shortly lead to a major investigation of the attendant circumstances.

CONCLUSIONS

From 1936 (and earlier) the Government's concept of the Air Corps had been as a source of technical personnel and expertise for the advancement of civil aviation. The priority given to the employment of pilots in civil ATC from 1936 onwards can be seen as a major aspect of that policy. Another was the conduct of three wings courses during the Emergency. With 63 students recruited and 43 qualifying, the output of the previous 17 years had duplicated in six while the post-War pilot requirements of civil aviation was more than adequately provided for. It is significant to note that the Air Corps second-in-command spent the Emergency period in a managerial capacity with Aer Lingus.

Given the lack of preparation and planning, that was, in effect, part of the Government's policy for the Emergency, it is easy to understand the quite unsatisfactory nature of the Air Corps' contribution during the first 12 to 16 months. Whatever the circumstances, with the decision to post an air detachment to the south-west, the campaign started badly and from there matters only got worse. It is fair to state that the R & MB Squadron detachment exchanged the aviation backwater of Baldonnell for the aviation wilderness of Rineanna. Nothing, whatsoever, in the aeronautical circumstances at Rineanna was conducive to the conduct of a successful military mission. The inadequacies included an ill-equipped and inadequately supported obsolete aircraft that was unsuited to the environment and to the mission. It could be convincingly argued that the loss of three aircraft early in the mission and the obvious lack of adequate spares were the main factors contributing to the degradation of the patrolling mission. However, the primitive nature of the aerodrome and facilities, including lighting, meteorology, communications, direction finding and other basic requisites, compounded by the absence of preparations and training, undoubtedly contributed in no small way to an outcome that proved, in the circumstances, inevitable.

At Baldonnell the position of the other operational squadron, from May 1940, was equally unsatisfactory. The composition of the 1940 Fighter Squadron, in effect, constituted an aeronautical absurdity. With 15 aircraft of six inappropriate and obsolete types, as the investigation report subsequently stated, it was fighter in name only. While it is not possible to adequately assess the likely affect, in practical operation, of GHQ's disjointed and unco-ordinated operations orders, the concept that Fighter Squadron would be the backbone of an 'Air Corps Interception Service' indicates a naivety on the part of the military leadership which defies belief. The terms in which Mulcahy indicated his acceptance of the suggestion that a training cadre might make a worthwhile contribution to the defence of Dublin strongly reveals how his great ignorance influenced those who should have known better. Subsequent events, culminating in the investigation of 1941–2, indicate that by the autumn of 1940 the pilots of the Air Corps was a demoralised and disillusioned group, and thus, was essentially the ineffective air element of a most inadequately prepared Army.

TEN

SERVICES RENDERED

—

The short-lived coastal patrol operation at Rineanna was the most substantial indirect assistance rendered by the Air Corps, on order, to the UK. In the wartime records this intelligence gathering activity was not acknowledged by the UK and was studiously ignored in the summary reports of the Emergency period. However, there were other air-related activities that contributed to the UK air war effort. In February 1945 the Dominions Office listed the 'facilities obtained from the government of Éire during the war'. The list, of 14 specific facilities granted to the UK, was supplied by the Dominions Office apparently to support 'the position of Éire in connexion with the question of inviting neutrals to attend the San Francisco Conference' considering the setting up of the United Nations. Among the facilities that the Éire Government considered 'would not be regarded as overtly prejudicing their attitude to neutrality' included the return of crashed Allied aircraft, the supply of meteorological reports, supplying reports of submarine activity, broadcasting details of aircraft sightings and providing direction finding services for UK ships and aircraft. The concessions included the internment, for the duration, of German aircrew while Allied aircrew were surreptitiously released to rejoin the battle. The use of the Donegal corridor by aircraft patrolling the Atlantic is the best-known and most often cited instance of the facilities granted by de Valera's Government. This allowed flying boats based at Castle Archdale on Lough Erne to fly out into Donegal Bay and enter the North Atlantic expeditiously to commence anti-submarine patrols.[1] While many facilities were policy matters negotiated at a political level, the co-operation of the military, north and south, was central to the provision of these and other concessions, particularly after May 1940.

MILITARY LIAISON AND CO-OPERATION

The Irish Army's co-operation with the UK, as well as the aviation related assistance directly or indirectly rendered to the RAF, took place in response to what appeared to be an unwritten mandate from Government. Initially this mandate allowed the Army meet and interact with their opposite numbers in British Troops Northern Ireland (BTNI), on matters mainly concerned with the defence of the 26 counties of Éire. The bones of this military-to-military co-operation are partially laid bare by Colonel Archer in his 'summary of contacts with foreign armies'. Archer's account, however, was greatly lacking in detail and, thus, had to be revised and greatly expanded. Lieutenant Colonel R. A. Childers, in his report of October 1947, succeeded in reviewing the documentation of the Emergency period and detailing much more fully, for the benefit of the post-War Army leadership, what had actually taken place.[2]

Archer observed that liaison had existed between the British War Office and GHQ on security and counter-espionage matters from September 1938. He did not mention the fact that he had represented de Valera at pre-war meetings dealing with military matters. He first visited London in October 1938 coinciding with a time (11th and 12th) when John Dulanty (High Commissioner) and Joseph Walshe (Secretary, Department of External Affairs) were having discussions with the Committee for Imperial Defence on defence and military matters of mutual concern. In December 1938 Defence wrote to Finance:

> I am directed by the Minister for Defence to state that, on the instructions of the Taoiseach, Colonel Liam Archer proceeded to London at very short notice on two occasions recently on business of a confidential nature. The periods of the visits were from the 10th to the 14th October 1938 and from the 4th to the 6th ultimo.[3]

The request was for authorisation to pay the rather modest costs arising out of two unexpected visits to London. While published accounts indicate that Archer was in London on intelligence business[4] it is suggested that this was not the only reason for his being there. The records do not confirm that Archer actually attended the CID meetings of 11 and 12 October. However, an unsigned memorandum records several observations and insights that indicate an intimate knowledge of the ethos and traditions of the Irish military, as well as assessments of Army officers. With Dulanty based in London (since 1930) and Walshe professing ignorance of military matters the presence of a serving officer is strongly indicated. The observations, thus, appear unwittingly to point to Archer as attending or being present on the periphery of the meetings that the CID had with Dulanty and Walshe.[5]

Archer also visited London on 25 August 1939, again on de Valera's instructions, but this time in connection with the setting up of the Coast Watching Service. As indicated above, records suggest this visit resulted in the Air Corps detachment to Rineanna. The appointment of a British naval attaché in October 1939 appears to have been directly related to the intelligence gathering effort on the west coast.

The Childer's Report cites the first military-to-military contacts as occuring in relation to the perceived threat of invasion that existed following the German invasion of Holland and Belgium. He states that Archer and Walshe had been sent to London to 'make contact with the British War Office in order to discuss with the British authorities steps to co-ordinate our respective defence measures against a German invasion of Ireland'.[6] The report went on to outline the visit of Archer and Walshe to London, their arrival into a conference at the Dominions Office being attended by senior officers of the British Army, Navy and RAF, and the resultant visit to Government Buildings in Dublin of Colonel Clarke and two other British officers on 24 May 1940.[7] Archer may have been in the UK for some time prior to the Dominions Office meeting of 23 May 1940. On 20 May 1940 Cecil Liddell, head of the Irish section of MI5 reported on events to CID:

> I have written to Archer as you suggested. In the meantime you may perhaps care to know what moves have been made recently. After the meeting the other day which Sir Vernon Kell attended, I saw Walsh[e] at Dulanty's office. He was quite unacquainted with the [illicit] wireless [interception] situation and asked me to discuss it with Archer when I saw him at Droitwich where he was undergoing a cure.[8]

At about five o'clock on 15 May 1940 Cecil and Guy Liddell arrived at Droitwich and met Archer at his clinic. While they had intended discussing wireless matters, the invasion of Holland had brought to their attention the possibility of something similar happening in Ireland. Encouraged by Archer's positive reaction to the suggestion of some form of staff talks to consider the situation where existing Irish forces might prove to be inadequate to repel an invasion, the Liddell brothers brought the idea of military staff talks to the Dominions Office. There they were informed that the two governments had been thinking along similar lines.[9] Arising from the military discussions, in London on 23 May and in Dublin on 24 and 25 May 1940, liaison contact between the military north and south was initiated on a quite informal basis. The Army's meetings with their northern counterparts were closely monitored by civil servants, while the more sensitive matters were referred for ministerial approval.[10]

A major aspect of the co-operation was the completion of a series of questionnaires provided by the British. The first concerned the technical aspects of the wireless broadcasts carrying details of foreign aircraft movements, as reported by Air Defence Command. A report accompanying a copy of the first completed questionnaire suggests that the meeting of 25 May 1940 concentrated on providing Squadron Leader Potter of Aldergrove with aero-nautical information on Baldonnell, Collinstown and Foynes, and other locations suitable for the operation of aircraft. It also provided armament and wireless details relating to Walrus, Anson, Lysander and Gladiator aircraft. The details provided on Foynes actually pertained to the Air Corps station at Rineanna. The Irish representatives, General McKenna and Colonel Archer, were unable to provide much detail without consulting others by telephone – a process that prolonged the Saturday afternoon meeting. The British were informed that Baldonnell was 'fully equipped'; but they remained sceptical. Of course, in reality, they had every right to be sceptical about Baldonnell as an airfield for possible use by modern fighter aircraft. Air defence consisted of the machine-gun posts just being installed. There was no anti-aircraft artillery and half the aerodrome was permanently staked to discourage aircraft landing. In addition, communication with aircraft was by wireless telegraphy only.[11] Aspects not noted include the absence of hard surface runways, not to mention the fact that the minute size of the aerodrome would have greatly precluded the effective dispersal of aircraft. It had no meteorological station or even a remote-reading anemometer. The most glaring inadequacy of all, however, was the absence of a direction finding service for military aircraft.

Subsequent questionnaires concerned the organisation, disposition and equipment of the Defence Forces as well as details of communications, infra-structure and resources, and other strategic considerations. The return of a completed questionnaire was occasionally delayed until political approval was given on matters the Army considered might be politically sensitive. Childers observed on the care taken when supplying military information to the UK:

> The answers were to be supplied on un-crested paper and were to be related to the questions only by paragraph number. Every care was to be taken that in the event of their capture by the Germans, their actual origin could not be proved.[12]

The chronology apparent in the layout of the Childer's Report suggests that the first formal north-south conference took place at Headquarters BTNI on 3–5 June, and the second in the Irish Army's Plans and Operations Branch on 1–3 July. The conferences agreed a plan for the evacuation of Dublin, a plan for military routes to be used by the British Army in a move southwards, and many other important aspects relating to a combined defence against a German invasion of Éire:

> The most valuable outcome was undoubtedly the fact that for the first time the two staffs had sat down together to consider a joint problem. They had oppor-tunities also, of informal talks together during reconnaissance and over meals.[13]

Although the Irish military supplied information relating to food, foodstuffs and other supplies, signal communications, roads, railways, hospitals, telegraph and telephone, and many other logistical and services matters, the more tangible benefits of the liaison included coastal artillery and, in particular, radio equipment supplied by the British to 'ensure the satisfactory and speedy transmission' of 'information collected by our air intelligence system' (LOPs, Garda stations and military posts).[14]

Notwithstanding the exchange of military papers, relationships between the military forces north and south were somewhat fraught during the spring and early summer of 1941, reflecting distrust emanating at government level and exacerbated by the presence of at least one representative of the Department of External Affairs at any time at military staff meetings in Dublin. Fundamentally, the level of military co-operation was restricted by a reserved and cautious approach insisted upon by the Irish Government. One main point of distrust arose from the Irish Army feeling that the British refusal to supply armaments was based more on the necessity to keep the Army weak, in an attempt to keep control of de Valera's administration, rather than on an inability to supply.[15]

Two aspects of the north/south co-operation took more tangible forms. At Carton House, Maynooth, a dump of 250,000 gallons of motor spirit was stockpiled for use by UK vehicles coming south to counter a German invasion. Another anti-invasion action was the preparation, for demolition, of the bridges across the Shannon. This action was to protect the British right flank as they moved south against an invading German Army. Surviving records indicate that some 490 other bridges throughout the country were prepared for demolition by and at the expense of local authorities. The blowing up of these bridges, including 18 in Dublin, was to be the final step to impede German progress towards and into the capital.[16]

The appointment of General Franklyn as GOC BTNI in June 1941 proved to be a positive turning point in the relations between the two armies and, indeed, the two countries. He visited Dublin on 16 to 18 June 1941 and, although the COS was not permitted by the Taoiseach to accompany him on a tour of the country, the Irish Army felt that they had convinced him that they would fight any invading force with determination and loyalty. This, and a further visit by Franklyn on 10 December 1941, helped to improve relations with the 18th Military Mission; and Franklyn is credited, in addition, with influencing a better supply of arms from the UK.[17]

MULCAHY AND THE AIR ATTACHÉ

The Archer Summary and the Childer's Report outline a succession of contacts, between British and Irish headquarters staffs, that were carefully monitored by officials of External Affairs. However the activities of the air attaché apparently were not subject to the same scrutiny. The appointment of a military (Army) attaché had been discussed by Colonel Clarke with Joseph Walshe in Dublin on 24–5 May 1940; the latter had indicated that if it was put forward by the UK Government it would be acceptable, provided that the appointment was suitably disguised and that the officer wore civilian clothes. Subsequently, the Air Ministry suggested that, if effective assistance was to be rendered by the RAF in an emergency, it was most desirable that an air attaché should also be appointed. By 29 May 1940 the appointment of both military and air attachés had been agreed by Dublin with Walshe insisting that neither attaché would wear military uniform or use military rank.[18] The UK having reluctantly agreed to the conditions, the air attaché, Wing Commander R. W. G. Lywood, left for Dublin via Holyhead on Monday 3 June 1940.

In the context of attachés and military liaison the matter of wireless communications was of primary concern. The UK authorities were prepared to supply a wireless set for direct communication between Dublin, Northern Ireland HQ and the Service Departments in the UK. However, as the three stations had to be operated by the same service, a decision had to be made as to whether the equipment and the British operating staff would be located at the UK representative's office or be attached to Éire's Defence HQ.[19] As a result, a wireless net, linking Army Headquarters, Belfast, the Air Ministry and station A. A. Dublin to the HQ of 75 Operations Wing (NI) was set up. The net was part of the organisation of an Irish fighter group being prepared for the air defence of Éire.[20] However there is some doubt about the actual use and the efficacy of the Dublin station. On 27 June 1940 a meeting was held at Kinnaird House regarding the communications with Ireland in case of an emergency situation. The meeting was informed that the air attaché possessed a wireless set which had not yet been used, as it was desired to keep its existence secret, and that it would be used should the normal landline (telephone and telegraph) direct to the Dominions Office break down. An outline plan was agreed:

> that an alternative set for Sir J. Maffey should be established in a friendly house in or near Dublin which Sir J. Maffey must arrange and that Col. Vivian [of MI6] should, as soon as possible, produce one set with two trained operators. The necessary arrangements with the Irish government should be made on the level of staff talks, and/or with Col. Archer.[21]

It appears that the air attaché's wireless was in place and in use as early as 11 July 1940. On that date 'Station AA' was sent a cipher message from the Air Ministry, directing certain actions to be taken by Lywood, and to which Lywood replied by telegram – presumably in order to avoid a wireless transmission that might be detected by the Irish Army.[22] This apparent early use of Lywood's wireless contradicts the claim that the equipment was not manned until August unless, as a qualified and experienced pilot, Lywood himself had received and recorded the coded message. Two Special Intelligence Service personnel, who were understood to be wireless operators, were attached to the air attaché's staff as domestic servants in August 1940. However, it has been put forth that subsequently it came as a surprise to the British representative's office to learn that the men considered themselves accountable to the Secret Intelligence Service and that the radio did not work.[23] Given the concern regarding good communications between Dublin, Belfast and London it might be assumed that this matter was soon put right.

On arrival in Dublin in early June 1940 'Mr' Lywood got down to business immediately. Arising out of his first meetings with External Affairs, GHQ and the Air Corps he submitted a detailed first report to his superiors:

> June 4th, 1940, I was introduced to Mr Walshe, . . . who subsequently arranged an introduction to Col. Archer, director of military intelligence. The discussion was of a very general nature, but I gathered that they wished my liaison with the Air Corps to be carried out as inconspicuously as possible.
> June 6th. [I was] Introduced to Col. P. A. Mulcahy, chief of Air Corps, by Walshe. It was suggested at this interview that I should be introduced to other members of the Air Corps as a civilian from the Air Ministry who was attached to the British Representative's Office to assist them in obtaining aircraft spares.[24]

It was explained that his identity should not be disclosed lest junior personnel with anti-British opinions might misconstrue his presence in Dublin, and deduce that the UK was exerting undue pressure or interfering in Irish affairs. In this respect he was, no doubt, referring to the four ex-IRA officers, at least two of whom were reputed to have retained trenchant anti-British views. After a very open and frank introductory meeting Mulcahy drove Lywood out to Baldonnell. Following a brief tour of the installation he had a lengthy discussion on Air Corps matters. Mulcahy supposedly regaled Lywood with accounts of his past activities in the service of Ireland – presumably his part in the Civil War on the pro-Treaty side. Mulcahy, commenting on the country's determination to resist invasion by any outside force stated that 'the country would rise up and tear limb from limb any invaders'. Mulcahy gave Lywood a briefing memorandum on Air Corps organisation. Lywood recognised the Air Corps squadrons as the training cadres they still were. He later got six attachments which dealt in detail with the matters of 'fuel', 'radio' and 'aerodromes' as they pertained to both military and civil air installations of the country, and 'ammunition', 'personnel' and 'aircraft' as particular to the Air Corps.[25]

On the following day Lywood resumed his familiarisation visit and observed basic flying and navigation training in progress. He remarked on the similarity with that conducted in the RAF but considered flying discipline to be more relaxed. In the afternoon he was brought on a reconnaissance flight of 'existing aerodromes and possible landing grounds' at the Curragh, Foynes, Rineanna, Ardnacrusha, Kildonan and the Phoenix Park. Mulcahy stood behind him in the Anson pointing out everything of interest. Lywood subsequently commented on the extent of the obstructions to aircraft landings that existed at the main military and civil aerodromes. He simply reported on the ground defences of Baldonnell and Rineanna but made no comment. He did, however, suggest that the air ammunition and bomb holdings were inadequate except for the briefest of aerial engagements.[26] In general it could be stated that Lywood received a most complete briefing on, and a comprehensive oversight of, the state of military aviation in June 1940. In this regard it is of interest that he had been introduced to Mulcahy by the Secretary of the Department of External Affairs and not, as might have been expected, by Archer or some other officer of the intelligence staff in GHQ. It could be surmised that both the Department of External Affairs and GHQ had little regard for the liaison and intelligence aspects of military aviation and, thus, placed no strictures on the attaché or, indeed, on Mulcahy. It is significant that neither Archer nor Childers, in their summary reports on the Emergency, gave any account of the activities of the air attaché while those of the military attaché were well noted.

Lywood met Mulcahy again at Baldonnell on 14 June 1940. The tone of his report suggests that Mulcahy understood that, in the event of invasion by Germany, the RAF would constitute the substantially greater part of air support to ground forces. Mulcahy indicated that he was anxious to know the nature and extent of assistance that the UK might give if the Germans invaded. Lywood indicated that, with the assistance then being given in France, he could not predict:

Regarding an aerodrome to be placed at the disposal of [the] RAF in [the] event of assistance being asked, for operating from and to be used as a possible storage for fuel, bombs and ammunition for RAF aircraft, Col. Mulcahy understands this to be Baldonnell, though I gather this was by no means definite.[27]

Lywood's comments on this question suggests a location other than Baldonnell but he indicated that the selection might depend on the type of

RAF aircraft and the balance, between Irish and British, of the eventual command structure.

There are several indicators that demonstrate that, unknown to Mulcahy, Gormanston was intended to be the first aerodrome to be used by the RAF. On 29 May 1940 the Air Corps recorded that the fuel tanks at Gormanston held almost 12,000 gallons, or 70 per cent of its capacity, of aviation fuel. On the same day Baldonnell held only about 60 per cent.[28] The obvious question arises as to why Gormanston, an unoccupied aerodrome on the east coast, should hold more fuel than Baldonnell at a time when a German invasion was expected. The fuel could not have constituted an Air Corps reserve as the fuel depot in the Dublin docks held 102 tons of aviation fuel about this time and was located much closer to Baldonnell.[29] Under Operations Order No. 1 of 29 May 1940 Gormanston was not designated for Air Corps use. In fact, surprisingly, the 'defence of aerodromes' annex that directed defence measures for Baldonnell, Collinstown and Tallaght, as well as three small private airfields in the Eastern Command area, did not specify any air defence measures to be taken at Gormanston, even though it was the most useful, accessible and vulnerable of all.[30] At the same time a list of 'Emergency Landing Grounds', designated for use in conjunction with ground troops, was distributed by Mulcahy on the instructions of the COS. Two landing sites, within a few miles of Gormanston were designated though the aerodrome itself was not included.[31] In December 1940 Gormanston was initially included in a list of 'Emergency Landing Grounds' but was later crossed off the list.[32] All the indications are that Gormanston was not to be used by the Air Corps and that it was designated for RAF use, for an unknown period from May 1940, if and when the Germans invaded.

Mulcahy also raised questions on matters he might well have dealt with before the outbreak of hostilities – such as the camouflage of Air Corps aircraft and of military aerodromes and the very limited supplies of 100 octane fuel available in the country. In regard to Army co-operation Lywood formed the opinion that, while some training had been carried out on Army manoeuvres, 'very little work of this kind' had actually been done. Lywood was interested in the conduct of reconnaissance of both land and sea areas, presumably in view of British suspicions about alleged IRA and German activities. He reported, without comment, on what was a mediocre capacity for general reconnaissance:

> Land [reconnaissance is] combined with training navigation flights over most of Éire. Any special information in the light of intelligence reports to hand are communicated to crews and are the subject of special attention on such flights. When necessary a special reconnaissance [flight] is ordered.[33]

Lywood was also brought up to date with regard to the coastal reconnaissance being carried out by the R & MB Squadron detachment at Rineanna:

> Sea reconnaissance has now been abandoned. 3 of 9 Anson aircraft were lost carrying out this duty. . . . Col. Mulcahy considers that the system of 'coast watches' . . . is carrying out effective work . . . in view of the small number of aircraft . . . he does not feel justified in using them for this particular duty.[34]

Mulcahy and Lywood discussed and agreed a system of visual and wireless telegraphy signals to be used by RAF aircraft crossing the Éire coast or land frontier, in the event of the Irish Government requesting air assistance. These were copied for the approval of the Air Ministry and RAF. The tone and content of Mulcahy's contribution to the discussion suggest that he understood that assistance from the RAF was practically guaranteed while he believed that he would get up to three hours' warning of a German attack. Arising out of his latest visit, Lywood was asked by Mulcahy to hasten the delivery of aircraft spares and equipment to the Air Corps. In view of his supposed role Lywood felt obliged to make these representations to the Air Ministry.[35]

In response to Lywood's first report, the Air Intelligence Division of the Air Ministry suggested that the Irish authorities be advised that the two most likely points for a German invasion were the Curragh and the Foynes/ Rineanna area. The Curragh was considered vulnerable, even if the Army reserve there was not committed elsewhere. This was because it had not been obstructed against aircraft landing, and, without question, German airborne troops had superior firepower to Irish infantry. Similarly, the Foynes/ Rineanna area was seen as being vulnerable because it was not adequately defended. It was considered that the capture of stocks of aviation fuel, at Foynes and Rineanna/Shannon, would be key German objectives.

Regarding the second report and the extent of assistance that might be expected, the Air Ministry suggested that few German aircraft would be intercepted en route, particularly if attacking by night. The extent of direct support would be limited by the fact that the enemy would be well ensconced before assistance was called for by the Irish authorities. They did not envisage occupying any existing aerodromes but suggested it would be necessary to identify aerodromes sites for RAF fighter squadrons on the south east and east coasts that Lywood might be able to collect information on suitable sites. The Air Ministry also put forth that Mulcahy was being extremely optimistic in believing he would get three hours' notice of invasion. The British expressed disappointment with the decision to terminate programmed coastal patrols out of Rineanna:

> It is felt that the abandonment of sea reconnaissance is a great error. The coast . . .
> contains many bays where a vessel might discharge personnel and small arms and
> even vehicles . . . suggest . . . that they should recommence coastal reconnais-
> sance . . . of bays and inlets for suspicious craft.[36]

In October 1940 Lywood arranged for Air Commodore T. N. Carr, AOC
RAF Northern Ireland, to visit Dublin specifically to meet Mulcahy and to
establish an understanding with him. Carr subsequently reported to his
superiors in London:

> The visit was a definite success and I was most cordially welcomed by Colonel
> Mulcahy. He showed me over the aerodrome at Baldonnel and also the head-
> quarters of the Observer Corps . . . discussed . . . with Colonel Mulcahy the state
> of his aircraft and the readiness of the Éire air force for active service. He agreed
> that as a factor in the defence of Éire it could . . . be ignored. The pilots are very
> keen but only half trained owing to lack of aircraft.[37]

Following Carr's visit to Baldonnell the RAF decided to capitalise on
Mulcahy's need of aircraft in order to further their own aims to better equip
the airfields.[38]

RAF PLANS FOR IRELAND

From the first days of planning by the RAF for air support of a British
defence of a German invasion of Ireland, the matter of aerodromes for the use
of several RAF squadrons was of primary concern. In the context of such a
defence it is considered that Gormanston was ideally placed for the initial
stages at least. As discussed above there is compelling evidence to show that
this aerodrome was, for an unknown period starting in May 1940, designated
for the use of the first RAF squadron or squadrons if and when the German
invasion took place. The concept of such a designation was so sensitive, in the
context of fears of a possible German invasion in the summer of 1940, that it
is almost certain that the arrangement was not put in writing.

Subsequent to the initial high alert period of May to July 1940 the RAF
commenced a planning process aimed at the air defence of Éire in relation to
a British response to a German invasion – if and when asked. An early warning
order directed that, in the event of German invasion of Éire or Northern
Ireland, immediate action was to be taken against the invading forces by the
air forces stationed in Northern Ireland.[39] Fighter squadrons were to be the
backbone of this defence:

when the situation in Éire permits, the need to establish fighter sector stations in the Dublin and Wexford areas with the object of affording protection to Éire and to British shipping in St George's Channel and the Irish Sea . . .[40]

This initial proposal provided for two fighter squadrons to operate from Baldonnell 'with an advanced landing ground at Wexford', and a further squadron located 'in southern or central Éire'.[41] Staff studies also considered the occupation of Collinstown, Curragh, and Rineanna and the posting to Ireland of five fighter squadrons and a servicing unit, in addition to the RAF Headquarters and seven squadrons already in Northern Ireland.[42] A later study projected as many as seven sector stations and two forward aerodromes in Éire. Also proposed were 28 RDF (later radar) stations – 14 Chain Home and 14 Chain Home Low stations – with priority given to the defence of Shannon, Queenstown (Cobh), Berehaven and Lough Swilly.[43] RAF planning concentrated on the concept of Baldonnell and Collinstown as fighter stations with an advanced or forward airfield in Wexford. This latter plan, which outlined the communications for the RAF in the event of operations outside Northern Ireland, put two Battle squadrons at Collinstown, three Hurricane fighter squadrons at Baldonnell and capacity for two fighter squadrons at or near Wexford. The same plan provided for No 11 Repair and Salvage Unit to be located at 'Gormanstown' and No 23 Workshop Service Unit at Baldonnell.[44]

FORWARD AIRFIELDS

As early as July 1940 the availability of existing aerodromes and the selection of forward landing grounds had concerned both the RAF and the Air Ministry.[45] Lywood was given direction on the matter:

It is desired make extensive reconnaissance Éire to ascertain landing grounds of possible use by enemy or ourselves . . . such reconnaissance might be conducted by two or three officers as tourists on instructions from Operations Department Air Ministry under your direction.[46]

Notwithstanding the extent of Mulcahy's co-operation with Lywood records suggest he did not provide information on the range of landing grounds, or other sites, he had identified in the command areas. Apparently operating alone Lywood, carried out a survey that did not fully satisfy his Air Ministry superiors. While they again considered sending a number of pilots, on leave, to survey more sites they reluctantly dropped the idea due to 'difficulties and objections'.[47] On 14 February 1941 Lywood was instructed by the Air Ministry to make representations to prevent the ploughing up of 13 sites to prevent German aircraft landing. The sites, at Newtownbarry (Bunclody), Ferns, Rosegarland, Tramore, Fermoy, Tralee, Newcastle West, Adare, Limerick, Rineanna, Oranmore and Tuam, were under consideration to establish 'the selection of suitable sites for [air defence] sector stations in Éire'. Lywood was also to ask Mulcahy for any detailed information he might have on sites in the sector areas.[48]

Following Carr's visit to Dublin the RAF considered that it might be possible to come to a mutually satisfactory arrangement with the Irish Government. It was hoped to get the Air Corps to develop aerodromes where the RAF required them at the price of some obsolete aircraft on the basis that Mulcahy would accept almost anything on which pilots could get some flying hours:

> It is necessary to make it clear that the question of the supply of these 10 Hector aircraft to Éire did not arise as the result of an official request from the Éire government. It has arisen in the course of a useful liaison which has grown up in the last few months between our Air Attaché in Dublin and Colonel Mulcahy . . .[49]

With Lywood supporting the supply of obsolete aircraft that had little offensive potential, Mulcahy had undertaken to ask his superiors for sanction for the construction of aerodromes in the Wexford and Cork sectors where the RAF needed them most.[50] In March 1941 Mulcahy gave instructions for the conduct of a survey to find airfield sites for RAF rather than Air Corps use. He issued guidelines detailing the characteristics required. Pilots were reminded that the surveys were to be kept secret and that in obtaining information about particular sites the real aim was not to be revealed.[51] Subsequently, Mulcahy was able to acknowledge receiving, from the Command Engineer, Eastern Command, reports and plans relating to nine selected sites. The sites included Gaybrook in Westmeath, Rathduff in Tipperary and Rosegarland in Wexford, the last being one of the RAF's preferred sites. Mulcahy forwarded copies of the reports and plans to the COS.[52] In August 1941 the COS made an announcement in relation to the preparation of emergency aerodromes:

> It has been decided that [only] two emergency aerodromes are to be prepared – one near Cashel and one near Mullingar . . . The selected sites are Rathduff, Co. Tipperary and Gaybrook, Mullingar.[53]

It was not explained why Rosegarland had not been considered further but it seems probable, in view of the parsimonious thread running through the file, that the projected development cost of £13,480 was the deciding factor.[54] It was specified that two runways at right angles with 'minimum dimensions of 1,000 x 50 yards', a capacity for further extension and 'capable of taking a total load of 7,000lbs' were required to be developed at each site. This represented a more demanding specification than the previous. The decision, to spend £1,020 on Rathduff and £530 on Gaybrook, was endorsed by the Minister a few days later.[55] Acknowledging the more demanding specification Mulcahy indicated that he would have the sites resurveyed by his civilian Aeronautical Engineer and an officer appointed by the Director of Engineers. Also, he warned that the change in the specification requiring greater runway length might cause difficulties in a particular case.[56]

Within the week he submitted a further report indicating that runways of the required length and at right angles could not be fitted in at Rathduff due to ploughed fields. The position in relation to Gaybrook was even less satis-factory, as the preparation of the site to the new specification would be a lengthy and expensive operation. It was recommended, in order to reduce expenditure, that a lesser runway length might be acceptable. Mulcahy considered that, before arriving at a final decision, 'certain interested parties' should be 'con-sulted and permitted to inspect the two sites'.[57] While there is no indication that Lywood actually visited the sites he most probably did. In early September the Department of Defence continued the charade when replying to a Finance query. Defence confirmed receipt of outline financial approval 'for the arrange-ments made in connection with the emergency accommodation of troops' and stated the landing grounds near Mullingar and near Golden were required for 'Air Corps purposes' and 'for use in certain eventualities'.[58] Subsequently, following further inspection by engineers, the development of Gaybrook was abandoned on the grounds of the potential expense resulting from the amount of grading required by the revised specifications.[59]

In requesting confirmation of verbal sanction previously given, the Department requested approval for agreements to be entered into with the three landowners at Rathduff, Golden, County Tipperary. It was proposed to pay annual reservation fees of £52 to Mrs D. H. Edwards, £12 to Thomas Burke and £6 to Denis Kennedy and to undertake to compensate for any damage done by removing fences. A rental payment was agreed in the event that the lands were actually used subsequently as an airfield.[60] By October 1941 it was reported that the aerodrome at Rathduff was being developed to the modified requirements of 21 August 1941, except that the runways were at 93 degrees to each other in order to fit them in with surface features. Outstanding works, including the levelling of banks, ditches and hollows, would take six weeks to complete with 100 men employed. The difficulty presented by the presence of a 200-yard strip of stubble that would not bear the required loading was easily resolved. Arrangements were made with headquarters,

RAF Northern Ireland to make available the necessary quantity of Sommerfeld track for emergency runways if and when required. Mulcahy had already gone to Fowlmere Aerodrome in August 1941 and inspected reinforced steel planking (RSP) used to stabilise soft ground. [61] Although Rathduff was therefore ready for use by the end of 1941, in fact, it was not destined to be used by the RAF at all. In reality, the only recorded Air Corps use occurred during the Army exercises of September 1942.[62]

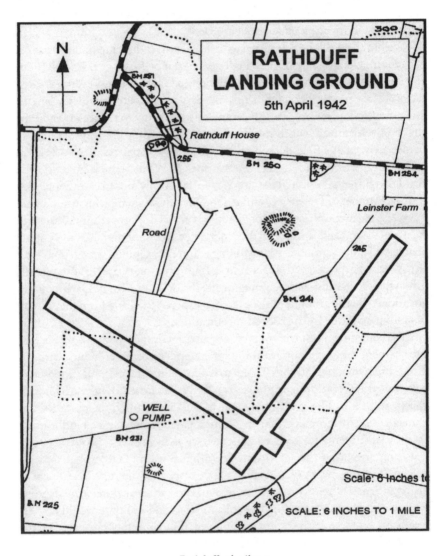

Rathduff, 5 April 1942

In May 1941 the Air Corps took delivery of ten ex-RAF Hawker Hectors and in January 1942 a further three. Accounts record that the machines were supplied at a notional cost of £200 each, plus £15 each for equipment.[63] On 10 February 1943 six ex-RAF Miles Master II training aircraft, of 1938 vintage, were supplied direct from RAF squadron service. Later, further dividends accrued to the Air Corps. In the latter part of 1943, the British supplied six Hawker Hurricane MK I aircraft in exchange for two MK II machines surrendered by the Air Corps. These were followed by four more in February /March 1944.[64]

Given what appeared to be an harmonious relationship between Lywood and Mulcahy it is not at all obvious why they did not co-operate more fully in the matter of selection and preparation of airfields for RAF use. In this regard Lywood seems to have functioned in a covert manner in gathering information from civilians[65] and supposedly selected several sites without consulting with Mulcahy, who, in the interest of getting more aircraft, was being most co-operative. While Lywood's main duty was to establish liaison and promote goodwill with the Air Corps he also had tasks on which he was directed not to refer reports 'to the authorities in Éire'.[66] O'Halpin's observations regarding the covert activities of Lywood would seem to confirm, if such confirmation is needed at all, that his role in Ireland was, in fact, primarily that of an intelligence officer.[67] Coinciding with the receding threat of invasion, Lywood left Ireland in 1942 and was replaced by Wing Commander Begg who, with Pryce's replacement – Brigadier Wodehouse – was appointed an attaché on an official basis.[68]

RECOVERY OF AIRCRAFT

Early in 1945 the Dominions Office acknowledged the fact that 'full assistance was given' by Éire 'in recovering damaged aircraft'.[69] The matter referred to was the operation, mounted mainly by the Air Corps under the direction of the intelligence branch of GHQ, to salvage and return repairable Allied aircraft to the border with Northern Ireland. Not mentioned by the Dominions Office was the not inconsiderable number of Allied aircraft that force-landed in Éire, due to lack of fuel or being lost, or both, and which were refuelled or otherwise helped to make speedy returns to their own jurisdiction.

It is not obvious how this process was initiated. As with other areas of co-operation, full records do not survive. Those available suggest that aircraft recovery was probably something that developed out of British necessity and an undeclared willingness on the part of the Irish Government to render assistance in covert ways. The precedent for allowing British aircraft landing in the State to depart again was set on the very first day of the War. On 3 September 1939 Royal Navy flying boats alighted at Skerries and Dun Laoghaire seeking shelter from particularly bad weather encountered while traversing the Irish Sea. After appropriate questioning of the aircrew and consultation between the COS and the Minister both aircraft were permitted to resume their journeys. Eleven days later an aircraft alighted at Ventry Harbour due to a broken fuel pipe. This machine was allowed to depart after effecting repairs with assistance from 'Sean Clancy's garage, Bridge St, Dingle' where 'the seaplane mechanic soldered the pipe himself'.[70]

Thereafter, the first land plane recorded as having been allowed to depart was a Hampden bomber that made a forced landing at the Curragh at about 05.00 hours on 16 May 1940 due to low fuel. The manner in which this forced landing was dealt with was probably typical of the many that followed. A young officer was woken early that morning in Baldonnell and was ordered to report to the airfield at the Curragh Camp. He was authorised by higher authority to supervise the refuelling of the RAF aircraft. Accompanied by the camp commandant, the officer supervised the refuelling of the aircraft with 200 gallons of the appropriate aviation spirit, from the stocks held in the Curragh, and received a receipt. The aircraft had fuelled and departed for Aldergrove by 09.30 hours the same morning.[71] While it is not explicit in the surviving records it would appear that, by this time, outline arrangements were in place, or at least being formulated, that would facilitate aircraft to be refuelled and depart so rapidly. The general position was later explained by Defence:

> During the period 1940–1945 aircraft of the British and American forces were forced down in this country as a result of fuel shortage, weather conditions, damage by belligerent aircraft, etc. Informal arrangements were made with the air attachés of these countries under which assistance was afforded by the Defence Forces in the rescue of crews and the salvage, repair, refuelling, etc. of any planes forced down to enable as many as possible to take off again.[72]

Although Wing Commander Lywood had not yet been appointed as UK air attaché in May 1940, it would appear that an informal agreement was sufficiently advanced to permit this aircraft to depart without delay. Subsequently, at least 29 British and 18 US aircraft were facilitated in a similar manner.[73] Colonel Keane recorded that the Air Corps rendered assistance in 31 of those cases, and that a total of 7,900 gallons of fuel was supplied.[74]

The next two RAF aircraft that force-landed in Éire were recovered to Baldonnell and, after repair, were pressed into service. The first of these, Hawker Hurricane P5178 of No 79 Squadron, RAF Pembrey (Wales), landed at Ballyvaldon near Enniscorthy, County Wexford on 29 September 1940.

Having landed with its undercarriage retracted, the aircraft sustained only minor damage to the underside. A local gentleman rendered assistance to Pilot Officer Paul Mayhew and was inclined to spirit him away and assist his return to the UK. However, a Local Security Force officer, Major Bryan, himself a former RAF pilot, intervened and made sure that the pilot was detained by Gardaí so that he could be subsequently interned in the Curragh. The aircraft was dismantled and brought to Baldonnell. It was repaired and entered service as Hurricane 93. A Miles Master which force-landed at Dungooley, County Louth on 21 December 1940 was also recovered to Baldonnell and, thereafter, entered service with the number 96.[75]

One of the largest and longest salvage operations taken on by the Air Corps followed the landing of RAF Lockheed Hudson No P5123 in a field near Skreen, County Sligo on 24 January. A salvage crew of an officer and 19 other ranks was dispatched the following day. On 30 January Mulcahy reported to the COS that the aircraft appeared to be in reasonable condition and that the question of making it serviceable so as to fly it to Baldonnell was being examined. As a modern reconnaissance aircraft, a generation ahead of the Anson and valued at about £30,000, there would have been a great desire within the Air Corps to acquire such a machine. The salvage operation, however, was hindered by the remote location, inclement weather and very soft ground conditions. Further complications would have been the lack of tools appropriate to American aircraft and lack of experience of aircraft of semi-monocoque construction.

After being raised onto its undercarriage the aircraft was moved to a firmer location and inspected for damage. The officer in charge of the salvage made a request for a considerable range of materials, tools, tarpaulins, duckboards, Wellington boots and other equipment. These were withheld pending the Minister's decision regarding salvage. Meanwhile the Civilian Inspector, Ted Hoctor, discovered serious cracks in the bottom members of the forward mounting on both engines. This matter having been reported, the COS authorised the sending of an officer to Northern Ireland to obtain materials required for the initial repairs – intimating that the Minister had approved the completion of at least the recovery phase of the salvage.[76]

The UK air attaché and RAF NI were very co-operative in facilitating the return of the Hudson to serviceability. Materials, spares and tools sourced in Northern Ireland were delivered by 'Mr Roberts' to the crash site via the Customs Post at Belcoo. All concerned with the delivery were instructed to keep the matter very quiet. The officer in charge in Sligo, Lieutenant Jim Teague, who had been instructed to tell his men to be discreet in regard to the origin of delivery, met the lorry at the border. The main items delivered to the crash site were two propellers, two engine bearers and an engine tool kit.[77]

Subsequently, engine mounting bolts, not available in Northern Ireland were procured in the UK through the good offices of the air attaché.[78]

In due course the aircraft was repaired to a condition that allowed it to be flown to Baldonnell. For this purpose the services of an officer of the Air Corps Reserve, Captain Ivor Hammond (Aer Lingus), were arranged by the Department of Defence. The aircraft was flown to Baldonnell on 27 March 1941, nine weeks after it had landed. While no financial calculation appears to have been made, there is little doubt that considerable resources were committed to the venture. Not least of these were a total of 1,609 man days of labour and the completion of some 5,690 miles by sundry Air Corps vehicles. A considerable inventory of spares and materials was used in the repair, while much equipment, tools and clothing were rendered unserviceable as a result of a difficult recovery operation.[79] After further inspection and more repairs the Hudson entered service with the Air Corps number 91. In a similar, but more straightforward manner a Fairey Battle light bomber was acquired on 24 April 1941. In June and August 1941 two Hawker Hurricane II aircraft force-landed, and were recovered to Baldonnell, subsequently entering Air Corps service.[80]

A system of skeleton crews, with five or six named individuals being nominated for the recovery of three different categories of Allied aircraft, was put in place in April 1941:

A. The repairing and servicing of aircraft that can be flown to an aerodrome in Éire.
B. The dismantling, packing and transporting to Baldonnel of aircraft that appear to be in a fairly good state of repair and are likely to be rebuilt.
C. The breaking up and transporting to a Military Post aircraft that are badly damaged.[81]

The information above suggests that previous to this an ad hoc system had been in place and that salvage crews were put together on a case-by-case basis. The recovery vehicles, possibly purchased in the 1940–1 financial year, comprised a single five-ton crane and two two-ton tractor and trailer combinations.[82] The records indicate that the first aircraft to be handed back to the UK was a Spitfire that force-landed at Clogher Strand, Donegal on 16 December 1941.[83] It is not obvious how this came about. Perhaps the air attaché had noted how the Air Corps had salvaged a number of aircraft, between May 1940 and August 1941, and had converted them to their own use. With aircraft always at a premium, and fighters such as the Spitfire particularly so, it would have made good sense to have repairable aircraft returned to Allied service. Defence's authority for providing this service was directed by Government:

the international situation existing during the Emergency was such that the state considered it politic at the time that belligerent aircraft landing on our territory should be removed . . . with all convenient speed.[84]

It must be noted that this assertion, suggesting that all foreign aircraft landing or crashing in Éire during the Emergency were repatriated, is not entirely true. In practice, the only aircraft allowed to depart, after refuelling or minor repair, were aircraft of the Allied countries. Similarly, only Allied aircraft that were repairable were salvaged and delivered to the border. Where the recovery of an aircraft was very difficult, secret or sensitive items of equipment, plus armament, were removed, and the wreckage left in place. Crashed German aircraft, if not already destroyed by the impact or by deliberate action of the crew, were usually blown up in situ after the removal of secret and sensitive equipment of intelligence value to the UK.

A notable exception was the case of the Junkers 88 (Ju88) that landed in Gormanston on 5 May 1945. This German aircraft had departed from a base near Aalberg in Denmark as many aircrew had been facilitated in escaping to a neutral country as the War ended. The aircraft made a safe landing at about 05.15 hours and the crew were detained in Gormanston before being interned in the Curragh.[85] About two weeks later the air attaché in Dublin contacted the RAF informing them that a Ju88, with interesting radar equipment, was located at Gormanston and that 'the Irish were prepared to let us collect' it. On 2 June 1945 a number of RAF personnel flew into Gormanston, now Fighter Squadron's new base. After a night of convivial hospitality, the RAF officers interviewed the German NCOs while the aircraft was made presentable for the flight to Farnborough. The swastika and other markings were painted over and crude RAF roundels superimposed before the machine was flown to the UK by Eric 'Winkle' Brown.[86]

With circumstances conducive to the quick dispatch of serviceable aircraft existing almost from the beginning, and the first repairable aircraft being handed back in December 1941, accounts suggest the aircraft recovery operation was put on a more formal basis in the first half of 1942. Defence subsequently explained the circumstances:

During the Emergency certain equipment was supplied to the Air Corps by both the British and American authorities under special arrangements made separately from the ordinary purchase channels. The supplies included equipment for the salvage of crashed aircraft together with equipment for general Air Corps use, e.g. spare parts and radio equipment. The total value of the equipment so supplied was £14,600 of which supplies to the value of £10,600 were expressed to be a free gift. . . .

The balance of £4,000 represents transport equipment of which £2,400 worth was received from the British and £1,600 worth from the American authorities.[87]

It seems that hand tools, sundry items of equipment and clothing, including Wellington boots, were included in the salvage equipment supplied.[88] It is not clear when this informal arrangement was strengthened or exactly when the donated equipment entered service. However, it appears that the main items, two 60-foot tractor and low loader combinations, were in service by May 1942. On 14 April 1942 a Hudson reconnaissance bomber had force-landed at Ely Bay, Blacksod, County Mayo. The following month it was transported by the Air Corps from there to Garrison, County Fermanagh. With an empty weight of 11,630lbs or more this aircraft could not have been moved on the small capacity Air Corps low loader, which strongly suggests that 'the heavy transport and equipment supplied from Northern Ireland' had been used.[89]

In the reference following, when hinting at the value of repatriated aircrew to the Allies, Fisk might have also referred to the return of aircraft:

> Of much greater material value was the collusion between the Irish and British Governments over the Allied air crews whose planes crashed in Éire and who should, under the rules of neutrality, have been interned for the duration.[90]

With 27 Allied aircraft being handed back at the border (in addition to some 47 aircraft that were permitted to take off again), the repatriation of experienced aircrew would also have been of great military value. The management of the repatriation function, to the extent that it can be assessed, would tend to suggest collusion at a political level. During the period, a total of 537 Allied aircrew survived crashes and forced landings in Éire. Of 273 RAF personnel in those categories only 45 were interned. Eleven RAF airmen escaped while the small number interned were released long before the end of hostilities mainly as a result of representations made at a diplomatic level. By way of contrast, all German aircrew (and sailors) were interned for the duration and remained so until 30 June 1945.[91]

The relatively small number of RAF internees is accounted for by the fact that all aircrew were encouraged to state that they were on training rather than operational missions or that they were involved in search and rescue. It appears that de Valera accepted such concocted stories, and that Colonel Archer had the authority to make decisions on the individual cases.[92] In early 1942, while Air Corps officers were handling the matter of a Hurricane that had landed at Collinstown, directions were handed down by GHQ:

Col. Archer, . . . phoned Comdt. Delamere to say he had decided to release the Hurricane and the pilot and that it was to proceed first thing on Thursday 29th. We were to ensure that the aircraft was checked and serviced . . . filled with petrol . . . the pilot given instructions to proceed straight to Aldergrove Aerodrome . . .[93]

The aircraft departed for Aldergrove at 10.28 hours on 29 January 1942. One pilot who could not make a claim to being on a training flight made a force-landing near Athboy on 21 August 1941. His Hawker Hurricane II had long-range fuel tanks (and 20 gallons of fuel), and no less than ten Browning machine guns with about 900 rounds of ammunition remaining.[94] On the following day the *Irish Press* carried a brief report under the headline 'British plane down in County Meath':

> The Government Information Bureau issued the following statement yesterday; 'A British plane made a forced landing in Co. Meath this afternoon. The pilot, who was uninjured, has been interned.'[95]

The Daily Mirror of 22 August 1941 carried the same standard announcement under the headline 'Éire interns RAF pilot'. Woolgar and Roberts Press Cutting Agency supplied their client, 'Éire', with the relevant cutting. While de Valera's Government was no doubt concerned to know how Irish affairs were being reported in the UK, they were probably more determined to give the impression to Irish people, home and abroad, that all such aircraft incidents resulted in the internment of the crew. This, as has been revealed of course, was not always the case.[96]

While the Department of Defence memorandum of May 1949 stated that the arrangements for the return of aircraft were informally agreed, evidence reveals that the Defence aspect of the matter – such as the salvage of aircraft as carried out by the Air Corps – was put on a regulatory basis in 1943. The main influence appears to have come from the British:

> Information has been received through W/C Begg that the incidence of forced landing next summer may be considerably higher than this year, which totalled 25. Workshops Branch will have to make provision for one permanent salvage party.[97]

This came about as part of the reorganisation of the Air Corps which took effect on 29 March 1943. In the Technical Workshops of the Maintenance Unit a 'Salvage' section, comprising a captain, six NCOs and 16 privates, was established. Listed under the heading of 'vehicles', in the Transport Section of the Air Corps Depot, were no less than five 'tractors, aircraft'.[98] A total of about 162 crashes and forced landings are recorded as having been dealt with by the armed forces during the period 1939–45.[99]

CONCLUSIONS

Archer's very general report of March 1944 only gives a vague impression of the actual extent of discussions that took place between the military of the north and south from May 1940 onwards. General agreement was reached on operational matters, logistics and services essential to the British countering a German invasion. However, Archer does not allude to his tasks and functions from October 1938 as he acted on behalf of de Valera in respect of military matters. Similarly we get no indication of the specific instructions received from de Valera with regard to the military co-operation that took place after May 1940. From May 1940 the great dependence of the Defence Forces on support from BTNI was evident in the degree of co-operation in the planning of matters relating to repelling a German invasion.

Though the actions of the military attaché and the military contacts, between the BTNI and GHQ Dublin, were closely monitored by External Affairs the duties and functions of R. W. G. Lywood were carried out largely unobserved. He was greatly facilitated in his pretend role as the man from the Air Ministry and successfully paid only lip service to his role as air attaché, while actively pursuing his functions as the intelligence officer he so obviously was. Mulcahy's co-operation with Lywood and the RAF paid off mainly in the number of well-used and harmless aircraft he managed to obtain from RAF service. He was less successful in obtaining spares for front-line aircraft such as the Anson, however. His relationship with Lywood suggests that, in intelligence matters, the Air Corps was, in effect, an open book. As a result it is likely that few aviation or army secrets escaped Lywood.

The services rendered by the Air Corps were clearly part of a general and unwritten plan of co-operation that evolved, after May 1940, within non-defined boundaries dictated by the Irish Government. In particular the refuelling and dispatch of serviceable aircraft, as well as the recovery and return to the Allies of crashed fighter aircraft, must have been invaluable help in the air battle in Europe. The Department of Defence memorandum of May 1949 indicates that the decision to provide such services was a major concession by the Government. There can be no doubt, however, that these services rendered were outside the accepted norms of neutrality.

ELEVEN

THE AIR CORPS' INVESTIGATION OF 1941-2

—

By the autumn of 1940 the operational capability of the reconnaissance detachment at Rineanna, never ever moderately effective, was at an extremely low point. In similarly degraded circumstances the Fighter Squadron at Baldonnell, which had an unsuited role in a scheme for the air defence of the Dublin area, was similarly impotent. The minuscule Coastal Patrol Squadron, also based at Baldonnell, had no operational function but acted as a training element for the reconnaissance detachment in Rineanna. It would be the events of the latter months of 1940 and the actions of two officers of CP Squadron that would finally demonstrate that relations between the Air Corps pilot officers and their Artillery Corps Commanding Officer, which should have been one of mutual respect, was substantially one of total distrust and that the morale and *esprit de corps* of the Air Corps, of the pilot officers in particular, was at rock bottom. It was to be through the genuine interest and the unprecedented intervention of two junior officers that the professional inadequacies of the commanding officer – Colonel Mulcahy – were to be highlighted, by exposing his ignorance of the technical nuances of the operation of military aircraft and the management of pilots. His failure to take appropriate action in such matters as aircraft modification and aerodrome support services, such as communications and direction finding services, were major areas of contention. The shocking action of junior officers in making unsolicited written submissions to the Minister for Defence was to give rise to a most unsettled and uncertain two year period while a committee investigated and GHQ and the Department of Defence considered the future of a corps the continued existence of which the General Staff would have had great difficulty in justifying.

THE COMMITTEE

The 'Committee of investigation into the effectiveness, organisation, equipment, training and administration of the Air Corps' was established by a convening order issued by the COS on 10 January 1941. The Committee consisted of Major General H. McNeill, Assistant COS, and three majors (then equivalent to lieutenant colonels). While the report states that the Committee first convened on 28 January, the taking of evidence actually began on 21 January 1941. Evidence was taken four days each week until 18 April, and again from 23 September to 21 November 1941. They began formulating their report and findings on 8 December 1941 and delivered the report to the COS on 10 January 1942, exactly one year after the order had been issued. Though not cited as such, the investigation process was in effect a court of inquiry as provided for by Defence Forces Regulation (DFR) A5 dated 10 April 1937. Under the terms of the convening order and DFR A5, the Committee was not required to follow the precise court of inquiry procedure. The regulation provided for the examination of witnesses and for rebuttal evidence in the event of a witness making remarks affecting the military reputation of an officer or giving evidence contradicting that of another witness. In practice the investigation examined each witness in private with all evidence being duly recorded. The preamble on procedure indicates that, while the conduct of the investigation was formal and on oath, the evidence was not necessarily spontaneous:

> While the evidence is recorded in the form of question and answer, it was found desirable . . . to permit witnesses to discuss with the committee and explain the points they desired to make. These discussions were then reduced to relevant and essential facts in the form of questions and answers and are so recorded.[1]

It is considered that this convenience the Committee afforded themselves may have provided scope for a degree of selectivity as to what was considered relevant or irrelevant. In effect the questions could be framed in such a manner as to elicit the answer required. The three volumes of witness evidence indicate instances where the Committee steered witnesses away from matters they might have preferred to pursue, but that the Committee might not. On occasion the Committee abruptly abandoned a line of questioning that was not producing the answers required, or that they did not understand. Forty-one witnesses, all Air Corps or Air Corps Signals personnel with the single exception of Colonel (later General) M. J. Costello, a senior officer on McNeill's operations staff in GHQ, were interviewed. A total of 588 pages, or approximately 265,000 words, of witness evidence was recorded while the report and findings, annexes and appendices added a further 274 pages of typed foolscap. Air Corps related correspondence, documents and reports, from the period 1937 to 1941, submitted in evidence were reproduced in 32 appendices. Due to his key roles as OC Air Corps and DMA, and to the amount of criticism expressed before and during the investigation, Colonel Mulcahy's evidence represented about 20 per cent of the total. Due to adverse comments, on aspects of his command, decisions and actions made by individual officers, he was recalled a number of times to explain points or give rebuttal evidence. The Committee also perused 44 Defence, GHQ and Air Corps files, the flying log books of 47 officers and sundry records and orders.[2]

THE CAUSE OF THE UNREST

The terms of reference, the evidence of witnesses and the report and findings of the Committee do not indicate, however, the exact circumstances that led to the investigation. It would appear that the condition of the Air Corps during what Mulcahy called the 'invasion nervous' months of the summer of 1940[3] and the manner in which Mulcahy exercised his command, and his functions as DMA, were central factors. From Mulcahy's final submission to the Investigation Committee on 21 November 1941 it transpires that, during the latter half of 1940 in particular, his command was under severe strain due to alleged irregular communications from junior officers to persons outside the Army, including the Minister for Defence. Mulcahy cited a visit to Baldonnell by the Minister and the COS on 23 October 1940 in connection with letters of complaint forwarded to the Minister. While it is not clear by whom complaints had initially been made it appears from the evidence that Captain T. J. Hanley and Lieutenant A. C. Woods were the two whistle-blowers. The exact nature of the allegations are not revealed but the evidence of witnesses confirms that standards of navigation and instrument flying, the standard of aircraft equipment – such as instrument panels, direction finding equipment, communications generally – and the failure to acquire vacuum pumps and loop aerial modifications for the Anson aircraft were the main matters of concern to the pilots.[4]

It is not at all clear what form the Minister's visit took but accounts suggest that the officers who had made written complaints were interviewed. Subsequently, with the Minister's permission, the two officers made further written submissions. Mulcahy's evidence to the investigation confirms that Hanley was one of those invited to write to the Minister – which he did on 4 November 1940.[5] One of the matters he complained of was the fact that vacuum pumps (for the more effective operation of gyroscopic instruments in Ansons) had been requisitioned by the Air Corps in June 1939, not ordered by Contracts Section, Department of Defence until June 1940, and had still not delivered in January 1941.[6] Although he possibly also mentioned the failure of the Air Corps to seek the purchase of loop aerials for Ansons (a modification that was available from A. V. Roe since November 1938), this was not explicit in his evidence to the Committee but rather was intimated in November 1940 in Mulcahy's initial defence of the allegations made against him.

On 15 November 1940 Mulcahy wrote to the COS in response to the matters contained in the two official letters to the Minister. One of the main planks of his defence against the allegations was to denigrate Hanley for his lack of experience of 'staff duties', stating that Hanley was in no position to criticise constructively the administration of the Air Corps. Mulcahy went on, in an oblique manner, to blame the procurement system for the failure to acquire modifications which Hanley saw as being of little importance to Mulcahy and his headquarters staff, but a matter of life or death to those who flew every day. He summed up Hanley as follows:

> Like many others he feels that every demand he makes for new or more equipment should be supplied without delay. . . . He forgets that these officers who built up the Corps flew for years without the aid of modern equipment which he now has and without the new instruments and equipment which he states are essential.[7]

While Mulcahy blamed the procurement system for the failure to acquire new instruments and equipment required by pilots, he avoided direct reference to the failure to purchase the loop aerial and vacuum pump modifications for the Ansons. In the case of the loop aerial modification Hanly's evidence to the Committee strongly suggests that Mulcahy had knowingly withheld authority to buy the equipment that constituted the newly developed modification.[8]

In denigrating Hanley's lack of administrative experience and knowledge of procurement and, in effect, stating that pilots never had it so good, Mulcahy attempted to deflect attention away from Hanley's fundamental point that Mulcahy's Air Corps was not keeping Anson aircraft up-to-date in terms of equipment conducive to good navigation. In addition, the failure to incorporate such modifications as vacuum pumps and loop aerials had rendered the reconnaissance operation more untenable than it might have been. It was in the context of this antipathy between Mulcahy and Hanley that the incident, concerning the irregular use of the civil DF Station at Baldonnell on 21 November 1940, had occurred.

On 24 October 1940 Mulcahy had felt obliged to issue orders prohibiting officers from visiting other offices in order to converse with fellow pilots except on official matters, and then only with the permission of their unit commander. When asked by the COS, Mulcahy explained that the necessity arose because of his belief, based on his observation of the casual movement of officers between offices, that 'the practice of officers consorting with each other' represented a waste of time and that it should be stopped. He considered that officers had deliberately misconstrued his order related to this matter and had reported it in an irregular manner.[9] The tone of Mulcahy's order about officers consorting with others, and the complaints made by some officers to higher authority, reveals the truly dictatorial nature of his command. His lack of appreciation of the technical nuances of the aviation of the day, together with the inevitable demoralisation caused by the impotence of the two main operational squadrons, all combined to result in great unrest amongst the flying officers. Cited the letters of complaint and other incidents Mulcahy subsequently stated that 'while these incidents were occurring, it was impossible to keep secret the fact that some disruptive element was at work and the effect on Corps morale and discipline will be appreciated'.[10] In this regard it might be considered that Mulcahy greatly mistook the symptoms for the cause.

A clear picture emerges of a demoralised and frustrated pilot officer body that was no longer prepared to grin and bear it. In Hanley, who qualified in 1928, the younger pilots had a spokesman who had the professional expertise and moral authority of a long-qualified pilot, having already seen service with Aer Lingus, probably saw his future career as being outside military aviation, who could highlight the inadequacies of the Director of Military Aviation. Notwithstanding, in the dictatorial atmosphere of the Army of the Emergency where higher authority was right by virtue of superior rank, Hanley's would have been a high-risk strategy. While it is possible that Hanley and others had secured their positions by keeping the Dáil opposition informed, it is more likely that their main safeguard lay in the fact that their ally, A. C. Woods, was well positioned in view of his allegiance to the ruling Fianna Fáil Party. He had originally been commissioned and appointed on the authority of the Minister, Frank Aiken, despite the objections of the Department of Finance. Had this not been the case, the pilots would most certainly have been subjected to disciplinary action by the Adjutant General.

THE INVESTIGATION

Accounts suggest that, irrespective of the nature of Mulcahy's defence against the written complaints, higher authority, in the name of the Minister, considered that a thorough investigation was warranted. In the circumstances the investigation might have been centred on Mulcahy's command and direction of the Air Corps which had resulted in 'demoralisation, inefficiency and stagnation'.[11] However, the terms of reference and the manner in which the Committee proceeded, was left to the discretion of GHQ which ensured that the spotlight was turned on the inefficiency of the Air Corps and the perceived inadequacies of the pilots individually and collectively. As a result the outcome presupposed that Mulcahy had little responsibility in such matters.

In investigating 'the effectiveness, organisation, equipment, training and administration of the Air Corps' the Committee addressed a number of standard questions, based on the nine main questions in the terms of reference, to the more senior witnesses in particular. More specific questions were put to individuals as appropriate to their appointments, responsibilities and evidence. The Committee reported their proceedings, findings and recommendations under nine broad headings and many subheadings.

EFFECTIVENESS OF THE PRESENT AIR CORPS

The effectiveness or otherwise of the Air Corps was assessed by the Committee in light of the first question:

> whether the Air Corps, as now organised and equipped, is capable of co-operating with other units of the Forces or of functioning usefully in any other capacity? To enable it to deal adequately with this question the Committee had to decide what type of co-operation our ground forces should expect from the Air Corps.[12]

The Committee, without taking evidence on the matter or citing existing planning or policy documentation, but presumably drawing on the operations backgrounds of Major General McNeill and Major J. J. Flynn, stated that the 'co-operation required by the Defence Forces of the Air Corps might be divided into war and peace missions'.[13] The peace missions broadly reflect the traditional roles of army aviation:

> To accustom the ground forces to the tactics of bombing, dive bombing and machine gunning aircraft, by means of exercises demonstrating these tactics.
> To test and examine the concealment of ground forces and thus perfect their technique in this important aspect of modern warfare.
> To test air discipline of ground forces in camps and on the move.
> To test alarm and evacuation of ground troops.
> To train ground forces and Air Corps, separately and in combination, in their War Missions.[14]

The peace missions listed were, by and large, the standard roles of army aviation deployed with, and in support of, ground formations. Had these principles been applied in training prior to the Emergency many squadrons of Army aviation, dispersed amongst the manoeuvring ground formations and devoted to the practice and simulation of wartime battle conditions, would have been required. However, in the context of totally limited resources, and the equally limited capabilities of the pre-War Air Corps, such principles were, of course, quite hypothetical.

The War Missions listed were in effect the roles of an independent air force, roles for which the Air Corps had not the organisation, structures, aircraft, equipment, training or other essential resources:

> Provision of information regarding strength, disposition and movement of hostile forces at sea en route to invade our territory.
>
> Provision of similar information of hostile forces which have invaded our territory and may be in contact with or moving against our forces.
>
> Provision of communications on a small scale such as message dropping and transport of commanders and staff officers.
>
> Interception of bomber and dive bomber formations.
>
> Limited attack on hostile ground troops.

The War Missions, in their broadest sense, could be reconciled with the missions implied in the nomenclature of the three operational squadrons, but could only have been feasible in Costello's ten-squadron Air Corps if properly equipped, manned and trained. In essence, the War Missions would have required an independent air force having an operational capacity many times that which existed during the Emergency. As such the War Missions were also utterly hypothetical.

Bearing in mind the fact that higher authority had not previously defined war and peace missions the introduction of such principles in regard to a major review of the effectiveness of military aviation might have unduly complicated the study. In the event the investigation was to concentrate on its perception of the effectiveness of the existing Air Corps and on the Corps' potential, as army aviation, to support ground forces. It should be noted that the Committee did not try to compare what it actually found with the stated ideal. The introduction of the concept of war and peace missions, in the report of 10 January 1942, only served to emphasise the fact that the Army had neglected to address such important matters at a more appropriate earlier juncture.

The effectiveness of Air Corps aircraft was assessed by the Committee with reference to the extent to which 'a heterogeneous collection of aircraft, service and training, having as many different characteristics as there are types' of aircraft could perform their war missions or, in a future reorganisation, be adapted to reconnaissance missions. There were no conflicts of evidence in regard to the manifest inadequacies of individual aircraft types. The Gloster Gladiator, of which there were only three in service, was described as a single seat fighter of limited range with a poor radio and no armoured protection for the pilot. 'In speed, armament and performance they would be completely outclassed by modern fighter aircraft.' It was described as having limited potential as a reconnaissance aircraft in that, as a single-seat machine, it did not have a rear gun and could not carry an observer.[15]

The Committee reported that the other service aircraft of Fighter Squadron consisted of five Lysanders organised in two flights. The sixth machine had been adapted with target-towing equipment to facilitate the training of anti-aircraft artillery crews.[16] This type was acknowledged as being a very suitable Army co-operation aircraft which, when used as fighters could use their low speed and manoeuvrability to avoid being shot down. In reality, however, this aircraft would have stood little chance in normal combat.[17] Thus, the Committee concluded that, with both fighter and Army co-operation aircraft, Fighter Squadron could fulfil neither role satisfactorily. While the bulk of its aircraft were Army co-operation machines that were unsuitable as fighters, the pilots were also inadequately trained in Army co-operation duties. The report did not state the exact position. During the pertinent period, the latter part of 1940, the Fighter Squadron had a total of 15 aircraft of no less than six different types – three Gladiators, six Lysanders, two Avros 636s, two Hawker Hinds, a DH Dragon and a Miles Magister – organised into four flights. The Squadron had only 11 pilots as against a notional war establishment of 27. 'B' Flight, Fighter Squadron had five aircraft of no less than four different types. Compared with the norms of the organisation and equipment of air squadrons of the period, the composition of the squadron was totally inadequate. [18] The Committee concluded that 'the fighter squadron is fighter in name only'. Its final assessment was brutally frank:

> The Committee considers that the employment of this insignificant unit would not be justified for fighter purposes. Such employment would, it is felt, be an unwarranted waste of life without any gain to the Army or the State.[19]

The Ansons of the R & MB Squadron were described by the Committee as 'twin-engined, slow, heavy and of limited manoeuvrability which renders them very easy prey to any type of enemy aircraft'. In this case the Committee seems to have seen qualified merit in the manner in which the Anson was, and could in the future, be used:

> The Anson machines can be employed on coastal patrol in normal weather during the present period of the Emergency. They have in fact been employed on such duties during the autumn and winter of 1939, operating from a base at Rineanna Aerodrome. . . . In the most favourable circumstances, they could be used to report whether hostile sea-borne forces were at sea, were approaching our coast and the location of such forces being put ashore.[20]

In considering a possible army co-operation role for this Squadron the Committee suggested that the Anson might be used 'over quiet sectors where hostile aircraft is [*sic*] not operating' or in lulls 'between periods of hostile air activity'.[21] As an operational unit the R & MB Squadron was assessed to have the deficiencies inherent in its aircraft. The CP Squadron was deemed to be similarly afflicted. It was cited as having two 'obsolete type Walrus aircraft', one Avro Anson and two Avro Cadet training aircraft. Due to the lack of spare floats the Walrus aircraft were not allowed to operate from water. The Committee summarised the operational capacity, and thus 'the effectiveness of the present Air Corps', in necessarily blunt terms:

> It will be seen from the foregoing that not alone is the Air Corps equipment obsolete, with the exception of the Lysanders, but is also totally inadequate. A so-called Fighter Squadron is maintained, possessing 8 service machines of which only 3 are fighters of an obsolete type. The Reconnaissance and Medium Bombing and the Coastal Patrol Squadrons have only enough aircraft to equip one flight each. In view of these facts the most that can be hoped for from the Air Corps under favourable conditions is intermittent [reconnaissance] information in limited areas subsequent to invasion. . . . Protection of the civil population and the Defence Forces is definitely not possible.[22]

The Committee found that the extent and nature of the co-operation that the Air Corps, as then organised and equipped, could offer to the ground forces to be 'so negligible that it can be discounted'. They considered two possible recommendations with regard to the future of the Corps. While they considered the disbandment of the Air Corps, with its personnel being formed into an infantry unit or transferred to other ground units, they recommended the other alternative 'to make the best use of personnel and equipment we have by a reorganisation of the Corps'. The main role of a reformed Air Corps would be 'assisting in the training of our ground forces in anti-aircraft measures and helping to overcome the psychological effects of aircraft bombing and machine-gunning attacks'.[23]

ORGANISATION AND EQUIPMENT: PREVIOUS POLICIES AND ORGANISATION SCHEMES

The second question addressed was that of the suitability of the current organisation and equipment of the Air Corps for defence purposes and consideration of the changes to both that might be required under the prevailing conditions of financial stringency and uncertain supply. The Committee prefaced its deliberations by stating that the 'organisation, equipment and training of the Air Corps, as in the case of any branch of the service, must be based on a definite policy', in turn based on the general policy of the Defence Forces. Before taking evidence they proceeded to 'examine all relevant and available documents' in order to review the historical position in respect of previous policies and organisation schemes. The Committee found that a conference of 17 January 1929 had been made aware, by Colonel Fitzmaurice, of the inadequacies of the aircraft then in service and of the poor level of technical expertise available to maintain them. Quoting from the same Defence file the Committee noted that the Minister, on 23 January 1929, had stated that he considered it important to have pilots and technicians trained before spending large amounts of money on aircraft and that, 'in whatever crisis that would arise' 'the machines could and would be found'. The Committee noted that no decision was taken as to whether pilots were to be trained for reconnaissance or fighter missions, or for both. 'In other words the defence role of the Air Corps was not adequately defined.'[24]

The Committee, quoting from another Defence file noted that Major Mulcahy had, on 16 September 1937, requested clarification from the COS regarding 'general aviation policy' and, pursuant to such a policy, the numbers of aircraft required for the following ten years. It was recorded in a minute of 18 October 1937, following discussions between the Minister and the COS, that 'the government cannot at the moment lay down the policy on which a decision could be reached'. On 28 September 1937 Mulcahy recommended, 'as the minimum number of Squadrons required', the establishment of five Fighter Squadrons and five Reconnaissance Squadrons to be dispersed to aerodromes in the vicinities of Dublin, Cork, Limerick, Sligo and Athlone. He further recommended that one squadron of each type should be maintained at full strength and the remainder at cadre strength strong enough to maintain all essential services and to carry out the required aircrew training.[25] The Committee considered that the Costello plan of 21 March 1938, which provided for the immediate raising of three squadron cadres and ultimately for a total of ten squadrons, appeared to have been the first time that the Air Corps was given a definite objective towards which to aim. 'From the nature and nomenclature of the Squadrons, their general role in the defence scheme can be judged.'[26]

The Committee noted that subsequently the Government had come to no definite decision on the ultimate development of the Air Corps.[27] On this point the evidence of Costello and Mulcahy differed. While Costello insisted that his plan of 21 March 1938 had been abandoned, Mulcahy stated that he had not been so informed. The Committee saw no point in resolving the matter. This may well have been because the problem was getting close to home. With the abandonment of the Costello plan the matter of air policy appears to have unknowingly devolved to Mulcahy by way of his advice to Colonel O'Higgins, while McNeill, Costello and Flynn, all of whom had occupied positions in the operations function of GHQ, could also be faulted for not taking action to make the position adequately clear.

The Committee noted that later, under the general scheme of organisation for the Army, war establishment tables were drawn up for one of each of the three types of squadron and financial sanction sought. Having received the approval of the Taoiseach on 10 December 1938, and of the Government on 31 January 1939, these 1938 tables became the War Establishment that eventually came into effect on 13 June 1940.[28] The Committee, however, noted that the approved Establishment included no provision for the expansion to ten squadrons, as favoured by both Costello and Mulcahy. The Committee indicated that, as the Air Corps' general role in the scheme of defence could be deduced from the nomenclature of these squadrons, this in effect, constituted an adequate statement of the Air Corps mission in wartime. They also considered that 'like the earlier proposals of March 1938, this organisation of December 1938, gives the Corps a definite, though more limited objective'; but stated that coming in January 1939 the decision was too late in terms of acquiring the numbers of the aircraft that would be required under a war establishment.

In assessing the 'form of organisation suitable for defence needs' the Committee considered that such a study should be carried out on the basis of 'what an Air Corps is required for', 'how it will be employed', and 'its size which must be governed by financial considerations'. It used the statement of missions as it had discussed earlier to suggest that 'close reconnaissance aircraft of the [army] co-operation type' were required to 'obtain information of enemy movement and disposition after he had gained a footing in our territory'. Two such squadrons would be required, one per Army division 'decentralised to provide flights to work in close co-operation with Brigades' and for 'occasional special missions'. It was calculated that the capital expenditure for 'new aircraft requirements and ancillary equipment for two squadrons' would amount to £290,000 with about £91,000 annual expenditure on personnel maintenance and spares.

In relation to coastal patrol aircraft the Committee considered that its primary function was to 'provide information of and on the approach of hostile forces to our shores'. The alternatives were summarised as:

In the case of invasion from the continent, it is possible that the other belligerent would be in a position to acquaint us of the movement by sea of hostile forces. In the case of invasion by the other belligerent, the main blow would almost definitely come overland and the need for long distance sea reconnaissance would not arise in an acute form.[29]

This obscure statement did more to demonstrate the paranoia of the military, who did not want to be seen to be supporting the Allies, than it did to clarify reconnaissance strategy. The Committee considered that long-range maritime reconnaissance could only be executed efficiently by modern multi-engine aircraft, as were in common use in Britain, costing over £30,000 each. They concluded that 'close-in reconnaissance of territorial waters' . . . 'sufficient to deal with an invader other than a continental one' could be done by close reconnaissance aircraft such as the Lysander. On the basis that long-range reconnaissance could be regarded as a passing phase, the existing Ansons could perform long-range reconnaissance prior to the outbreak of hostilities. Ridiculously, having earlier highlighted the inadequacies of the Anson, the Committee, influenced by the cost of re-equipping with Lockheed Hudsons, for example, recommended that no financial provision be made for more appropriate aircraft and that the Anson could now be considered adequate![30]

Acknowledging that 'the maximum size of the force maintained must be determined by our financial resources considered in relation to our commitments for other elements of the Defence Forces', the Committee then proceeded to embark on a study of a fighter force of outlandish proportions. The study of the employment of fighter aircraft considered that, although it would be impossible 'to estimate accurately the strength of an adequate fighter force' in order to be reasonably safe, 'a force of 30–40 squadrons would probably be required.' It calculated that the capital cost of a force of 40 squadrons would be £6,400,000 based on a 'fighter aircraft of the Hurricane type' while the recurring annual expense per squadron would amount to £48,000:

> To this must be added the cost of the necessary ancillary services required to enable a fighter force to function efficiently, including observer system, radio detection system, direction finding system, central control, provision of aerodromes and accommodation.[31]

The latter facilities were, in essence, the essential facilities that were absent from the authorised War Establishment, the absence of which, along with inappropriate and inadequate numbers of aircraft, rendered the Squadrons ineffective. In attempting to cut their cloth to suit the State's financial measure, the Committee, recognising 'the necessity of affording some degree of protection for Dublin, Cork and Limerick' considered 'that a force of five fighter squadrons is the absolute minimum required' for which capital expenditure of £800,000 and recurring costs of £240,000 per annum, exclusive of ancillary services, would be required. While acknowledging that it was for the Government to decide whether the degree of protection which would be afforded by such a force would justify the level of expenditure, the Committee recommended that the minimum number of operational squadrons required would be five fighter and two reconnaissance squadrons.[32]

FACTORS AFFECTING ORGANISATION AND EQUIPMENT

The Committee considered the time, personnel, and financial aspects of the implementation of this proposed expansion and made equally significant recommendations. The two reconnaissance squadrons were to be equipped with a total of 42 Lysanders, at a capital cost of £432,000, and were to be located at Rineanna and Collinstown. These Squadrons would be manned by existing pilots. The Flying School would need to be organised and re-equipped to train an additional 73 officer and NCO pilots for the Fighter Squadrons. It was recommended that 70 modern fighter aircraft (with another 35 in reserve) were required for the five fighter squadrons which would be dispersed to separate locations – Collinstown, Cork, Rineanna, Curragh and Gormanston. They proposed capital expenditure of over £1,184,000 spread over four years at an annual cost of £296,000. In relation to fighter aircraft the Committee suggested that 'nothing but the most modern aircraft should be considered and that the 'complete equipment for one fighter squadron should be purchased every year *ad infinitum*'.[33]

The 'recurrent annual expenditure' for two reconnaissance and five fighter squadrons, as well as an 'administrative and training organisation' that would entail the recruitment of no less than 742 more personnel, was put at £440,000.[34] While recommending the decentralisation of service squadrons – thus 'throwing them on their own resources' and making 'them more self reliant,' no provision was apparently made for the new aerodrome facilities that would be required.[35] The decentralisation of five squadrons should have been seen to be totally impractical except in the context of a substantial investment in infrastructure and camp staffs. As had been demonstrated in the case of the Rineanna detachment squadrons had no resources on which to rely if removed from an established aerodrome.

In the context of the investigative review, and of the actual annual expenditure on the Air Corps (£176,644 for 1940–1), the capital and recurring costs of the proposed reorganisation could only be described as alarming.[36] Having regard for the parsimony of Finance, even in the Emergency circumstances of the time and the very real threat of invasion not passed it is not clear how the Committee could justify such an ambitious expansion. The financial circumstances and the Government policy of the time should have indicated to them that such a scheme was not feasible. It should have been obvious, based on the known opinions of the Minister and An Taoiseach, that the Government saw no necessity for other than a token level of military aviation. The heterogeneous collection of aircraft purchased in the years prior to the outbreak of war was all the Government was prepared to fund and was, in effect, appropriate to its neutrality stance as well as in keeping with the level of co-operation with the British in defence matters. Perhaps the Committee felt it their duty to identify the extent of fighter defences required irrespective of the State's ability to fund such forces.

The most radical recommendation regarding reorganisation was that '[Air] Corps Headquarters be abolished and replaced by a directorate of military aviation located at the Department of Defence'. The reason for this was explained:

> It will bring the head of the Air Corps into closer touch with the General Staff; it will relieve him of many of the duties of administration and interior economy which seem to occupy so much time at the moment; it will give him greater freedom to concentrate on the inspection and training of the Corps; by removing him from so much close contact with junior officers in our principle air station and placing him on the same basis as any other Director, his prestige would be enhanced.[37]

One would have to see the above argument as being totally spurious, reflecting, as it does, the belief that a director could function better if the functions of his appointment were moved to GHQ/Department of Defence. This proposal appears to reveal a desire to remove the current Director who clearly had such technical and professional deficiencies and lacked the fundamental qualifications to satisfactorily perform the functions of OC Air Corps or DMA. The recommendation, thus, seems to have had the aim of rehabilitating Mulcahy by minimising his contact with turbulent pilots.

ADAPTATION OF THE EXISTING ORGANISATION

Realising that their grand plan for an Air Corps, expanded to seven squadrons and some 1,440 personnel, would require Government approval and, if authorised, would take a considerable time to implement, the Committee recommended that, as an initial step, the existing organisation and equipment should be adapted to form the basis for the establishment of two reconnaissance squadrons. It was suggested that existing aircraft could be reorganised to form two 'provisional Squadrons' – No 1 Squadron, Rineanna (with

Lysander, Anson and Walrus aircraft) and No 2 Squadron, Collinstown (with Lysanders, Ansons and Gladiators). It was argued, that while the grouping of the Lysanders and Gladiators at Collinstown and the Ansons and Walrus at Rineanna would have been more logical from organisational and maintenance points of view, both squadrons should have some of the most useful aircraft, the Lysander. It was considered that the Rineanna Squadron could not perform co-operation training with the 1st Division without Lysanders. Therefore, while the primary role of the two composite squadrons was to be Army co-operation, they could also do 'coastal missions'. It was suggested that, if each squadron had a 'properly equipped' Anson, training in navigation would be facilitated. This proposal was to be modified if and when 13 Hurricanes 'on order for a considerable time' were delivered and if the Government was prepared to proceed with the programme for five fighter squadrons.[38]

TRAINING OF THE AIR CORPS

In the overall context of the Committee's investigation into the state of the Air Corps in 1941 the module that addressed the questions relating to the training of the Air Corps and suggested changes in personnel, administration and training was possibly the most crucial and most telling in terms of higher authority's attitude to the Air Corps, in general, and flying officers, in particular. This section sought to define the efficiency of individual officers as service pilots and to decide whether or not 'flying practice' was 'properly organised and carried out by flying personnel'.[39] This emphasis, implying that Mulcahy was not one of the 'flying personnel' clearly suggests that the Committee intended to place responsibility for the perceived poor state of flying training and efficiency on squadron commanders and the pilots in general, rather than on the commanding officer.

Notwithstanding, the Committee experienced considerable difficulty in arriving at definite conclusions on the abilities of individual pilots due to the conflicting nature of the evidence given and the fact that, as they saw it, no standards were laid down for service pilots. This difficulty arose because of the contradictory evidence of, on the one hand, the commanding officer and the squadron commanders who contended that the pilots were 'capable of carrying out any service mission using the aircraft available' and, on the other hand, of the more junior flight commanders and younger pilots who claimed that they had 'insufficient training in one aspect or another', a situation that undermined their confidence to execute service missions under difficult conditions.[40] The Committee's overall impression was summarised:

> On the evidence of the pilots . . . of the school . . . and service squadrons, the
> interrogation of individual officers, the absence of prescribed standards of training
> for service pilots and the nature and methods of training in the service squadrons
> provides . . . cumulative proof that the pilots of the service squadrons have not
> attained as high a standard of training . . . as should be possible with existing
> aircraft . . .[41]

The Committee observed that, from an operational point of view, no
standards of proficiency, in flying or ground subjects were laid down by Air
Corps Headquarters or by the Department of Defence for either the flying
school or the three service squadrons.[42] They did not comment upon the fact
that the General Staff equally had not laid down such standards. It might
surely have been considered appropriate that the 'first assistant to the chief
staff officer', as designated in 1924, who was the 'technical officer responsible
for inspecting the Air Corps', or his current equivalent, would have had some
responsibility to ensure the setting of flying standards.[43] The stark fact
remained, however, that no aviation expertise existed outside the Air Corps
pilot group. It was, after all, at the insistence of GHQ that the drafting of the
syllabus of flying training for officers and cadets, that became DFR 7/1927,
had been carried out by C. F. Russell for the 1926–8 'wings' course.

It was noted that the last satisfactory training directive had been issued as
early as 1936. This was presumably drafted by a flying officer on behalf of his
newly appointed and uninitiated superior. It was considered that those direc-
tives issued by Mulcahy in later years could not be regarded as having been
adequate guides as to the exact nature and standard of flying expected of pilots
in the operational squadrons. Mulcahy's evidence to the Committee on the
matter was received without comment:

> There are no definite standards laid down, but unit commanders are sufficiently
> conversant with their duties and with what would be required of their officers to
> bring their units to a satisfactory standard.[44]

As the above quote reflects Mulcahy accepted no responsibility in the matter
of setting flying standards. Apparently satisfied with the hands-off policy
adopted by Mulcahy, the Committee proceeded to cross examine the three
squadron commanders with particular reference to their respective undefined responsibilities in the matter of flying training and proficiency.
Notwithstanding the lack of direction from Air Corps HQ, GHQ and
Defence there was no cause for the Committee to question the effectiveness
of *ab initio* pilot training in Air Corps School:

It should be noted that not a single witness had any adverse criticism to offer of the school training, which training, in the opinion of the committee is generally satisfactory, except that advanced training is not catered for. In the school, the standards of training to be reached by the pupils in each subject are clear-cut and definite.[45]

The Committee did not comment on why this should be so. If it had done so it might have confirmed that the School's current training syllabus was fundamentally sound having been based on DFR 7/1927, probably brought up to date as a result of the RAF instructors' courses attended by Lieutenant Keane in the early 1930s. Having been redrafted in 1935, it was probably further refined on the basis of the visit to RAF training establishments by W. P. Delamere and K. T. Curran early in 1939.[46] The accumulated experience of the flying instructors, who were adjudged by the Committee to be efficient and painstaking, would also have contributed to this satisfactory situation.

The original syllabus of flying training should be considered the single most important document relating to the aviation history of the period in question. Essentially, it was the only substantive regulatory instrument relating to the flying of military aircraft, which laid down the standards required of pupil pilots of the Air Corps and, in effect, underpinned standards generally. Notwithstanding, not only did the Committee neglect to connect the syllabus with the satisfactory state of Schools' training, but they failed to include DFR 7/1927 of 18 March 1927 among the list of files and other records examined.[47] Whether this indicates that the Committee failed to consult the regulation, or – while not wishing to highlight Mulcahy's manipulation of regulation – merely chose to ignore its importance, is not clear.

The Committee did, however, interrogate the three service Squadron Commanders, at some length, to examine the contention of many of the flight commanders and junior pilots that they had inadequate training in various aspects of their profession. The Committee found 'that the service training in the Fighter Squadron was of a haphazard type lacking in organisation, control and direction' with, for example a course for three young pilots started in February 1940 likely to take two years instead of six months, while essential ground school subjects had not started by 20 November 1941. Shockingly it also found that there was no organised training for older pilots. While many squadron pilots criticised, OC Fighter Squadron contended that training in aerial combat and formation tactics was carried out to the best of his ability; nonetheless, the Committee found that 'individual training in aerial combat had not been as efficient as it should be' and that there was 'a definite lack of training in formation combat tactics'. The Committee recognised, however, that squadron formation could not be taught when there was only approximately a flight of three fighter aircraft available.[48] The criticism of Commandant Mick Sheerin can be understood on the basis that he was about 40 years old in 1941. This ex-IRA officer is remembered in Air Corps folklore more for his proficiency with handguns rather than his enthusiasm for flying!

In regard to the R & MB Squadron the Committee found that during 1940, and up to the spring of 1941, training was carried out in an unco-ordinated manner that prevented progress being measured, however, that the lack of organisation had since been remedied to the extent that pilots got more regular and useful flying. Given the inadequate level of manpower, poorly equipped obsolescent aircraft, primitive airfield and inadequate support services, not to mention the total lack of preparation prior to the occupation of the 'aerodrome', it could be argued that the adverse comments on training in the Rineanna detachment were unfair. It was found that training in the CP Squadron, (that functioned as the training element for the Reconnaissance Squadron detachment in Rineanna), was found to be conducted in a satisfactory manner. This included 'elementary aerial observation and elementary navigation instruction for other-rank aircrew members aimed at making them more efficient'.[49]

The overall comment by the Committee on training was to the effect that the 'majority of the officers of the Air Corps are not as efficient and capable of carrying out the duties of their appointments' as the available aircraft would permit. The finding detailed the many shortcomings perceived:

> The most important subjects in which the officers are backward are – Navigation, Signals, Night Flying and Service Flying in general, including operating from improvised flying fields. . . . flying training is not properly organised in service squadrons in as much as it is not designed to ensure the systematic progress of pilots towards acquiring and maintaining a definite standard of service proficiency. With the exception of the pupil pilots in the School and the young officers in the Coastal Patrol Squadron, such flying training as is engaged in could be described as flying without an objective.[50]

These comments make no allowance for the fact that the rather basic navigation training carried out in June/July 1939 was totally inadequate for the squadron detachment that was later dispatched to Rineanna. Similarly the ground and airborne signals (or communications) equipment, as well as aircraft flying instrumentation, were inappropriate to the task. In effect the squadron commanders were blamed for inadequate standards when, in fact, Mulcahy, as Commanding Officer and Director, had overall responsibility for equipment and training standards and for the standard of service support.

With the exception of a mild rebuke in the matter of his failure to adequately direct training standards, Mulcahy did not come in for the adverse comment he surely deserved. On the basis of the accepted military principle that the commanding officer is wholly responsible for all his formation does or fails to do, Mulcahy should have been found responsible for the unsatisfactory state of flying training. However, the Committee placed most of the blame on the commanders of the two operational squadron, Commandant Mick Sheerin (Fighter Squadron) and Captain W. J. Keane (R & MB Squadron, Rineanna) whom Mulcahy had considered were 'sufficiently conversant with their duties and with what would be required of officers' regarding standards of proficiency – a convenient escape clause for Mulcahy.

In view of his 'limited technical training', however, the Committee did find that Mulcahy 'could not be expected to supervise and inspect' all aspects of Air Corps training 'without having to rely, to an undesirable extent, on his subordinates'.[51] There is adequate proof that Mulcahy did not welcome advice on such matters as navigation, meteorology, navigation and signals even though he functioned as Director of Military Aviation in relation to all such matters. However, the Committee appears to have accepted that his limited technical training allowed him to devolve responsibility for training standards to the Squadron Commanders, in effect absolving him from the responsibility for those substantial functions he purported to perform from 3 June 1935.

The Committee's main recommendation was that definite standards of flying proficiency should be laid down and that 'all standards should have the force of regulations'. The standards 'to be reached and maintained by service pilots' were to be appropriate to 'the peculiar conditions under which the Air Corps must operate'. The Committee put major emphasis on the development of co-operation with ground forces including having 'a sound knowledge of the tactics, technique and organisation of such forces including practical experience in operating with these forces', in effect, therefore, recommending a return to the anachronistic Army co-operation role largely abandoned before the Emergency.[52]

Under a sub-heading of training the 'efficiency of Air Corps officers' was assessed on the basis of the verbal evidence given. The Committee put the pilots (excluding Mulcahy) into four categories, reflecting their assessment of individual standards. The first group included a number of experienced pilots who were considered to have failed to keep up to service standards due to lack of flying practice and instruction. It was considered that the majority of the pilots that made up the second group had completed a relatively good initial flying course but had not received progressive training since qualifying. The third group was made up 'of very keen and efficient officers' while the last group were 'a few officers whose ability as Air Corps officers is in question'.[53]

FLYING QUALIFICATIONS OF THE COMMANDING OFFICER

Although not portrayed as such by the Committee, the questions as to whether the commanding officer should or should not be a flying officer, over the flying qualifications of Mulcahy and over the receipt by him of the flying pay appropriate to a duly qualified pilot, collectively represented possibly the most contentious issue to be examined by the Committee. Paradoxically, of the seven substantive questions that it addressed the Committee appears to have devoted least attention to what the pilots would have considered to be the most important issue. The apocryphal accounts of the early Emergency, still frequently recalled during the author's service in the 1960s, indicate that those pilots who qualified by successful completion of the standard flying course greatly resented the fact that Mulcahy was in receipt of flying pay at the rate appropriate to a fully qualified pilot and, more importantly, wore the flying badge or pilot's 'wings' on the basis of an abbreviated flying course authorised by himself. In fact, many officers of that era strongly believed that this matter had in fact been the main reason for the demoralisation and, thus, the investigation. Mulcahy had completed a flying course of less than 15 flying hours before putting up his wings. Subsequently, having had his certification of entitlement accepted and, being paid as a qualified pilot, he was not permitted, by the CFI, to fly except when accompanied by a fully qualified pilot. In carrying out a somewhat superficial examination of the question as to whether the commanding officer should be a fully qualified pilot or not, it is a fact that the Committee chose to ignore the historical situation even though they had available, and presumably examined, the file 'Additional Pay – Major Mulcahy' (2/47557). The Committee in effect glossed over these matters claiming that 'the term "flying officer" was somewhat lacking in precision in as much as it had no particular meaning in the Defence Forces':

> It does not lend itself to an exact definition which will convey precisely a standard of proficiency or degree of knowledge. The committee decided that the term must have been intended to mean an officer who is fully qualified to take off and land service aircraft and to perform service missions under all conditions'.[54]

In adopting this attitude the Committee ignored the glaring fact that Mulcahy, who was barred from flying solo, could not, and did not meet the above criteria. In arriving at a loose definition of the term 'flying officer' the Committee, therefore, overlooked the approved syllabus, the satisfactory completion of which qualified pilots for the appropriate rate of flying pay. Reference to the 'Young officers' syllabus of flying training' would have provided a more than adequate definition of the term 'flying officer' but would have identified Mulcahy as having qualified for flying pay without meeting the qualification standards laid down by himself.

It is certain that the Committee was well aware that Mulcahy was drawing flying pay at the higher rate of eight shilling per day, as against the five shillings paid to those who qualified after him in accordance with the syllabus proper.[55] However the Committee seemed prepared to accept, as the Department of Defence had previously, that Mulcahy had undergone flying training in accordance with the current regulation, DFR 40/1936. In accepting Mulcahy's qualifications the Committee chose to ignore the fact that the DFR alone did not provide for the award of 'wings' and certainly did not provide for qualification for receipt of flying pay. The Committee instead addressed the 'problems of commanding officer':

> It was brought home to the Committee at a comparatively early stage of the proceedings that the task of the present commanding officer of the Air Corps is a particularly difficult one for the following reasons: –
> (a) Numerous problems of a highly technical nature are constantly coming up for solution.
> (b) The long absence of a clearly defined policy for the Corps together with the lack of adequate up-to-date equipment and the difficulties of its procurement.[56]

In defence of Mulcahy, the Committee also cited as a difficulty the fact that the younger officers, who were highly critical of the Commanding Officer and his staff because of the small amount of flying the senior staff had engaged in, also blamed the headquarters staff for the 'present lack of equipment and weakness of the Corps in general'. The junior officers did not understand that 'financial considerations and the attitude of foreign powers in respect of supplies are insuperable factors', not to mention that administering the Air Corps, curtailed the amount of flying Mulcahy and his headquarters staff could engage in.[57]

The report stated that the Commanding Officer should be a fully qualified pilot. Among the reasons cited were that he might 'have the necessary prestige in the Corps' and 'set an example to the older as well as to the younger officers'. It was considered that such qualifications would ensure the CO had 'the necessary knowledge to fully appreciate the practical problems involved in flying, navigation and aerial operations' and to 'successfully guide training in Squadrons and Schools'. He would also be able to 'give satisfactory decisions on the many technical matters' that arise and appreciate modern developments. It was further observed:

> The committee does not considerate it absolutely essential, though undoubtedly it is desirable, that the commanding officer should fully undergo the course, as a

pupil pilot, prescribed for personnel qualifying as pilots in the school. He should however undergo such instruction as is necessary for him to get the qualifications required to fit him for his appointment. In the event of the committee's recommendation regarding the appointment of director of military aviation and consequent abolition of the appointment of commanding officer, being accepted, the director should possess the qualifications outlined above as being essential to the commanding Officer.[58]

In adopting the above position the Committee appears to have contradicted their previous stance, in effect endorsing Mulcahy as having qualified to receive pilot's flying pay as per DFR 40/1936 while, at the same time, intimating that he was not actually qualified as a service pilot. A majority of the Committee subsequently recommended that Mulcahy be appointed Director of Military Aviation (in GHQ) but that he 'should be required, at an early date, to undergo the additional training to obtain the qualifications which the committee' considered absolutely essential for the officer holding the appointment. [59]

It is obvious that the Committee's position on the reappointment of Mulcahy as DMA was quite contradictory. At one level they had no apparent difficulty with Mulcahy's flying qualification and receipt of flying pay, in affect considering him to have been a duly qualified pilot. At another level the Committee accepted that Mulcahy lacked the flying qualifications required, did not have the respect of his subordinates and generally lacked sufficient knowledge of flying to make aviation decisions or to direct and inspect flying training. His most glaring deficiency, as actually stated by the Committee, was that he was unable to cope with the numerous problems of a highly technical nature that kept coming up. However, these accumulated short-comings were cited as mitigating circumstances that meant that he should undergo necessary additional training to qualify him to undertake the duties that the Committee considered he had been performing satisfactorily since 3 June 1935. The contradictions in the Committee's position suggest that they had wrestled unsuccessfully with their collective consciences in order to endorse the decision of the General Staff and the Department of Defence to make the original appointment back in 1935 and the decision to grant him flying pay in questionable circumstances in 1936.

FLYING PAY

In considering the question 'is the present system of pay and additional pay satisfactory, and, if not, what changes are considered necessary, and is additional pay for flying personnel at all desirable?' the Committee mainly considered the case of the eight pilots of the 1937 class. These had commenced flying training just prior to the publication of DFR 7/1937 of 8 February 1938 that had reduced the flying pay for newly qualifying pilots from eight to five shillings per day. In brief the Committee recommended that the officers affected should get the higher rate of pay.

Another cause of concern to the younger pilots, who did the major part of service flying, was that the more senior officers, by virtue of their appointments, did little or no training or service flying but received the flying pay at the higher rate. Similarly some pilots and observers, who were in effect ATC officers with the Department of Industry and Commerce, did little or no military flying, and yet continued to receive eight shillings per day. In this matter the Committee recommended 'that flying pay should not be paid unless flying is being properly engaged in' and 'be payable only on certification'.[60] As in other aspects of their investigations the Committee applied a different set of rules to Mulcahy's receipt of flying pay at the higher rate.

TURNOVER OF PILOTS

The Committee discussed how a reserve of pilots might be built up in such a manner as to have sufficient pilots available for an emergency. Although it was not so stated, the position that had existed immediately prior to the Emergency was one of stagnation with active flying appointments filled by relatively old pilots while the number of younger pilots was inadequate for the 1939 Peace Establishment and particularly so for the War Establishment of June 1940. It was considered that newly qualified pilots, after a number of years of service flying with a squadron, might revert to an appointment in some other Army corps and complete a short period of refresher flying training with the Air Corps on an annual basis. This idea, however, was discarded on the basis that an officer could not develop a career in two Army corps at once. Once properly trained a pilot would have to function as such in any emergency thus depriving the other corps of an officer at a time of need. On a practical level, it was recognised that a trained and motivated pilot would not easily settle down in any other corps.

The Committee considered the existing short service scheme as a basis for a turnover. The fact that promising young officers could be retained in the Air Corps was felt to be a considerable advantage as was the fact that a reserve could be built up without affecting any other units. However, the scheme was seen to have one major disadvantage that on passing on to the reserve if officers could not get employment in the State there would be a temptation to seek employment abroad and, thus, devalue the reserve. In terms of the strength and composition of the reserve it was considered that the number would depend on the number of squadrons to be organised and on the basis of having three pilots per aircraft – one pilot in permanent service and two on the reserve. In effect, therefore, the Committee succeeded in endorsing the short service scheme already in use.[61]

NON-COMMISSIONED PILOTS

The Committee then considered the question of training non-commissioned personnel as pilots in the context of the formation of an active reserve of pilots under the short service scheme. Without investigating the matter in any detail the Committee made what they saw as a pertinent recommendation on the subject of NCO pilots reflective of the practice in other countries. In this regard they had the unsolicited comments and recommendations of Commandant P. J. Hassett.[62] In accepting the suggestions they were very specific as to the main condition to be met:

> The Committee is satisfied that there is a case for non-commissioned officer pilots in one circumstance only, and that is if it is proposed to build up five fighter squadrons. In that it is recommended that non-commissioned officer pilots be recruited in the proportion of two to each flight of three aircraft.[63]

In this manner it was foreseen that NCO pilots could replace short service officers on the basis that 24 NCO pilots would require to be trained for each fighter squadron, of whom 16 would be maintained on the reserve. It was considered that NCO pilots were not required for reconnaissance squadrons because such pilots required a particularly high standard of training and a good knowledge of the tactics and techniques of ground forces, and had to exercise command over non-commissioned aircrew members.[64]

GENERAL MATTERS

The Committee noted that various matters outside the terms of reference had been introduced in evidence and commented, generally very briefly, on some. Several complaints regarding aspects of the responsibilities of the Air Corps Company, Signal Corps, had been raised in evidence. In commenting on these matters the Committee demonstrated that it had understood little of the evidence relating, in particular, to aircraft wireless telegraphy and radio telephony sets and their uses. Similarly, their grasp of the communications requirements of operational squadrons in general was not the best. In particular, they did not understand the necessity for radio telephony sets for fighter aircraft

> In future . . . if possible fighter sets should be capable of operating on the medium wave-band as in the case of the TR 1082/83 [wireless telegraphy set], thus obviating in normal operation the need for a multiplicity of ground stations.[65]

It had been adequately demonstrated in evidence, and by the demonstration of Thomas Murphy's transmitter in conjunction with Lieutenant Woods's receiver, that fighter aircraft required a short wave radio telephony set, operated by the pilot, for the effective two-way voice communication demanded by the role. However, the Committee recommended, quite ill-advisedly, that fighter aircraft should operate with W/T sets compatible with those of reconnaissance aircraft so as to reduce the number of ground stations.

The Committee's pronouncement on loop aerials was similarly lacking in perception. Suggesting that loop aerials were only required on longer-range aircraft they stated that 'except for the existing Ansons the problem does not call for any serious consideration'. No mention was made of the fact that loop aerials could and should have been fitted to Ansons from November 1938 and that such action would have been a boon to safe navigation and would have contributed to making the reconnaissance mission more effective. To have drawn attention to this point might have implied criticism of Mulcahy and his command but this was something the Committee desired to avoid.

While the Committee acknowledged that 'the system by which ground direction finding facilities were [not] made available until recently' (late 1941) had been the subject of adverse comment, they found no fault with the manner in which the matter of direction finding stations in general had been handled. They surprisingly made no comment on the fact that it had taken so long for the Air Corps to be granted control and unrestricted use of the DF station at Baldonnell or to the fact that Mulcahy and his staff had contributed to the delay by conceding that civil aviation should have prior call on DF services ahead of military pilots who needed, and demanded, more and better aids to navigation. In addition, shockingly, they made no comment on the fact that, as their report stated, 'two short-wave direction finding sets' had been 'lying in the stores of the Signal Corps' for two years while no military DF stations had been installed anywhere in the country.[66]

The Committee commented favourably on, in particular, the communications available at Baldonnell after the acquisition of Thomas Murphy's short-wave transmitter. This comes as no surprise, as although this was a considerable improvement on the abysmal situation pertaining on 3 September

1939, it only affected three aircraft and, thus, improved the general situation very little. There is no evidence that either the Committee or the Signal Corps witnesses were aware of the excellent communications and direction finding services provided for a modest level of civil aviation by the Department of Industry and Commerce. Similarly the Committee demonstrated no appreciation of what the Air Corps expected of the Signal Officer in the matter of aviation communications. From January 1929 the duties of the Signal Officer had been laid down:

> He will be responsible for all types of signal communication in the [Air] Corps . . .
> He will be responsible for keeping in touch with all new designs and improvements
> in the different types of wireless apparatus used in aircraft from time to time.[67]

In the above regard the evidence, mainly of Signal Corps personnel, adequately demonstrated that the Signal Corps had not kept abreast of developments and had served the Air Corps very poorly at a critical time. However, the Committee had identified no short-comings in Signals matters and expressed satisfaction with the Corps' air related functions.[68]

PERSONNEL MATTERS

In its appraisal of individual officers, separate from their functions as pilots, the committee was generally complimentary. It commented favourably on the service of many of the more senior officers – P. A. Mulcahy, W. P. Delamere, P. Quinn, D. V. Horgan, T. J. Hanley, W. J. Keane, F. O'Cathain and K. T. Curran. However, while promoting the retention of P. A. Mulcahy as Director of Military Aviation, the majority of the Committee expressed a major reservation about possible successors:

> Whilst there is within the Corps a number of promising officers, none of them, in
> the opinion of three members of the committee concerned is fitted at this stage to
> effectively direct the Corps in its present condition.[69]

A further five officers were considered to have performed to a lesser level of satisfaction and, in effect, having stagnated in their appointments, were recommended for transfer within the Air Corps in the interest of efficiency. A further four officers were recommended for transfer out of the Air Corps. Two of these were so recommended on the basis that they were ATC officers seconded to the Department of Industry and Commerce. A single officer, Lieutenant F. F. Reade, was identified as being indisciplined, unsuited for instructional flying and unfit for appointment as a Flight Commander and therefore unsuited for the duties of an Air Corps flying officer. His transfer to another corps, or his dismissal from the service entirely, was recommended. The last officer, Lieutenant A. C. Woods, who had been active in his distrust of Mulcahy, was effectively identified as a disruptive influence and recommended for transfer out of the Corps. The Committee considered, without explaining the circumstances, that his action of procuring possession of official documents in an irregular manner, irrespective of his motivation, was reprehensible. In addition to those senior personnel favourably mentioned, the Committee also identified six very keen and efficient junior officers and recommended that they be considered for promotion when opportunities arose.[70]

The appraisal of P. A. Mulcahy by the majority of the Committee set out to ensure his reappointment as Director of Military Aviation. However, the assessment does not amount to a fulsome endorsement of his performance and record as officer commanding. To a certain extent it highlighted the shortcomings and failings identified earlier though, as he was only in a minor way the subject of adverse comment:

> Whilst Colonel Mulcahy bears responsibility for the low standard of training in the Air Corps, the mitigating circumstances mentioned . . . must be taken into consideration.[71]

It might be considered that the mitigating circumstances referred to, including the problems of the commanding officer already referred to, in effect, made Mulcahy unsuitable for the functions of his appointment. The impact of this endorsement was further reduced by the proviso that Mulcahy undergo training to fit him for his appointment. While the majority recommendation of the Committee was fundamentally in favour of maintaining the status quo in terms of the command and direction of the Air Corps, the minority opinion of the Chairman, Major General Hugo McNeill was to totally undermine that position:

> I am satisfied that no one other than a fully qualified flying Officer possessing considerable practical experience should be placed in charge of the Air Corps. The fact that in the past several such non-qualified officers have been from time to time placed in charge of the Corps is, in my opinion, one of the causes of the condition of affairs this committee was set up to investigate.[72]

McNeill went on to state that Mulcahy was not qualified for the appointment and did not enjoy the confidence of the officers under his command. 'With the best will in the world I do not see how this state of affairs can be completely set right.' He recommended that Mulcahy be relieved of his appointment and that 'Major G. J. Carroll be recalled to active duty as director of military aviation'.[73] He was clearly stating, therefore, that Mulcahy was a major part of the problem and could not be part of the solution. This was a very radical position for McNeill to adopt. As a very senior GHQ staff officer since 1922 and, latterly, as ACS, Operations, he had been central to the decision making processes that had appointed Mulcahy in 1935 and that had, in effect, given him the free rein that had resulted in the demoralisation of the Air Corps. More recently, as ACS Operations, he had issued the various operations orders that had specifically tasked Fighter Squadron to the defence of Dublin. Thus, he probably realised more than anyone the extent to which GHQ was culpable for the mismanagement of the aviation functions of the Army.

CONCLUSIONS

The general belief, handed down by successive generations of flying officers, is that the matter of Mulcahy being in receipt of flying pay and wearing pilot's wings, was the main cause of the unrest that resulted in the investigation. However, the evidence, both written and verbal, as presented to the Committee confirms that various complaints regarding the technical and professional competence of Mulcahy were actually the main factors that caused the inquiry. When charged with instituting an investigation it appears that the GHQ, rather than investigate and highlight its own role in the affair, had sufficient leeway to place Air Corps flying officers under the spotlight while generally playing down the affects of Mulcahy's command and direction of the Corps that were the major factors in causing demoralisation and unrest.

The main complaints from flying officers were in respect of the failure to keep aircraft modified with the latest equipment, the failure to ensure the availability of appropriate communications and direction finding and the failure to ensure best practice in navigation. Notwithstanding the inadequacies elucidated before and during the investigation, the Committee, mainly from a position of not understanding such technicalities, failed to address these matters properly and were unable, or unwilling, to identify the failings of both Mulcahy and the Signal Corps in their respective areas. The terms of reference, the conduct of the investigation and the contents of the report suggest that the Committee placed greatest emphasis on finding fault with the Air Corps and was, thus, not inclined to find fault elsewhere.

To a great extent, when investigating the various matters, the Committee was selective. While they had little difficulty in deciding that the Air Corps was ineffective and inefficient they appeared to accept Mulcahy's plea that the UK dictated the type of aircraft and the supply, or not, of spares. In addition, they did not comment on Mulcahy's judgement in the matter of the selection and purchase of aircraft, or his role in committing a hapless squadron to the defence of Dublin that put pilots' lives needlessly at risk.

In the matter of training standards the Committee found Mulcahy to be at fault to a limited extent, but in a contradictory manner, allowed his lack of qualifications and expertise in aviation matters to constitute mitigating circumstances excusing his inability to perform many aspects of the duties of his appointment. The role of GHQ, whose functions included directing training standards for the whole Army and coordinating the conduct of defensive operations, was ignored while the Committee found it appropriate that the two Squadron Commanders should shoulder the major part of the blame for poor flying standards.

The Committee, in trying to come to terms with the impotent state of the squadrons, researched much of the background of the lack of planning and the indecision that resulted in an unprepared Air Corps being assigned to what soon became impossible tasks. The Committee did not consider, let alone adjudicate on, the wisdom of the decision to send a detachment to Rineanna on a wartime mission. In contrast, however they found Fighter Squadron's mission, in defence of Dublin, to be so futile as to be an unacceptable risk to the lives of pilots. Shockingly, they made no comment on the fact that the tasking of the Air Corps Interception Service had resulted from operations orders made or endorsed by the COS and his staff officers.

In view of the fact that the Air Corps had been found wanting, the medium term solution which recommended two reconnaissance and five fighter squadrons is difficult to understand. However, the Committee, recognising that whatever establishment was put in place, its size, role and equipment would be dictated by financial constraints without reference to perceived defensive needs. The fact should have been obvious to the Department of Defence and the Government long before the inability of the Air Corps to perform any worthwhile defensive role was proved without doubt with the submission of the report of January 1942. Though the Air Corps was engaged in the recovery of aircraft and had co-operative contacts with the RAF, as well as being a support to civil aviation, the reality was that it had no real function in the defence of the country before the investigation and would have even less after its limited and probably selective circulation. As Fighter Squadron was to have only a notional role in the defence of Rineanna/Shannon for the last two years of the Emergency, the training of a large class of sergeant pilots in 1943–5 was to become the main function, and achievement, of the latter anti-climatic years.

Although the investigation might have been expected to presage a new beginning for the Air Corps, the demoralisation and uncertainty of the year due to the long drawn out inquiry and a further 16 months of anxious anticipation before the implementation of recommendation was to do little for the Corps' morale. While command would devolve to a flying officer in late 1942, a much reduced Establishment and the purchase of many more obsolete aircraft would do little to restore a Corps' aviation culture that had been stagnating for too long.

TWELVE

RE-EQUIPPING, REORGANISATION
AND DEMOBILISATION

—

By May 1940 the role of the R & MB Squadron detachment at Rineanna had been rendered ineffective due mainly to the loss of three Avro Ansons and the lack of spares to maintain the remaining six. At the same time at Baldonnell the Fighter Squadron, designated by the Army as the Air Corps Interception Service, was equally ineffective and would later be deemed by the Investigation Committee to be fighter in name only with the potential for wasteful loss of life. Whatever credibility the Air Corps retained at the end of 1940, the completion of the Investigation Report in January 1942 and its subsequent consideration in GHQ and the Department of Defence would have reduced it to naught. While its not clear that a conscious decision was taken, the commencement of Air Corps demobilisation appears to have coincided with completion of the report. The Air Corps had reached a maximum strength of just under 800 all ranks in late 1941 and decreased slowly thereafter while the number of squadrons was to be halved by March 1943. Notwithstanding the manifest inability to maintain and operate the aircraft already in service, aircraft numbers were to display a significant and contradictory rise from 1942, a rise that was in inverse proportion to the operational capacity of the Corps.

The 1940 War Establishment, for the first (and last) time, had specified the number of aircraft to equip individual squadrons, but the totals in each never reached that specified. At the same time there was a total mismatch between the numbers and types of aircraft available and in service and the theoretical operational role of each squadron, as implied by the nomenclature. The greatest mismatch was in the Fighter Squadron of December 1940 which had only 15 aircraft but six different types – none of them appropriate to the designated operational role.[1] The number of aircraft and their distribution had changed little by the end of March 1941.

After the reorganisation of March 1943 the distribution of some 63 aircraft was to bear no relation to good practice in terms of either operations or maintenance. The Fighter Squadron in Rineanna had 12 aircraft and four types, three Hurricanes, Gladiators, Masters and Lysanders. With the disbandment of Reconnaissance and Medium Bomber and Coastal Patrol Squadrons in March 1943 all the remaining aircraft were attached to the Air Corps School – 51 aircraft of eight separate types.

Table 12.1 **Types and numbers of aircraft on charge**

Aircraft type	16 Aug. 40[2]	31 Mar. 41[3]	5 Nov. 43[4]	3 Jan. 45[5]
Avro Anson	9	8	8	7
Walrus	3	3	3	1
G. Gladiator	4	4	3	3
Lysander	6	6	4	4
Avro 636	2	2	–	–
Avro 626	1	3	–	–
Avro 631 Cadet	5	5	3	1
Miles Magister	15	15	14	13
DH Dragon	1	1	–	–
Hawker Hind	5	5	4	4
V. Vespa	–	1	–	–
H. Hurricane	–	–	5	12
L. Hudson	–	–	1	1
H. Hector	–	–	10	10
Miles Master I	–	–	1	–
Miles Master II	–	–	6	6
F. Battle	–	–	1	1
Total	51	53	63	63

This disorganised situation came about due to the manner in which aircraft had been acquired before and during the Emergency. Prior to September 1939 Mulcahy acquired the aircraft of his choice though the numbers supplied (Gladiators and Ansons in particular) were restricted by the Air Ministry. After the appointment of the air attaché Mulcahy used his relationship with Lywood and exploited the value of the services rendered to the UK to elicit supplies of aircraft. Mulcahy was happy to accept aircraft of any type just to have machines for pilots to fly with little if any consideration for operational considerations or of his corps' ability to maintain such a large number of aircraft. Early efforts to buy Hurricanes failed due mainly to the reluctance of the Air Ministry to supply operational aircraft to de Valera's suspect administration. While details relating to the supply of Hawker Hinds in 1940 are not readily available, a report from May 1941 records that Mulcahy and the Department of Defence were endeavouring to buy 15 advanced training aircraft. In due course, the Air Ministry authorised the release of only six

Hawker Hinds from RAF squadron service and these were delivered on 1 June 1940. Five Miles Magister elementary trainers, delivered on 7 June, were apparently released by the Air Ministry in lieu of more Hinds.[6]

In February 1941, at a time when 'the Éire government' was seeking, and had been refused, '10 Harvards, 13 Hurricanes and 3 Battles', Churchill decided that '10 Hector aircraft could however be supplied in order to maintain [the] present improved atmosphere'. This approval was in keeping with the policy of 'no arms for de Valera till he comes in [to the war] except a few trifles by the RAF in return for conveniences'.[7] The minute recommending Churchill's approval had cited 'the increasingly helpful attitude of Colonel Mulcahy', and the advantages to the UK resulting from his co-operation.[8] The delivery of six Miles Masters in February 1943 resulted from 'separate negotiations with the Éire Air Corps' in recognition of 'services rendered in respect of RAF internees and force landed aircraft'.[9]

While the acquisition of three Hurricanes, a Hudson bomber, a Fairey battle and a Miles Master as a result of RAF aircraft crash landings in Ireland may appear to have been beneficial, this was not necessarily so, as no doubt maintenance and supply of spares for four new types must have been difficult. With the reduced threat of the invasion of either country, and the mellowing of political relations between the two countries, greater numbers of obsolescent aircraft became available from RAF squadron service. Between February 1942 and June 1945 a total of 29 ex-RAF aircraft were delivered. These more than replaced those aircraft that had been formally boarded (or written off charge) at the end of service lives.

With over 60 aircraft and up to 13 types the spares and maintenance situation must have been made particularly difficult, which is demonstrated by the fact that, in 1943, the average flying time for an individual aircraft was less than 35 flying hours. The annual average, for a single aircraft – for the years 1943 to 1945 – was less than 50 flying hours.[10] With about 45 squadron pilots each would have managed to get less than 60 hours flying per annum in the same period. By any yardstick the aircraft figures indicate great inefficiency as a result of the dilapidated condition of the aircraft procured from RAF service and by a chronic lack of spares. Similarly, 60 flying hours per year would not have been conducive to maintaining and improving flying skill or gaining valuable flying experience. The only saving grace perhaps arises from the fact that the aircraft were supplied at nominal prices. In effect, the unstructured and chaotic acquisition practices of the period built up a huge fleet of used and worn-out aircraft that could not be maintained using the technicians and spares available. The situation was such that on 3 January 1945 the Air Corps had 63 aircraft in service as compared to a notional establishment of 40. Such was the condition of the machines that no less than 22 were earmarked for disposal.[11] This number included the last ten of 13 Hectors which, according to the records of flying times kept by Lieutenant Colonel Jim Teague, did no flying after the total of 65 hours flown in 1943.

The reorganisation of 29 March 1943 came about as a result of interpretation of the Investigation Report, findings and recommendations. Bureaucratic inertia apparently delayed the implementation process by 15 months. The 'Report and Findings of the Committee and Annexes thereto' together with the 'Proceedings of [the] Committee of Investigation. . .' was completed on or about 10 January 1942 and were submitted to the COS immediately. GHQ summarised the condition of the Air Corps as action on the Report was awaited.

> It should be noted . . . that the unavoidably protracted sittings of the board were bound to have an adverse affect as pending the issue of that report all promotions were held up and a general spirit of uncertainty prevailed in the Corps.[12]

The air of uncertainty extended into the area of morale and *esprit de corps* during the period of deliberation that followed the completion of the Report. The demoralisation, evident in Baldonnell in 1940, was mirrored in Rineanna in 1941–2 in CP Squadron that had replaced the R & MB Squadron there in May 1941.[13] The main evidence for this comes as a result of an incident that happened the day before the restricted circulation of the Report. The incident mainly involved Lieutenant A. J. Thornton, a graduate of the first short service course of 1939–40. He had been posted to Rineanna on 10 September 1941. His log book shows that he flew three different aircraft types at Rineanna, Anson, Walrus and Avro Cadet, and completed a total of 31 hours flying between 10 September and 21 November 1941. He flew ten hours on the Avro Cadet mainly doing aerobatics, formation and 'general flying practice'. While the squadron's notional role was the patrolling of the 12-mile limit, the single patrol on which Thornton flew on 13 November 1941 did not extend beyond the Kerry coastline. His other four reconnaissance flights were inland, in Cork and Kerry, apparently in conjunction with local army exercises. The rest of his flying in Walrus and Anson consisted of approaches and landings, instrument flying and flights between aerodromes. Examination of Thornton's log book reveals how a young and enthusiastic pilot could get bored with Rineanna and its unstructured and aimless flying regime. These conditions no doubt contributed to the events of 9 January 1942.[14]

On 9 January 1942 Thornton and three non-commissioned personnel left Rineanna on an unauthorised flight in a Walrus aircraft. Thornton had previously been in touch with the German Embassy in Ireland and intended flying to a German base on Jersey. However, the aircraft was forced by the RAF to land at St Eval in Cornwall. The aircraft and crew were subsequently returned by the UK authorities and the pilot was charged and found guilty of taking an aircraft, arms and ammunition without authorisation.

Late in his retirement Alphonsus J. (Alan) Thornton, and his short service classmates, explained the circumstances, without in any way trying to justify a serious offence. They described Rineanna as a forlorn spot 'four miles down a dirt lane' in the middle of nowhere. The only public house, 'The Honk', was a primitive facility only accessible by trudging two or three miles across rough fields. The Squadron's flying, using the limited Anson and the more limited Walrus, was aimlessly harmless. When they encountered a German aircraft, which was not often, all the crews could do was wave sheepishly and withdraw. There was, in effect, 'nothing to do' and the pilots were 'completely bored' and frustrated. The pilots, particularly Thornton, longed to fly modern fighter aircraft like the Hurricane and to see real action. He was a young, keen and clearly impetuous flying officer who decided to take matters into his own hands. He had little difficulty in getting three non-commissioned personnel, for whom the conditions at Rineanna were, according to Thornton, much worse than those of the officers, to join him.[15] Air Corps folklore and the personal comments of pilots and others who served in Rineanna also testify to this account of the demoralising nature of service there.

Notwithstanding the demoralisation and unrest caused by the strained relations between pilots and their commanding officer, and the continuing uncertainty caused by the delay, it was almost another year before any action was taken on the report and findings of the Committee. It is probable that consideration of the Committee's two main options – to reorganise the Air Corps or to disband the Corps and form the personnel into a ground combat unit or transfer them to other units of the forces – was lengthy and contentious. The length of time taken to arrive at a decision suggests that the disbandment option was seriously considered. Ironically, the Air Corps of the period was better equipped to function as an infantry formation than in an air defence role. This situation derives from the fact that while the 1940 War Establishment did not specify the number or types of weapons to be carried by any aircraft, reconnaissance or fighter, the number of revolvers, rifles and machine guns allotted to each unit or squadron was so specified. Although only 236 rifles, 157 revolvers and 32 light machine guns were to be specified in the 1940 War Establishment already, in March 1939, the Air Corps units had a total of 485 rifles, 47 revolvers and 13 machine guns. By April 1944 the units were to hold a total of 565 rifles.[16]

In May 1941 the flying squadrons held a total of 50 aerial machine guns. However, as early as December 1941 it was directed that some 34 machine guns, some recovered from Allied aircraft, be sent to the Ordnance Depot to be converted to ground use.[17] The situation being such, with the Air Corps better equipped in infantry weapons juxtaposed with impotent and ineffective operational squadrons, it is not surprising that the disbandment of air units was contemplated. Air Corps folklore reflect the fact that, when threats of invasion were more acute, the aircraft were placed on the aerodrome as a deterrent to uninvited landings, while officers (with pistols) and other ranks (with rifles) were dispersed on the aerodrome to defend the aerodrome against paratroop attack.

With no military records available to indicate the nature of the discussion and study of the Report, confirmation of the possibility of disbandment came from an unusual quarter:

> The Air Ministry should, however, be allowed to provide sufficient equipment to Éire at their own discretion in exchange for certain useful concessions which they are able to obtain through the Éire Army Air Corps, whose disbandment would not be in our interest.[18]

It is probable that this reading of the situation reached the UK Chiefs of Staff through intelligence gleaned from Mulcahy by Lywood.

Though the Investigation Committee had made over 40 recommendations, of greater and lesser import, the vast majority were not acted upon. The major decision was that the Air Corps was to be retained but reorganised. The new organisation was not to take the form of five fighter squadrons and two reconnaissance squadrons as had been recommended by the Committee. Prior to the reorganisation a new commanding officer had been appointed. Despite the opinion of the Committee that none of the serving air officers were ready for the top leadership role, Commandant W. P. Delamere was promoted to acting Major and appointed Acting OC Air Corps on 11 December 1942. At about this time he was given a copy of the Investigation Report, to be read and returned without delay. Delamere recorded that he found it 'impossible to read and digest all the evidence sufficiently quickly to make suitable recommendations'. However, having just been appointed Officer Commanding, he readily agreed with the main trust of the recommendations.[19] Delamere appears to have been the only Air Corps officer to see the Report. There is no folk memory, within the Air Corps, of the findings or recommendations of the Committee.

With the new establishment of 29 March 1943 Delamere was made substantive in his new rank. The appointment is considered significant in that he was the last ex-RAF pilot in permanent service. The rate of turnover of officers commanding in the early years and the brevity of the periods of command does not give the impression of a good rapport between the ex-RAF officers and their largely infantry superiors in GHQ. However, in appointing Delamere GHQ and the Department of Defence had ignored the credentials of the remaining four ex-IRA officers who had been advantageously placed in 1928. The senior of this group, Commandant P. Quinn, might have been considered as suitable as Delamere though he lacked the latter's broad experience, particularly in flying appointments. It is probable that Delamere's record in charge of Air Defence Command was a deciding factor. In due course, however, Quinn would get his turn when Delamere was headhunted by the Department of Industry and Commerce and took the position of General Manager of Dublin Airport, with effect from 2 October 1946.[20]

The Committee's recommendation, that two reconnaissance and five fighter squadrons be established, was completely ignored as a considerable scaling down of the organisation and a reduction in personnel was put into effect. The June 1940 War Establishment had provided for a notional 1,295 all ranks (less the 366 other rank appointments that were 'not to be filled'). As up to 200 more remained unfilled at times the Air Corps strength in the early years of the Emergency did not reach 800 all ranks.[21] The 1943 Establishment provided for 48 officers, 176 NCOs and 438 privates (a total of 662 all ranks), dispersed in six main elements: Air Corps Headquarters (6), Depot, Air Corps (186), Maintenance Unit (169), Air Corps Schools (148) Central Control (Air & Marine Intelligence) (13) and Fighter Squadron, Rineanna (148). The Establishment appears to have been tailored to absorb the strength in officers, NCOs and men serving in early 1943. With the disbandment of two squadrons and the dispatch of 12 aircraft to Rineanna with Fighter Squadron all the remaining aircraft, about 51, were placed on charge to Air Corps Schools, the only other flying unit.[22]

Notwithstanding the fact that five fighter squadrons, as recommended by the Committee of investigation, were not going to be established, a sergeant pilot course was undertaken as a continuation of the short service scheme started in 1939. The course in Air Corps Schools commenced with 31 pupils and ran from December 1943 to December 1945. In the absence of operational flying this wings' course was to be the main preoccupation for the Air Corps for the latter years of the Emergency.[23]

THE 1942 ARMY EXERCISES

The future direction of the Corps had already been set by the nature of its participation in the Army exercises of September 1942. Aircraft operating from Rineanna and Rathduff supported, respectively, the 1 and 2 Divisions.

With no fighter support on either side the style of air reconnaissance conducted was more appropriate to the early stages of the Great War. The main air task was the observation and reporting of the movements of the opposing forces. Operation below 1,500 feet was prohibited except for message dropping and for a final small river defence exercise.

A major aim of the ground troops was to avoid observation by proper use of camouflage. All manner of aircraft, whether suited to the task or not, were committed to the exercises. Included were several low-winged monoplane types that afforded very poor observation of ground troops. The Avro Ansons, which had been noted by the Investigation Committee as being unsuited for such a role, were used. Also used were the Miles Magister training aircraft. These, without a wireless, were even less suited to the task and so the pilots had to resort to the archaic practice of dropping handwritten messages.[24] While this anachronistic use of aircraft appears to have done little for the advancement of the Air Corps, General M. J. Costello was well satisfied:

> I am more than pleased with the work of the [blue] air component . . . They supplied a stream of information which was in the main much more accurate than that supplied by the ground forces. Their reports gave an excellent picture of most of the various crossings and attempted crossings of the Blackwater. They gave phase-by-phase reports of the movement of 4th Brigade at the last stage of the first exercise and the movement of the 2nd Brigade at the last stage of the second exercise.[25]

This testament to unopposed air observation might appear to endorse the outmoded use of aircraft and the artificial air situation of the exercise. However Costello, who had a greater appreciation of Air Corps matters than had Mulcahy, was probably acknowledging the direction of his air component by Commandant W. J. Keane and his judgement in the evaluation of all reconnaissance reports, in an effort to raise the morale of the pilots and the Corps. It is highly unlikely that aircraft were used in the 1942 exercises just to lend realism to an exercise in modern warfare. It is much more likely that the exercises were intended to bring the Air Corps back to its Army co-operation roots and to remind the pilots that they were still part of the Army.[26]

Another exercise that may have had a similar aim took place in May 1944. Colonel Liam Archer led a team of no less than 35 officers in an inspection of the Air Corps' basic skills with infantry weapons. The other ranks of the Corps were divided into five companies of approximately 80 each and were tested, mainly in accuracy of aim and rapid fire. Other aspects of infantry training examined included syllabi, programmes and training diaries, coaching, range duties and zeroing of weapons.[27]

A CLASH OF CULTURES

The manner in which the Army regime dealt with the matter of aircraft accidents demonstrates how such incidents were perceived as serious disciplinary matters requiring admonition and, in some cases, punishment by means of penal deductions from pay. By way of contrast, the aviation centred command that replaced Mulcahy in late December 1942 saw aircraft accidents as requiring study and conclusions leading to improvements in airmanship and flight safety.

As court of inquiry files are not available it is not possible to examine individual accidents in any detail. However, sufficient detail, of the findings of courts of inquiry for the period 1938 to 1945 exists to allow examination of the matter of disciplinary action taken against allegedly negligent pilots. The Department of Finance file dealing with 'cost of repairs' arising from 'damage to army aircraft from accidents during exercises' summarises the administration of some 54 accidents and indicates any disciplinary action taken. It will be recalled that, from at least as early as 1930, courts of inquiry were considered by GHQ as important aspects of the administration of flying discipline. Courts of inquiry were used to assess the negligence of pilots and to precipitate disciplinary action. Such was the concern of GHQ in such matters that Commandant G. J. Carroll was, in effect, replaced as OC Air Corps for failing to be sufficiently diligent in the conduct and administration of such inquiries. The summary reports of the accidents of the Emergency period suggest that Mulcahy, GHQ and the Department of Defence shared a similar disciplinary mindset when dealing with the findings of courts of inquiry.

The 54 courts of inquiry conducted in the period investigated accidents reflecting damage to military aircraft amounting to over £42,000. Individual cases varied from as little as £3, for damage caused by an aircraft striking an aerodrome obstruction stake, to £9,114 for the total loss of an Avro Anson and its equipment. The final assessment, by the General Staff and the Department of Defence, of courts' findings were generally expressed in terms of 'negligence' or 'no negligence'. In the vast majority of the cases (42 out of 54) the accidents were not attributed to pilot negligence, thus attracted no punishment. The remaining 12 accidents gave rise to deductions in pay, as the Minister was permitted to authorise, in accordance with Section 130(D), Defence Acts 1923–1945. In three of the cases 'penal deductions' of £50 were made from the pay of junior pilot officers by the Department of Defence on the recommendation of the General Staff and Mulcahy. These were made on the basis of damage to public property which, 'after due investigation', it appeared to the Minister had been 'occasioned . . . by negligence on the part of the officer'. The final decisions were made by departmental officials acting on behalf of the Minister. At the time, £50 represented a lieutenant's regimental pay for 120 days. As such, a summary reduction of one third of a yearly salary, without due process and without appeal, constituted a most severe punishment. It is, however, of note that no penal reductions were made in respect of any accident that occurred after Delamere took over as Officer Commanding late in 1942.[28]

In contrast to a regime that treated individual flying accidents as possibly indicating ill discipline or negligence, the Air Corps under W. P. Delamere saw the number and frequency of accidents as a matter to be studied and from which to derive lessons. In November 1943 Delamere circulated a 'memorandum on flying accidents for the information of senior officers of the Air Corps'. The study of some 67 accidents that had occurred in the period 1936 to 1942 showed that the flying time between flying accidents had reduced from 950 hours in 1936 to 200 hours in 1942, and that some 67 per cent of accidents were caused by pilot error. The contribution of unauthorised low flying to the incidence of accidents was particularly noted. The study also identified the fact that there were three points in a pilot's career when he was particularly susceptible to pilot error. At about 100 hours of flying experience a pilot tended to consider himself to be competent but actually tended to be foolhardy. At 300 hours pilots were considered to be overconfident and prone to taking unjustified risks, while at about the 600 hour point a pilot tended to become careless and to forget the airmanship principles of his training. Though junior pilots were not permitted to read the document, squadron commanders were urged to make them aware of the lessons contained in the study. There was an immediate and positive response to the study's findings. Although 28 aircraft had been written off due to accidents during the seven years of Mulcahy's command, only ten were lost in the following seven years under the command of flying officers while the number of flying hours between accidents increased dramatically.[29]

FIGHTER SQUADRON AND HAWKER HURRICANES

In what was probably a welcome break from the constant standby at Baldonnell the Squadron's three Gladiators were based at Ballinter House near Navan, County Meath during the early summer of 1941. As Ballinter was only ten minutes flying time from Gormanston, an aerodrome with adequate hangars and more than enough fuel, the decision to operate from an unprepared field appears odd. However, as suggested elsewhere, Gormanston was almost certainly designated as the first base for RAF squadrons coming to assist in the event of a German invasion and the indications are that the aerodrome was retained for the exclusive use of such forces.

The aircraft were tasked to support the 2 Brigade's operation, monitoring the alleged suspicious movement of British troops along the border. Special reconnaissance of the Boyne and Navan areas started about 29 April while Ballinter was first visited by Captain D. K. Johnston on 7 May. On 8 May 1941 Johnston flew Mulcahy to Ballinter in order to carry out an 'examination of the landing ground' which was the parkland in front of the grand house. The three Gladiators apparently moved to Bellinter on 12 May and, when not flying, were parked under the trees. The bulk of special reconnaissance was confined to May and June 1941. Miscellaneous flights, including exercises with brigade troops, were flown while the reconnaissance operation petered out by early July.[30] With the withdrawal from the Boyne Valley, the practice of maintaining a flight of aircraft on 'Stand-to' at Baldonnell, for the purpose of intercepting belligerent aircraft infringing the country's neutrality, was resumed. In January 1942 a single aircraft was still being detailed for the duty on a daily basis. In view of the futility of the operation, however, the Committee recommended that the practice be discontinued.[31]

As has been discussed, from as early as October 1940 the Air Corps had been trying to obtain more advanced fighter aircraft (see chapter 9 and 10 for more detail). Mulcahy used the occasion of Air Commodore Carr's visit to Baldonnell to indicate his 'urgent requirement' for 'one squadron of fighters, preferably Hurricanes, and ten advanced trainers'.[32] The RAF indicated that it was well disposed:

> In these circumstances it would surely be a mistake not to follow up the discussions which took place between Colonel Mulcahy and Air Commodore Carr? Probably we could get the Éire air force to build aerodromes where we want them at the price of some obsolete aircraft [for training], perhaps with the addition of a promise that we will give them Hurricane Is some time next month when we shall be replacing them with Hurricane IIs. The political effect of such an agreement would be wholesome.[33]

As a result of Mulcahy's co-operation with the RAF ten ex-RAF Hawker Hectors had been delivered in May 1941 and a further three in January 1942. However, Hurricanes were not immediately forthcoming. The Air Corps had already acquired a Hurricane I that had force-landed on 29 September 1940 and acquired two Hurricane MK IIs in similar circumstances in June and August 1941.[34]

With the selective implementation of the recommendations of the Committee of Investigation, Fighter Squadron was relocated to Rineanna in April 1943. On 15 April 1943 the main body of personnel left Baldonnell at 07.00 hours, marched to Lucan South Station and took a train to Limerick.

From there they marched to Rineanna. The Squadron consisted of a HQ and three flights. Initially it had 12 aircraft of four different types. The squadron, though designated a fighter squadron, initially at least, took the form of one of the two provisional reconnaissance squadrons as recommended by the Committee. Eighteen pilots were provided for – eight officers and ten sergeant pilots – though the latter had yet to be trained.[35]

At Rineanna the Squadron was attached to 8 Brigade, Southern Command – notionally as part of the defence of Rineanna/Shannon Airport.[36] Gradually, however, it began to assume the form of a fighter reconnaissance squadron. In July 1943 four Hurricane I aircraft were received in exchange for two Hurricane MK II machines which were returned to the RAF. The Gladiators were returned to Baldonnell, with two being boarded in late 1943 and the third the following year. By November 1943 the three Lysanders, Nos 61, 63 and 66 had been returned to Baldonnell. With the receipt of three Hurricane I aircraft in November and a further four by March 1944, followed by the return of the three Masters to Baldonnell, Fighter Squadron finally became a single aircraft-type squadron for the first time.[37]

Notwithstanding its notional role in defence of Shannon the maintenance and operation of aircraft was not a priority with 8 Brigade. In July 1943 it was reported that the number of aircraft unserviceable on a monthly basis was increasing rapidly, 25 per cent in May, 42 per cent in June and 58 per cent in July, with a prediction that it would as high as 80 per cent in August. This was put down to the fact that only 22 of the 77 technical personnel were available to work on aircraft on any given day. This in turn was put down to the number of personnel, by direction of OC 8 Brigade, who were distracted by other duties in fatigues, guard duties, infantry training, kit inspections and cutting turf in the bog.[38] A pilot recalled his feelings at the time:

> I was only down there 4 days and I was sent off to the bog and I was the adjutant.
> I was out in the bog cutting turf! . . . and a lot of the aeroplanes were up on stilts in
> the hangars because the fitters . . . were on guard duty, out cutting turf and on
> fatigues . . . it was appalling, appalling.[39]

The situation did not improve with the delivery of the ex-RAF Hurricanes in November 1943. Soon after arriving in Rineanna it was found that aircraft had various components that were excessively worn. One machine was in such poor condition that its continued service was in doubt. A major factor contributing to poor aircraft serviceability at this time was the fact that the Hurricanes, like practically all aircraft acquired during the Emergency, came direct from active service with RAF squadrons. It would have been axiomatic that the RAF would get rid of the worst aircraft first and that many 'hangar queens' would be pawned off on the Irish. R. W. O'Sullivan recognised the fact that 'certain parts' of the Hurricane I and the Merlin III engine 'which, though within permissible limits of wear at the time of delivery, very rapidly exceeded these limits' were in dire need of overhaul, replacement or repair. He also noted that the technicians at Rineanna had not been made familiar with the aircraft and engine and were 'not able to acquire the experience necessary for efficient maintenance'. He commented that inexperienced technicians were often caught unawares by defects 'which might' in more favourable circumstances 'have been anticipated', resulting in aircraft, even during routine inspection, being grounded for longer periods than would be usual or desirable.[40] Another major difficulty that was noted was that adequate spares were not available for aircraft like the Hurricane that was still in active RAF service. Tony Kearns summarised the situation:

> During 1943–4 very little flying was done due to a chronic lack of spares, especially tyres. Day after day a Hurricane would be taken out of the hangar; its engine run up for five minutes and then silenced as it was pushed back into its stable.[41]

As an indicator of serviceability the return of flying hours show that the Hurricanes flew an average of 33 hours each in 1943 and less than 50 in 1944. None of the Hurricanes acquired in 1943–4 did more than 170 hours flying in four years of Air Corps service. With about ten pilots in the Squadron they would have averaged less than 50 flying hours each in 1944.[42] The COS's report for the year ending 31 March 1945 gives the misleading impression that Fighter Squadron had been carrying out a worthwhile defensive role at Rineanna for the previous two years:

> The general improvement in training, discipline and morale [in the Air Corps] . . . has been well maintained. Towards the end of the period it was decided to move the Fighter Squadron to Gormanstown. This decision was made possible by the lessening danger of any sudden invasion.[43]

In reality the main reason for this move actually was that the authorities at Shannon (and the Department of Industry and Commerce), who had wanted rid of the Air Corps from very early in the Emergency, saw the squadron as a hindrance to civil aviation. The fact that precious Hurricane tyres were wearing out too rapidly on the concrete runway was also cited as a reason. Relief, for Shannon and the Squadron, eventually came when the unit, with its 95 personnel, (64 per cent of the approved establishment), and nine Hurricanes finally moved to Gormanston on 1 May 1945.[44]

DEMOBILISATION

It is a moot point as to whether a concept of demobilisation existed in the Air Corps towards the end of the Emergency. The strength returns strongly suggest, in fact, that the Air Corps was subjected to a slow demobilisation from late 1941, as the Investigation Committee was still putting the finishing touches to its report. While the Air Corps War Establishment of June 1940 provided for some 929 personnel, except for an initial burst of recruitment in 1940, no effort appears to have been made to bring units up to the authorised establishment. With a maximum strength, of 791 all ranks, attained on 1 October 1941, the Air Corps subsequently became increasingly under strength. Just before the reorganisation of March 1943 the Corps was at 75 per cent of the authorised figure. On 31 December 1943 the Air Corps consisted of 686 all ranks or 85.5 per cent of a new permitted figure of 802. By the end of 1944 this was down marginally to 84.5 per cent but was down to less than 72 per cent at the end of 1945.[45] This reduction was mainly due to natural wastage as non-commissioned personnel and short service officers completed their terms of enlistment and were discharged. This reduction was achieved despite an annual intake of 30 apprentices each year from 1941 to 1945 – a total of 150.[46] As in the Army generally, desertion contributed to decreasing strength during the Emergency. By 26 May 1944 100 Air Corps NCOs and men had been 'declared deserters' or absentees at Baldonnell.[47] Those Air Corps personnel who deserted from Rineanna would have been accounted for by Southern Command. All Defence Force personnel who were declared deserters at the end of the Emergency are listed in a bound volume held by Military Archives. The list contains about 5,723 names:

> Deserters – List of the Defence Forces personnel dismissed for desertion in times of national emergency pursuant to terms of Emergency Powers (No 362) Order 1945 or Section 13 Defence Forces (Temporary Provisions) Act 1946 (No 7/1946).[48]

It is generally believed that many of the deserters had joined the British forces. The main provision of the above instruments was a prohibition, for a period of seven years, on the possible employment of deserters in the public service. While many deserters were arrested and charged in the legislative and regulatory process, the State apparently largely abandoned the traditional military imperative of pursuing, arresting, charging and punishing those guilty of desertion.

There was no dramatic exodus of pilots in 1944–5. Towards the end of the Emergency there were 48 flying officers in the Air Corps. By the end of 1944 four pilots had taken up permanent positions in Civil ATC (Department of

Industry and Commerce), while four others were still on loan to ATC in 1945. Thereafter, as civil aviation picked up slowly, about half the short service officers moved into civil aviation with Aer Lingus and other airlines.[49]

CONCLUSIONS

The post-investigation Air Corps was in a very similar condition to that which pertained after the Civil War. In considering the future of the Air Corps the General Staff and the Department of Defence would have had to consider – as the Department of Finance had done in 1923–4 – whether such an ineffective Corps was worthy of continued existence. It is of considerable note that personnel, whose task it was to salvage Allied aircraft, were provided for in the 1943 Air Corps Establishment. In effect the salvage section was established by a regulation issued by the Minister for Defence in accordance with the Defence Act 1923–1943. The placing of this service on a regulatory footing emphasises the importance of the salvage operation in the overall scheme of things. Similarly, the Department of Defence Memorandum of May 1949 makes it abundantly clear that co-operation with the UK, particularly the recovery and return of Allied aircraft, was a major aspect of Anglo-Irish co-operation from 1940 onwards. The circumstances suggest that these various services rendered by the Air Corps were central aspects that helped to justify the Corps' continued existence after the Investigation.

While the Air Corps under the new establishment retained the Fighter Squadron, equipped eventually with Hurricanes, its operational capacity was largely notional. In addition, the poor serviceability of the machines would have had a major demoralising affect on an already under strength unit. In the circumstances, the Air Corps of 1943 to 1945 was little better than the care and maintenance organisation of 1924. As Mulcahy had doggedly pursued the acquisition of (mainly) training aircraft from the RAF, in return for services rendered, a total of over 50 dilapidated ex-RAF aircraft were acquired before the end of the Emergency. It must be noted that it was a futile pursuit, as they added little to the Air Corps other than a modicum of experience for demoralised young pilots.

With the change in command of the Air Corps at the end of 1942 a new emphasis was placed on airmanship and flight safety. The appointment of W. P. Delamere as Officer Commanding brought a welcome return to aviation values. Despite the poor mechanical condition and serviceability of aircraft and the small amount of flying available to individual pilots, the number of flying hours between accidents showed a significant increase. Aircraft losses were proportionally fewer while accidents were no longer viewed as disciplinary matters. In effect the poor airmanship perceived by Mulcahy to constitute indiscipline was soon eliminated thanks to Delamere's professional understanding of pilots and flying discipline.

With no worthwhile operational tasks to perform the aircraft recovery operation and the ATC assistance rendered to Industry and Commerce represented the only real contribution to the State during the latter years of the Emergency. As these employed few personnel it was inevitable that the Air Corps generally would stagnate as was demonstrated in Rineanna in the latter years.

POSTSCRIPT

It is not easy to retrace the evolution of the Air Corps and what it had become by 1945. This was due to the fact that most of the key decisions were made by higher authority, in the absence of defence policy. Though greatly influenced by C. F. Russell, the main force behind bringing the Military Air Service into being was Michael Collins. A modest yet significant expansion of the Air Service, authorised by Collins immediately before his death, ensured that an Army Air Corps would survive the demobilisation and reorganisation processes of 1923–4, even though W. J. McSweeney could not make a sufficiently cogent case. In the ten year period after Collins's death successive Commanding Officers – McSweeney, Maloney, Russell, Fitzmaurice and Carroll – appear to have had little real influence, as the direction of military aviation was dictated by the Council of Defence and GHQ. During this period a token organisation managed to survive Army indifference to slightly expand and in time evolve into an army co-operation organisation. With no military role of major consequence throughout the 1920s and 1930s the Government saw that Air Corps main potential lay in its aviation expertise and as a source of trained personnel for civil aviation.

The training of ex-IRA officers as pilots, along with a constant tinkering with the intake system, ensured that the pilot body was made up of several disparate groups. It is highly probable that tensions between, and within these groups, contributed to the unpleasantness that resulted in Captain P. J. Hassett's sudden departure of the Corps in 1936. The command of the Corps by Liston and Mulcahy while supposedly primarily focused on disciplinary considerations, in retrospect, could be viewed as having had a contrary affect. In particular, the exercise of air command and the functions of Director of Military Aviation by P. A. Mulcahy – who assumed the qualifications and pay of a flying officer – had a major detrimental impact on the Air Corps, especially on the morale of many pilots. In addition, the manner in which

Mulcahy influenced and made decisions resulting in an ineffective Corps being given improbable operational tasks during the Emergency, contributed directly to a further demoralisation of the Air Corps. In the folk memory Colonel P. A. Mulcahy is probably the most notable Commanding Officer of the period – but for all the wrong reasons. In command for almost one third of this period under review, Mulcahy had many negative effects, while the spectre of the Investigation Committee hung over the Corps for many years.

After the investigation it appears that the Air Corps only survived on the basis of the pilot training for civil aviation, services rendered to the Department of Industry and Commerce and, most importantly of all perhaps, services rendered to the RAF and the British administration. The appointment of W. P. Delamere facilitated a welcome return to aviation values centred on the need for good airmanship. This renewed emphasis was greatly influenced by a young and very talented electrical engineer who was commissioned into the Corps in 1939. From an early date Lieutenant (later Lieutenant Colonel) Jim Teague's study and statistical analysis of pilots' flying and aircraft accidents illustrated the way to foster flight safety and prevent aircraft accident. This marked the initiation of a much more professional approach to the flying and maintenance of aircraft – a development that would serve the Air Corps well in future years.

The Air Corps of 1945 was still the small training element it had traditionally been, and with a turnover of personnel providing pilots and mechanics for civil aviation, would remain so for many years. Defence spending would be such for the next 20 years that minimum numbers of training aircraft would remain in service for as long as possible. The short service pilot training scheme that ran during the Emergency resumed in 1953 and would run for nine years. It was to be as late as 1963, however, before the Air Corps would be given the genuine responsibility for and the resources to perform a practical operation task that was of service to the State – search and rescue operations with Alouette III helicopters.

Appendices

—

Appendix I: Summary of Expenses, Captain C. F. Russell[1]

Received from	Date	£ s d
Art O'Brien	19 Oct. 1921	1,500. 0. 0.
Art O'Brien	12 Dec. 1921	1,300. 0. 0
Art O'Brien	30 Dec. 1921	250. 0. 0.
'Other sources'	Date unknown	60. 0. 0.
Total received		3,110. 0. 0.

	£ s d
Purchased from Messrs. Martinsyde, Ltd. One five-seater aeroplane	2,300. 0. 0.
One set of floats for same	300. 0. 0.
One Avro machine	130. 0. 0.
Alterations to five-seater machine to increase accommodation	100. 0. 0.
To dual instruction by Company pilot	17. 10. 0.
Lt McSweeney, IRA, Expenses before the purchase	25. 0. 0.
Maps, helmet and Compass Box	10. 0. 0.
Petrol, oil and mechanic's fees	25. 0. 0.
Travelling expenses Brooklands	40. 0. 0.
Travelling expenses Woking	34. 0. 0.
Irish travelling expenses	130. 0. 0.
Hotel expenses – London & Dublin	65. 0. 0.
Miscellaneous expenses	10. 0. 0.
Expenses in connection with two machine guns – London	5. 0. 0.
Expenses (to Capt. Clarke)	10. 0. 0.
Insurance & garage for aeroplanes	20. 0. 0.
Report on Haulbowline as an Air Station, supplied by Director of Handley Page London – Paris service	25. 0. 0.
Total expenses	£3,247.10.0.
Total amount received	£3,110 .0. 0.
Balance due to C. F. Russell	£137.10.0.

1 Statement of expenses, C. F. Russell to M. Collins, 27 Feb. 1922 (NAI, DT S.4002); London Office accounts 1 Oct. 1921 to 31 Dec. 1921 (NLI Mss 8431–2, Art O'Brien Papers); MFD to MFF, 7 Apr. 1922 (NAI, DT S.4002).

Statement of immediate financial requirements, 7 April 1922

	£ s d
Packing & shipping of Martinsyde, London to Dublin	150. 0. 0.
Packing & shipping of Avro 504K, London to Dublin	100. 0. 0.
Balance due to Martinsyde for Avro 504K	130. 0. 0.
Balance due to Martinsyde, garage & insurance	10. 0. 0.
Salaries of staff for one month	40. 0. 0.
Special – expenses for mechanics to watch disassembling of aircraft	40. 0. 0.
Miscellaneous	50. 0. 0.
	£ 520. 0. 0.
Total costs associated with the purchase of two aircraft:	**£3,767. 10. 0.**

Appendix II: *Telegram Received in the Irish Office*[1]

Date: 4 July 1922

Handed in at <u>DUBLIN CASTLE</u> at Received here at <u>11.39 am</u>
From <u>Cope</u> To <u>Curtis for Mr Churchill</u>
Collins wants two aeroplanes one with undercarriage for bombing and one without. Reasons for request are

(1) McSweeney has not brought over his plane yet due to inclement weather.

(2) Telegraph and telephone communication is interrupted and particulars of the surrounding country are not available.

(3) Reports come in of concentrations of irregulars in Dublin County and neighbouring Counties. Troops and transport are sent out on these reports and search country for hours for these concentrations but fail to find them and men and time are wasted.

(4) Collins is satisfied he could clean up the Country districts if he could get early information of concentrations and keep up communications. As an example of (2) there were reports yesterday that irregulars were doing well in Drogheda. At P. G.'s request I got through to Gormanstown by wireless for information but wires were down between Gormanstown and Drogheda and no information could be obtained.

It would be most undesirable for P. G. to use our pilots owing to the dead set which is being made by republicans on P. G. receiving assistance from us. Each issue of the Republic of Ireland mentions either Mr Churchill, General Macready or myself as giving assistance in the fight and the mainspring of the republican propaganda is that British forces are prompting and assisting in the killing of Irishmen.

I suggest one aeroplane being handed over at once. Can this be done please. The handing over should be at Baldonnell. The P.G. have one or two efficient airmen – of this I am certain.

1 Cope to Curtis, 4 July 1922 (TNA, Air 8/49).

Appendix III: **Statement of Expenditure, Major General McSweeney: £3,800 advanced by the Ministry of Defence**[1]

1922		£ s d
20 June	Received from Chief of Staff	1,300. 0. 0.
	Received from Chief of Staff	2,500. 0. 0.
21 June–4 July 1922	McSweeney – misc. expenses Dublin/London/	
	Dublin/London/Dublin	43. 1. 2.
24 June	Aircraft Disposal Co.	400. 0. 0.
26 June	C. Baker	3. 7. 0.
26 June	Gamages	2. 2. 0.
1 July	Aircraft Disposal Co.	400. 0. 0.
1 July	G. Adams	4. 17. 0.
2 July	C. Baker	9. 0. 0.
13 July	Yeates	1. 10. 0.
15 July	T. S. Harris	86. 12. 6.
15 July	Aircraft Disposal Co.	1,100. 0. 0.
17 July	Royal Air Force	4. 18. 4.
30 July	Col. Russell	9. 17. 4.
30 July–14 Aug.	McSweeney – misc. expenses Dublin/London/Dublin	37. 1. 2.
31 July	Dixon Hempenstall	2. 2. 0.
1 Aug.	Burberrys	3. 10. 0.
4 Aug.	Gieves	2. 10. 0.
	Wages	21. 2. 3.
	C. Baker	9. 0. 0.
	G. Adams	6. 5. 6.
	Col. Russell – Expenses	10. 18. 0.
	Advance – Mr Piercey, ADC	15. 0. 0.
11 Aug.	Wages	18. 7. 10.
18 Aug.	Wages	19. 2. 5.
25 Aug.	Wages	25. 14. 1.
30 Aug.	Cox Shipping Co.	32. 6. 6.
	Lieuts. Crossley and Maloney	35. 0. 0.
1 Sept.	Wages	17. 16. 3.
9 Sept.	Wages	29. 13. 0.
16 Sept.	Wages	3. 15. 0.
	Wages	31. 5. 6.
	Mr Piercey [Mono engine]	100. 0. 0.
22–27 Sept.	McSweeney misc. expenses Dublin/London/Dublin	15. 14. 8.
3 Oct.	L. B. Fitch	1. 10. 0.
10 Oct.	Dairy Engineering Co.	6. 6.
14 Oct.	Fox, carter (Wages)	12. 0. 0.
28 Oct.	Fox, carter (Wages)	8. 0. 0
27 Jan. 1923	Jacob's [second hand flying suits]	67. 10. 0.
1 Nov.	Allowed as expenses – McSweeney	129. 12. 1.
	Total expenditure	2609. 11. 1.
13 Sept.	Refunded to DOD (National Land Bank)	829. 10. 7.
13 Sept.	Refund to DOD (Munster & Leinster Bank)	360. 18. 4.
1 Nov. 1923	Account balanced	3,800. 0. 0.

1 Statements of expenditure, Maj. Gen. W. J. McSweeney, 23 July 1923; 30 Oct. 1923 (MA, AC/2/2/1).

Appendix IV: Department of Civil Aviation, 20 July 1922[1]

Name	Duties	Salary	Commenced
Chas. F. Russell	Director, Civil Aviation, Sec. Aviation Council	£300 p.a.	1 April 1922
Miss McLoughlin	Typist, Civil Aviation Department	£2. 10s. p.w.	1 April 1922
A. J. Russell	Junior Clerk	£1. 10s. p.w.	23 April 1922
W. J. Guilfoyle	Engineer, Baldonnell & Tallaght	£6. 10s. p.w.	30 April 1922
Frederick Laffan	Switch Board Attendant	£3. 10s. p.w.	30 April 1922
A. Conmee	Switch Board Attendant	£3. 10s. p.w.	30 April 1922
J. Byrne	Engine Driver, Clondalkin Pumping Station	£2. 10s. p.w.	30 April 1922
L. Nelson	General Labourer	£2. 16s. p.w	6 May 1922
Vol. G. Dunne	Labourer, Sewage & Fire Hydrants	£2. 10s. p.w.	10 June 1922
Vol. M. Horan	Fitter & Turner	£3. 13s. p.w.	10 June 1922
P. Condon	Store Keeper/Caretaker of Aerodrome fittings	£2. 18s. 4d. p.w.	10 June 1922
Chas. O'Toole	Aero Ground Engineer	£5. 0s. p.w.	14 June 1922
H. Mathews	Labourer, cleaning duties	£2. 3s. p.w.	16 June 1922
M. Perkins	Fitter	£3. 13s. p.w.	17 July 1922
M. O'Gorman	Electrician	£3. 10s. p.w.	17 July 1922
E. Broy	Accountant & Clerk	£5. 0s. p.w.	19 July 1922

1 P7/B49/38 (UCDA, Mulcahy Papers).

Appendix V: *Department of Military Aviation, 22 July 1922*[1]

Rank	Name	Duties	Appointed	Pay per Week
Lt	G. Dowdall	Adjutant	25 May 1922	£4. 0s. 0d.
2/Lt	T. Nolan (in Hospital)	Observer	7 July 1922	£2. 0s. 0d.
2/Lt	J. McCormac	Pilot (Dismissed)	11 July 1922	£2. 0s. 3d.
S/Capt.	W. Stapleton	Observer (Acting)	11 July 1922	
Capt.	Mills	M. O.	11 July 1922	Not paid through Aviation
Sergt	J. McCarthy	Rigger	1 Feb. 1922	£2 10s. 0d
Cpl	J. Curran	Rigger	30 Mar. 1922	£1 6s. 3d
Cpl	A. Hughes	Fitter	30 Mar. 1922	£2 14s. 3d
Cpl	H. White	QM & Discipline		£2 14s. 3d
Vol.	F. Kerrigan	Fitter	7 June 1922	£1 4s. 6d
Vol.	M. Lawler	Rigger	20 June 1922	£1 4s. 6d
Vol.	T. McGee	Fitter MT	20 July 1922	£1 4s. 6d
Vol.	Behan	Fitter MT	Attached from Garrison	
Vol.	J. Stephenson	Fitter MT	20 July 1922	£1 4s. 6d
Vol.	Gerard	Rigger		£1 4s. 6d
Vol.	T. Clarke	Rigger	20 July 1922	£2 12s. 6d
Vol.	J. Reid	Fitter	20 July 1922	£
Vol.	Hussy	Fitter MT	19 July 1922	£
Sergt.	Sean Waldron	Medical	⎫	Not Paid
Vol.	W. Winters	Medical	⎪	By
Vol.	M. Adamson	Medical	⎬	Aviation
Vol.	J. O'Leary	Medical	⎭	
Miss	M. Kiernan	Typist	24 Mar. 1922	£2 10s. 0d.
Mr	W. Keogh	I/C MT Repair		£4 10s. 0d.
Mr	H. Cleary	Cook	12 July 1922	£4 10s. 0d.
Mr	F. Sullivan	Cook	12 July 1922	£3 10s. 0d.
Mr	M. Hennebry (Survey)	Carpenter	10 July 1922	£3 0s. 0d.
Mr	J. Hennebry	Carpenter	10 July 1922	£3 0s. 0d.
Mr	A. Fay	Carpenter	10 July 1922	£3 0s. 0d
Mr	Doyle	C / Labourer	21 July 1922	£2 10s. 0d.
Vol.	P. Kelly	Telephone operator	22 July 1922	£1 4s. 6d.
Vol.	D. Kelly	Telephone operator	22 July 1922	£1 4s. 6d.
Vol.	M. Campbell	Rigger	22 July 1922	£3 0s. 0d.
Vol.	J. Daly	Rigger	22 July 1922	£2. 12s. 6d.
Vol.	E. Sutcliffe	Rigger	22 July 1922	£1. 4 s. 6d.

1 P7/B49/37 (UCDA, Mulcahy Papers).

Appendix VI: *Aviation Department of the Army, 18 October 1922*[1]

CABHLACH EITILEACHT ÉIREANN

OFFICERS

Rank	Name	Remarks
Comdt-General	W. J. McSweeney	Commanding Officer
Comdt	E. Broy	Adjutant
Captain	J. Arnott	Acting 2nd in command of flying
Captain	T. Donnelly	Quartermaster
Lieutenant	W. Delamere	Pilot
Do.	G. Dowdall	A/Adjutant
Do.	W. Hardy	Pilot
Do.	W. Keogh	I/C Transport
Do.	J. Crossley	Pilot
Do.	J. Maloney	Pilot
Do.	J. Fitzmaurice	Pilot
Do.	A. Russell	Observer
Do.	T. Nolan	Observer
Do.	L. Gogan	Observer
Do.	T. J. Conba	Observer
Do.	C. Flanagan	

NON-COMMISSIONED OFFICERS

Rank	Name	Remarks
Sgt	J. McCarthy	Rigger
Sgt	J. Curran	Rigger
Cpl	T. Doyle	Batman
Cpl	F. Kerrigan	Fitter
Cpl	A. Hughes	Fitter
Cpl	J. Maher	Rigger
Cpl	? Tracey	Fitter
Cpl	R. Reid	Fitter and storeman
Sgt Major	[H] White	Discipline

1 C. Hogan to CIC, 18 Oct. 1922 (MA, A/07279).

MEN

Henry Lalor	Rigger	Richard Maher	Carpenter
Leo Beahan	Fitter	Patrick McNulty	Blacksmith
Geo. Garrard	Fitter	Rd. Spittle	Armourer
Thos. McGee	Fitter	R. Long	Fitter
James Reid	Fitter	L. Stafford	Fitter and turner
Thos. Clarke	Rigger	D. Monaghan	Fitter
Jas. Stephenson	Armourer	G. Goulding	Cook
James Tracey	Fitter	G. Parker	Orderly
John Daly	Rigger	J. Kelly	Orderly
Ml. Campbell	Rigger	H. Brennan	Photographer
Ed. Sutcliffe	Rigger	J. O'Keeffe	Blacksmith
Patk. Soohan	Orderly	J. Doyle	Carpenter
Wm. Kelly	Armourer	J. Ryan	Rigger
Henry Marsh	Fitter	H. Gregan	Fitter
J. O'Reilly	Labourer	J. Grassick	Fitter
– McDermott	Labourer		

CIVILIANS EMPLOYED IN MILITARY AVIATION

Jas. Hennebry	Carpenter	J. Moran	Labourer
Ml. Hennebry	Do.	C. Ellis	Do.
P. Condon	Storeman & Timekeeper	P. McGrath	Labourer
Chas. O'Gorman	Electric Wiresman	J. Downey	Labourer
		J. Dowling	Labourer
Ml. Horan	Fitter	Ml. Doran	Fitter
L. Nelson	Sewage	Patk. Kavanagh	Labourer
A. Fay	Carpenter's Helper	D. O'Brien	Switchboard
N. Matthews	Stores	F. J. Laffan	Do.
J. Stynes	Labourer	J. Byrne	Labourer
T. Coates	Do.	C. J. O'Toole	Ground Engineer
Wm. Nolan (boy)	Do.	Wm. J. Guilfoyle	Engineer
R. Matthews	Do.	Jas. Connolly	Labourer
G. Dunne	Do.	Jas. Mullally	Do.
P. Flynn	Do.	Thos. Birmingham	Do.
G. McNulty	Do.	C. O'Reilly	Motor transport
Ml. Coates	Do.	S. Clancy	Accountant
C. Fitzsimons	Do.	Miss Kiernan	Typist
S. Ellis	Aero Carpenter	P. Downey	Labourer

Appendix VII: Colonel P. A. Mulcahy's Pre-Invasion Address, 4 July 1940[1]

To all officers at Baldonnel, through Unit Commanders.

Although there is abundant evidence that an invasion of our country is contemplated, it is not possible for us to be told when or where the enemy will strike.

It is our duty to be at our posts, ready to take our part at the moment of attack.

We may get a few hours warning. We may get no warning. Nevertheless, it is our duty to be ready at the precise moment.

If we fail to get into the air, if we loose our aircraft on the ground, we have failed utterly in our duty to our people. It is, therefore, necessary that the crews of the Service Squadron and detachment at Baldonnel be readily available to their aircraft at all times.

Until further notice, the crews of service aircraft will occupy quarters in Camp. Married personnel whose families live out of Camp will be granted permission to visit their families during the afternoon or evening, dependent on military exigencies and such personnel must return to camp before 23.59 hours or earlier, if required.

All officers sleeping in Camp must be in their quarters before 25.59 hours and strict quiet will be maintained in quarters after that hour.

If any married officer should consider this order harsh because other Army units are not on active service let us remember that an officer of the ground forces may be able to make up for a few lost hours but an Air Corps officer who fails to get into the air to carry out his allotted task, has betrayed his trust.

Let us, therefore, bear inconveniences cheerfully now so that we will be standing by to perform whatever the task and whatever the hour.

[Signed] P. A. Mulcahy, Colonel, Officer Commanding, The Air Corps, 4 July 1940.

1 Col. P. A. Mulcahy to AC Investigation, 21 Jan. 1941 (MA, ACS 22/23).

Appendix VIII: *Colonel P. A. Mulcahy, Flying Training*[1]

Date	Time	Pilot	Observer/pupil
21 May 1936	.30	2/Lt Reade	Maj. Mulcahy, Cadet 1[2]
3 June 1936	.30	Lt McCullagh	Maj. Mulcahy, Cadet 5

Mulcahy was paid four shillings per day, the pupil pilot rate, from 3 June 1936

3 June 1936	.25	Lt McCullagh	Maj. Mulcahy
4 June 1936	.20	Lt Reade	Maj. Mulcahy
10 June 1936	.45	Lt Hanley	Maj. Mulcahy
12 June 1936	.55	Maj. Mulcahy	Lt Hanley
12 June 1936	.35	Maj. Mulcahy	Lt Hanley
15 June 1936	.30	Maj. Mulcahy	Lt Hanley
18 June 1936	1.10	Maj. Mulcahy	Lt Hanley
21 June 1936	.35	Lt O'Cathain	Maj. Mulcahy
6 July 1936	.40	Maj. Mulcahy	Lt O'Cathain

Mulcahy was paid eight shillings per day as a qualified pilot, from 7 July 1936

16 Sept. 1936	.40	Lt Hanley	Maj. Mulcahy
17 Sept. 1936	.45	Lt Hanley	Maj. Mulcahy
25 Sept. 1936	.55	Lt Hanley	Maj. Mulcahy
5 Oct. 1936	.25	Lt Hanley	Maj. Mulcahy
5 Oct. 1936	.15	Maj. Mulcahy	Solo
6 Oct. 1936	.20	Lt Hanley	Maj. Mulcahy
6 Oct. 1936	1.00	Maj. Mulcahy	Solo
7 Oct. 1936	.45	Maj. Mulcahy	Solo
8 Oct. 1936	.35	Maj. Mulcahy	Solo
8 Oct. 1936	.30	Maj. Mulcahy	Solo
12 Oct. 1936	.55	Maj. Mulcahy	Solo
12 Oct. 1936	.45	Lt Hanley	Maj. Mulcahy.
3 Nov. 1936	.35	Maj. Mulcahy	Solo
10 Nov. 1936	.20	Lt Hanley	Maj. Mulcahy
13 Nov. 1936	.25	Maj. Mulcahy	Solo
14 Nov. 1936	.50	Maj. Mulcahy	Solo
16 Nov. 1936	.40	Maj. Mulcahy	Solo
18 Nov. 1936	2.45	Lt Hanley	Maj. Mulcahy
13 Jan. 1937	.05	Lt Hanley	Maj. Mulcahy
13 Jan. 1937	.30	Maj. Mulcahy	Solo
19 Jan. 1937	.45	Maj. Mulcahy	Solo
21 Jan. 1937	.40	Lt Hanley	Maj. Mulcahy Cadet 5

March 1938 – 2 hrs 20 mins instruction from Lt McCullagh – no more solo.

Totals: .30 mins Pass; 13 hrs 20 dual, 7hrs 45 Solo.

Total time in the air to 31 March 1939 – 135 hrs 30 mins.

1 Log book, Avro Cadet 5 (in my custody). While it appears that Mulcahy did most of his flying in Avro Cadet No. 5, in the absence of his log book it cannot be confirmed that the above is a complete record of his flying training.

2 Probably a passenger flight.

Appendix IX: Damage to Army Aircraft from Accidents During Exercises, Cost of Repairs 1938–45[1]

The details on pp. 299–301 relate to accidents in relation to which courts of inquiry were held. In each instance the findings were examined by the Department of Defence in order to assess the degree of negligence on the part of the pilot. Where a pilot was considered sufficiently negligent a stoppage from pay was authorised by the minister in accordance with Section 130 (d) Defence Act 1923.

1 NAI file S.008/0029/39

Date	Aircraft	Pilot	Location	Circumstances	Cost	Penal deduction
20 Oct. 1938	Gladiator 23	Lt M. Higgins	Near Baldonnell	[Engine failure, heavy landing]	£4800	Nil
8 Sept. 1939	Anson 45	Lt W. Ryan	Ballyferriter	Forced landing, fuel problem no negligence	£1645	Nil
2 Oct. 1939	Magister 40		Johnstown	Loose prop, F/landing	£68	Nil
10 Oct. 1939	Anson 44	Capt. W. J. Keane	Nenagh	Forced landing due bad weather. No negligence	£1245	Nil
8 Nov. 1939	Avro 636 14	Lt J. O'Brien	Baldonnell	Landing incident	£36	Nil
9 Feb. 1940	Avro 636 15	2/Lt O. L. Toole	Baldonnell	Engine failure	£1400	Nil
4 June 1940	Anson 21		Baldonnell	Hit fence, 'error of judgment'	£77	Nil
5 June 1940	Hind 69		Baldonnell	Known aircraft defect	£160	Nil
14 June 1940	Magister 33	Lt M. T. Cregg	Rineanna	Hit obstruction, 'Want of reasonable care'	£29	£3
20 June 1940	Hind 72		Baldonnell	Tipped over on engine run. 'Pilot incompetent'	£70	Nil
23 June 1940	Magister 40		Baldonnell	Aircraft hit rough ground, no negligence	£34	Nil
28 June 1940	Magister 32		Rineanna	Hit obstruction, 'pilot incompetence'	£315	Nil
27 July 1940	Hind 70	Lt M. J. Ryan	Laytown	Low flying, pilot error	£282	Nil
9 Sept. 1940	Magister 40	Lt M. P. Quinlan	Baldonnell	Stalled, hit obstruction, no negligence.	£101	Nil
18 Sept. 1940	Walrus 19	Lt D. V. Cousins	Baldonnell	Aircraft damaged during taxying.	£10	Nil
27 Sept. 1940	Hind 71		Baldonnell	Engine failure, heavy landing	£50	Nil
28 Oct. 1940	Magister 73	Cadet K. P. Loughran	Baldonnell	Stall at 30/40 feet	£110	Nil
19 Dec. 1940	Anson 43	Lt P. O'Sullivan	Galway Bay	Engine failure, No negligence	£9,144	Nil
6 Mar. 1941	Magister 36	Lt T. P. McKeown	Baldonnell	Hit obstruction	£10	£5
23 May 1941	Anson 19	Lt O. L. Toole	Rineanna	Hit obstruction	£40	Nil
25 May 1941	Cadet 6	Lt D. V. Cousins	Rineanna	Hit obstruction	£3	£1. 10s
29 May 1941	Hector 83	Lt C. F. Cagney	Baldonnell	Tipped over, no negligence	£6	Nil
30 May 1941	Hind 68	Lt M. T. Callaghan	Baldonnell	Propeller hit ground, no negligence	£26	Nil
4 July 1941	Lysander 62	M. T. Cregg	Ballybunion	Aircraft struck beach	£148	£50
13 July 1941	Lysander 65	Lt B. C. Peterson	Edgesworthstown	Heavy landing	£4793 + £750	Nil
21 July 1941	Hector 87	Lt D. J. Healy	Baldonnell	Unauthorised simulated forced landing	£30	Reprimand

Date	Aircraft	Pilot	Location	Circumstances	Cost	Penal deduction
26 Aug. 1941	Magister 39	D. J. Healy	Baldonnell	Hit obstruction	£10	£5
7 Oct. 1941	Magister 76	Lt B. M. Flanagan	Baldonnell	Stall, hit obstruction stake.	£360	£10
16 Jan 1942	Cadet 5	Lt W. Ryan	Rineanna	Hit obstruction, no negligence	£181	Nil
23 Mar. 1942	Cadet 6	Lt T. P. O'Mahoney	Croom	Simulated attack, hit gun crew	£288	Nil
25 Mar. 1942	Gladiator 26	Lt D. J. Healy	Baldonnell	Hit obstruction stake	£45	£10
1 July 1942	Magister 33	Lt M. T. Calaghan	Laytown	Low Flying, pilot killed	£953	Nil
26 July 1942	Cadet 2	Lt H. F. Howard	C'knock	Low flying, hit tree	£333	£50
28 Aug. 1942	Hector 83	Lt S. Kelleher	Rathduff	Engine failure, forced landing, no negligence	£216	Nil
3 Sept. 1942	Magister 34	Lt M. Cusack	Glenville	Poor landing in the dark. Hit boundary fence	£101	Nil
3 Sept. 1942	Walrus 20	Lt T. P. O'Mahoney	Castletownroche	Engine failure, no negligence	£2000	Nil
4 Sept. 1942	Hector 81 Hector 83	Lts J. O'Connor and B. Crawford	Gormanston	Ground collision, 'pilots admit not complying with regulations'	£263	£50
5 Sept. 1942	Magister 35	Lt D. V. Cousins	Kilmurray	Stalled at low level	£439	£35
28 Sept. 1942	Anson 22	Lt D. V. Cousins	Boher, Co. Limerick			Nil
2 Oct. 1942	Lysander 64	Lt D. J. Healy	Near Guinness Aerodrome	Low flying, 'unnecessary zeal' hit trees	£4033	Nil
16 Dec. 1942	DH Dragon	Lt J. O'Brien	Baldonnell	Engine failure, forced landing	£5561	Nil
24 Mar. 1943	Anson 21	Lt W. Ryan	Rineanna	Took off with control locks in place	£779	£50
26 Mar. 1943	Magister 99	Lt J. B. O'Connor	Baldonnell	Struck wheel-barrow & survey pole.	£10	Nil
16 Apr. 1943	Anson 21	Capt. T. J. Hanley	Rineanna	Nosed over on to propeller	£30	Nil
15 Feb. 1944	Magister 37	Pte Casserly	Scribble-stown	Undercarriage collapse	£59	Nil
7 Mar. 1944	Magister 77	Pte N. K. Brennan	Baldonnell	Low flying	£572	Nil
25 Feb. 1944	Magister 73	Pte J. Gibney	Templemore	Ground loop, undercarriage collapse, no negligence.	£40	Nil
25 Feb. 1944	Magister 38	Capt. P. Swan	Templemore	Lost, Forced landing, fuel	£45	Nil
13 July 1944	Fairey Battle		Gormanston	Hit hedge on take off	£30	Nil
21 Nov. 1944	Master 102	Pte R. O'Keeffe	Baldonnell	Forced landing, engine failure	£350 +	Nil
				Brake failure	£8	Nil

Date	Aircraft	Pilot	Location	Cause		
8 Jan. 1945	Master 100	Cpl J. Gibney	Baldonnell	Heavy landing	£98	Nil
3 Apr. 1945	Hurricane 108	Lt J. B. O'Connor	Rineanna	Nosed over	£30	Nil
3 Apr. 1945	Master	M. P. Quinlan	Baldonnell	Undercarriage problem	£39	Nil
27 May 1945	Lysander 66	Sgt. Gleeson O'Connor, Toomey	Rineanna	Defective workmanship. (during maintenance)	£190	£5 £4 £4
1 June 1945	Anson 21	Capt. D. K. Johnston	Baldonnell	Engine failure	£1350	Nil

Appendix X: Majority Report on Colonel P. A. Mulcahy Air Corps Investigation[1]

COLONEL P. A. MULCAHY

Three members of the Committee favour the appointment of Colonel Mulcahy as Director of Military Aviation. The fourth member dissents and is submitting a separate recommendation. In recommending Colonel Mulcahy, these three members do so on the grounds that:

(i) Colonel Mulcahy took over the command of the Air Corps in 1935 when it was in a very bad condition. He was seriously handicapped by not having a policy for the Corps and by inadequate equipment.

(ii) He has a good conception of the requirement and role of the Air Corps and endeavoured to obtain a decision on policy. The supply of equipment has been to a large extent outside his control.

(iii) In a small Unit such as the Air Corps with officers having grievances about their flying pay, the inadequacies of equipment and lack of policy, discontent was bound to arise, thus making Colonel Mulcahy's task very difficult. In such circumstances, criticism is always rife.

(*a*) That Colonel Mulcahy succeeded in maintaining a high standard of discipline in such circumstances rebounds to his credit.

(*b*) Whilst Colonel Mulcahy bears responsibility for the low standard of training in the Air Corps, the mitigating circumstances in Section V, paragraph 24, must be taken into consideration.

The above mentioned considerations render it necessary that any Officer nominated to replace Colonel Mulcahy requires to be a good administrator, have technical ability and possess strong character and personality. Whilst there is within the Corps a number of promising officers, none of them, in the opinion of the three members of the Committee concerned is fitted at this stage to effectively direct the Corps in its present condition.

In order to satisfactorily fill his appointment, however, the three members in favour of his appointment consider that Colonel Mulcahy should be required, at an early date, to undergo the necessary additional training to obtain the qualifications which the Committee have recommended in Section V of the Report as being essential for the Officer holding the appointment of Director of Military Aviation.

1 Report and findings, p. lxi (MA, ACS 22/23).

Appendix XI: Minority report on Colonel P. A. Mulcahy

I regret that I cannot agree with my colleagues regarding the advisability of retaining Colonel P. A. Mulcahy in the Air Corps, even on the conditions set out in paragraph 39 (b).[1] This opinion is based on two factors, neither of which should be taken as reflecting in any way on Colonel Mulcahy personally. These factors may be summarised as follows:

(*a*) I am satisfied that no one other than a fully qualified Flying Officer possessing considerable practical experience should be placed in charge of the Air Corps. The fact that in the past several such non-qualified officers have been from time to time placed in charge of the Corps is, in my opinion, one of the causes of the condition of affairs this committee was set up to investigate. As already stated in this Report, such officers must of necessity rely on their subordinates to an undesirable extent in matters connected with the organisation, training and administration of the Air Corps.

Colonel Mulcahy does not possess these qualifications and I firmly believe that no amount of training at this stage could bring him up to the required standard. Furthermore, if the continuance of Colonel Mulcahy in the Air Corps is made conditional upon his attempting to qualify as a Flying Officer as set out in Section V of this Report, I believe that this would have a very adverse effect on his prestige and upon the discipline of the Corps in general.

(*b*) The confidence of a large number at least of the junior officers in Colonel Mulcahy has, through one cause or another, been hopelessly been undermined. Furthermore the confidence of Colonel Mulcahy in the loyalty of a large number of his officers has similarly been undermined as a result of the existing situation. With the best will in the world, I cannot see how this state of affairs can be completely set right while Colonel Mulcahy and the Officers concerned are required to serve together in the same Corps.

I, therefore, recommend that Colonel Mulcahy be relieved of his present appointment and posted to some other appointment commensurate with his rank, qualifications and experience and that Major G. J. Carroll be recalled to active duty as Director of Military Aviation. This officer is a very experienced pilot, as far as I know he has not been connected with any of the existing factions in the Air Corps, and should, therefore, enjoy the full confidence of the Officers of the Corps.

[Signed] Major General.
(Aodh MacNeill)
President, Committee of Investigation

---

1 Should read 41 (b).

Notes

—

ONE: EARLY AVIATION IN IRELAND

1   Karl E. Hayes, *A History of the Royal Air Force and the United States Naval Air Service in Ireland 1913–1923* (Irish Air Letter, 1988), pp. 3–5.

2   Ibid.

3   Ibid., pp. 29–39.

4   Ibid.

5   Ibid.

6   Ibid., pp. 7–19. Many early sources, primary and secondary, cited here use the spelling Baldonnel. Baldonnell is the Ordnance Survey spelling and that used in the first RFC/RAF documents.

7   OPW to Sec DF, 4 July 1939 (NAI, S.007/0002/35).

8   Lt Col. M. O'Malley, *Gormanston Camp 1917–1986* (Defence Forces, 1986), p. 2.

9   McLaughlin & Harvey materials ledger (MA); Site map, Gormanston, 1917 (NAI, OPW 4/6/28).

10   Copy RAF photograph (in author's possession).

11   Lt Col. Michael C. O'Malley 'Baldonnell aerodrome 1917–1957', in *Dublin Historical Record* LVI: 2, Autumn 2003; Walter McGrath, *Some Industrial Railways of Ireland and Other Minor Lines* (Cork, 1959), pp. 23–5; *Irish Industrial and Contractors Locomotives* (London, 1962), p. 13.

12   Copy of RAF photograph (D. McCarron).

13   Hayes, *History of the Royal Air Force*, pp. 7–17.

14   Patrick J. McCarthy, 'The RAF and Ireland 1920–1922', *Irish Sword* XVII: 68 (1989), pp. 174–88, passim.

15   Charles Townshend, *The British Campaign in Ireland 1919–1921* (Oxford, 1973), p. 171.

16   Ibid., pp. 170–1.

17   [RAF, List of aerodromes], SO Book 122; Capt. C. H. Pixton, 'Complete List of Landing Grounds – Ireland' (in author's possession). The mainly undated references in these books suggest that the data was collected in the period from late 1917 to late 1922.

18   Townshend, *British Campaign*, p. 171. See also McCarthy, 'The RAF and Ireland, 1920–22'.

19   Hayes, *History of the Royal Air Force*, pp. 50–7.

20   Madeleine O'Rourke, *Air Spectaculars: Air Displays in Ireland* (Dublin, 1989), p. 17; Fergus D'Arcy 'Flying high in Foxrock', in Fergus D'Arcy and Con Power (eds), *Leopardstown Racecourse, 1888–1988* (Privately published, 1988), pp. 44–9.

21   Air Navigation Act, 1936.

22   *Baldonnel: Dublin's Civil Airport 1919 to 1939* (Irish Air Letter, 1989), passim.

23 Air Council minutes, 6 Apr. 1922 (NAI, DT S.4002).

24 Hayes, *History of the Royal Air Force*, pp. 60–5.

25 *Freeman's Journal*, 4 May 1922.

26 Michael O'Malley, 'Military aviation in Ireland 1921–1945', doctoral thesis, NUI Maynooth, 2007, p. 41.

TWO: CIVIL AVIATION: DEVELOPMENTS IN SAORSTÁT ÉIREANN

1 Defence Act 1954; *Irish Defence Forces Handbook* (1968), p. 1.

2 *Irish Defence Forces Handbook* (1968), p. 1; John P. Duggan, *A History of the Irish Army* (Dublin, 1991), p. 1.

3 Donal McCartney, 'From Parnell to Pearse (1891–1921)', in T. W. Moody and F. X. Martin (eds), *The Course of Irish History* (Cork, 1984), p. 311.

4 Annex to Articles of agreement for a treaty between Great Britain and Ireland, 6 December 1921.

5 Liam Byrne, *History of Aviation in Ireland* (Dublin, 1980), p. 52; Donal MacCarron, *Wings over Ireland: The Story of the Irish Air Corps* (Leicester, 1996), p. 11.

6 Enclosure, David Lloyd George to Eamon de Valera, 20 July 1921, in Ronan Fanning et al. (eds), *Documents on Irish Foreign Policy*, 1, 1921–2 (Dublin, 1999), p. 237.

7 Memorandum by Erskine Childers, July 1921, in Fanning, *Documents on Irish Foreign Policy*, 1, 1921–2, p. 239.

8 Enclosure, David Lloyd George to Eamon de Valera, 20 July 1921, in Fanning, *Documents on Irish Foreign Policy*, 1, p. 242.

9 Memorandum by Erskine Childers, July 1921, in Fanning, *Documents on Irish Foreign Policy*, 1, p. 239.

10 Ibid.

11 Conference of Ireland, Committee of Defence, 13 Oct. 1921 (UCDA, MP, P7/A/73/32).

12 Ibid., 17 Oct. 1921 (UCDA, MP, P7/A/73/53).

13 Jim Ring, *Erskine Childers* (London, 1996), passim.

14 Conference of Ireland, Committee of Defence, 17 Oct. 1921 (UCDA, MP, P7/A/73/53).

15 Sir Philip Joubert de la Ferté, *The Third Service: The Story Behind the Royal Air Force* (London, 1955), pp. 72–3.

16 Michael Armitage, *The Royal Air Force: An Illustrated History* (London, 1993), Appendices 1, 2 and 3.

17 Minutes of Committee of Defence, Conference of Ireland, 17 Oct. 1921 (UCDA, MP, P7/A/73/53).

18 Minutes of conversation, Tom Jones/Erskine Childers, 28 Oct. 1921, in Fanning, *Documents on Irish Foreign Policy*, 1, p. 296.

19 Amendments by the Irish representatives to the proposed articles of agreement, 4 Dec. 1921 (NAI, DE 2/304/1).

20 Frank Pakenham, *Peace by Ordeal* (London, 1962), p. 265.

21 Ibid., p. 372.

22 Emmet Dalton to Lt Col. W. J. Keane, 23 Oct. 1951 (MA, PC137).

23 Summary of expenses, C. F. Russell to M. Collins, 27 Feb. 1922 (NAI, DT S.4002).

24 In the Irish Defence Forces the rank of commandant equates to the more conventional major. In the 1920s and 1930s the rank of major equated to that of lieutenant colonel.

25 Emmet Dalton to Lt Col. W. J. Keane, 23 Oct. 1951 (MA, PC143).

26 Sean Dowling to Lt Col. W. J. Keane, 12 May 1965 (MA, PC143).

27 Enclosure 2, 21 Oct. 1921, Emmet Dalton to Lt Col. W. J. Keane, 23 Oct. 1951 (MA, PC143).

28 Officer's history sheet, 16 Feb. 1924 (MA, SDR 3718).

29 Enclosure 1, Emmet Dalton to W. J. Keane, 23 Oct. 1951 (MA, PC143).

30 A. J. Jackson, *Avro Aircraft Since 1908* (London, 1990), p. 68.

31 Enclosure 1, Emmet Dalton to Lt Col. W. J. Keane, 23 Oct. 1951 (MA, PC143).

32 DOD special expenditure, 1 July 1921 to 31 Dec. 1921 (NAI, DE 3/4/10); Irish Self-Determination League, London office accounts, 1 Oct. 1921 to 31 Dec. 1921 (NLI, Art O'Brien Papers, MSS 8431–2); C. F. Russell to M. Collins, 27 Feb. 1922 (NAI, DT, S.4002).

33 Michael Hopkinson (ed.), *The Last Days of Dublin Castle: The Mark Sturgis Diaries* (Dublin, 1999), p. 60.

34 C. F. Russell to M. Collins, 27 Feb. 1922 (NAI, DT, S.4002).

35 Enclosure 2, Emmet Dalton to Lt Col. W. J. Keane, 23 Oct. 1951 (MA, PC143).

36 Ibid.; Ray Sanger, *The Martinsyde File* (Tunbridge, 1999), p. 181.

37 Enclosure 2, E. Dalton to W. J. Keane, 23 Oct. 1951 (MA, PC143).

38 Irish Self-determination League, London office accounts, 1 Oct. 1921 to 31 Dec. 1921 (NLI, Art O'Brien Papers, MSS 8431–2).

39 Russell to Collins, 27 Feb. 1922; Civil Aviation, 7 Apr. 1922 (NAI, DT, S.4002); Quartermaster General's account, 1 Jan. 1922 to 1 Oct. 1923 (NAI, DE 3/4/7).

40 Ronan Fanning, *The Irish Department of Finance, 1922–58* (Dublin, 1978), pp. 13–23.

41 Aircraft log book, Avro No 1; Log book Martinsyde Type A, MK II (in my custody).

42 Aviation Department memo., 2–3 Mar. 1922 (MA, PC143).

43 C. F. Russell to Michael Collins, 27 February 1922 (NAI, DT, S.4002).

44 MFD to MFF, 7 Apr. 1922 (NAI, DT, S.4002).

45 Ibid.

46 M. O Coileain to MFD, 12 Apr. 1922 (NAI, DT S.4002).

47 Extract, Cabinet minutes, 18 Apr. 1922 (NAI, DT, S.4002).

48 Refund to MFD, 1 Nov. 1922 to 15 Aug. 1923 (NAI, DE 3/43A).

49 E. Dalton to Lt Col. W. J. Keane, 23 Oct. 1951 (MA, PC143).

50 Extract, cabinet minutes, 27 Feb. 1922 (NAI, DT, S.4002).

51 Civil Aviation minutes, 23 Mar. 1922 (NAI, DT, S.4002).

52 C. F. Russell to M. Collins, 20 Feb. 1922 (NAI, DT, S.4002).

53 Ibid.

54 Ibid.

55 Ibid.

56 Ibid.

57 Civil Aviation Department minutes, 23 Mar. 1922 (NAI, DT, S.4002).

58 Ibid.

59 Air Council minutes, 6 Apr. 1922 (NAI, DT, S.4002).

60 Ibid.

61 *Freeman's Journal*, 4 May 1922.

62 E. Dalton to C. F. Russell, 12 May 1922 (MA, Liaison office file).

63 Department of Civil Aviation, 20 July 1922 (UCDA, MP, P7/49/38).

64 C. F. Russell to M. Collins, 25 Apr. 1922 (NAI, DT, S.4002).

65 Ibid., 2 May 1922 (NAI, DT S.4002).

66 Unsigned report, Ernest Mills to Chief liaison officer, 6 May 1922 (MA, Liaison office general file).

67 Ibid.

68 General Routine Order No 9, 20 Dec. 1922.

69 Department of Civil Aviation, C. F. Russell to W. J. McSweeney, 20 July 1922 (UCDA, MP, P7/B/49/38).

70 Capt. M. Dunphy to COS, 8 Apr. 1922 (NAI, DT, S.4002).

71 Ibid.

72 Maryann Gialanella Valiulis, *Almost a Rebellion: The Irish Army Mutiny of 1924* (Cork, 1988), p. 24.

73 Civil Aviation Department minutes, 6 Apr. 1922: Air Council agenda, 15 May 1922 (NAI, DT, S.4002).

THREE: MICHAEL COLLINS, THE MILITARY AIR SERVICE AND THE CIVIL WAR

1 Unsigned memo. dated 13 Feb. 1922 (MA, Liaison office file). This document appears to be a typed copy.

2 Aviation department to COS, 4 Mar. 1922 (MA, Liaison Office file).

3 Military Aviation, 20 July 1922 (UCDA, MP, P7/B/49/37).

4 Sgt J. Curran, statement to W. J. Keane, June 1944 (MA, PC143).

5 Michael Hopkinson, *Green Against Green: The Irish Civil War* (Dublin, 1986), p. 52.

6 Ibid., p. xix.

7 Ibid., pp. 58–72.

8 Ibid., pp. 72–3.

9 Ibid., p. 121.

10 AM note to CAS, 30 June 1922 (TNA, Air 8/49).

11 Log book, H.1585, 29 June 1922 (in my custody).

12 Receipt, 20 June 1922 (MA, AC/2/2/1).

13 AFO to W. J. McSweeney, 12 July 1923 (MA, AC/2/2/1).

14 Hopkinson, *Green Against Green*, pp. 72–3.

15 'Statement of expenditure', Comdt Gen. McSweeney, 28 July 1923 (MA, AC/2/2/1).

16 COS to Martinsyde, 30 June 1922 (TNA, Air 8/49).

17 Copy by Lt Col. W. J. Keane, 21 June 1954 of Comdt Gen. McSweeney to ADC, 1 July 1922 (MA, PC143).

18 Aircrew service record, W. J. McSweeney (TNA, Air 76/329).

19 Expenses of Major General McSweeney during year 1922–1923; 'Statement of Expenditure', 28 July 1923 (MA, AC/2/2/1); Chaz Bowyer, *Bristol Fighter F2B: King of Two-seaters* (Shepperton, 1985), p. 124.

20 Log book, BF II (in my custody); Bowyer, *Bristol Fighter*, p. 124.

21 Copy by Lt Col. W. J. Keane, 21 June 1954, W. J. McSweeney to ADC, 4 July 1922 (MA, PC143).

22 Log book, B II, p. 5 (in my custody).

23 Message A/179, 7 July 1922 (UCDA, MP, P7/B/106/266).

24 Telegram, Cope to Curtis for Mr Churchill, 4 July 1922 (TNA, Air 8/49). The word 'undercarriage' should, more correctly, read 'bomb racks'.

25 Ibid.

26 Cipher, War Office to McCready, 4 July 1922 (TNA, Air 8/49).

27 AM to Bonham-Carter, 4 July 1922 (TNA, Air 8/49).

28 Microfilm P7/B/49 (MP, UCDA).

29 Ibid.; Log books, B I, B II and B III (in my custody).

30 Log book, B III (in my custody).

31 W. J. McSweeney to Vol. T. Nolan, 7 July 1922 (MA, PC143).

32 Log Book, B I (in my custody); Military Aviation, 22 July 1922 (UCDA, MP, P7/B/49/37).

33 McSweeney to COGS, 24 July 1922 (MA, A/06886).

34 McSweeney to AG, 17 July 1922 (UCDA, MP, P7/B49/41); Log book B III (in my custody).

35 Aircrew service record, W. J. McSweeney (TNA, Air 76/329).

36 *Royal Air Force Flying Training Manual, Part 1* (Air Ministry, 1923), pp. 158–63.

37 Reconnaissance report, 17 July 1922 (UCDA, MP, P7/B/107/69).

38 *Freeman's Journal*, 18 July 1922.

39 Report, 28 July 1922 (UCDA, MP, P7/B/10/40).

40 Civil Aviation, 20 July 1922 (UCDA, MP, P7/B/49/38).

41 Officer's personal file (MA, SDR169).

42 Civil Aviation, 20 July 1922 (UCDA, MP, P7/B/49/38).

43 Military Aviation, 22 July 1922 (UCDA, MP, P7/B49/38).

44 Baldonnel return, 12–13 Nov. 1922 (MA, Army census).

45 MFD to McSweeney, 17 and 21 July 1922 (UCDA, MP, P7/B49/43–4).

46 DMA to MFD, 22 July 1922 (UCDA, MP, P7/B/49/36).

47 Ibid.

48 Baldonnel return, 12–13 Nov. 1922 (MA, Army census).

49 W. J. McSweeney to COS, 19 July 1922 (UCDA, MP, P7/B/49/34).

50 These books were presented by Broy to the Officers' Mess, Baldonnell in the early 1950s. The flyleaf of the earlier book was annotated 'E. O'Broite, 15 Cadogan Gdns., Sloan Sqr. 28/X/1921'.

51 Unsigned file memo., 13 Nov. 1925 (MA, SDR169).

52 Officer's history sheet (MA, SDR169).

53 Padraic O'Farrell, *Who's Who in the Irish War of Independence and Civil War 1916–1923* (Dublin, 1997), p. 145.

54 W. J. McSweeney to COGS, 24 July 1922 (MA, A/06886).

55 COS to General Dalton, 24 July 1922 (UCDA, MP, P7/B/49/25).

56 E. Dalton to COS, 26 July 1922 (UCDA, MP, P7/B/49/24).

57 Report, 30 July 1922 (UCDA, MP, P7/B/10/28).

58 Unsigned, undated memo., Telephones at Baldonnell, *c.*Aug. 1922 (UCDA, MP, P7/B/10/7).

59 Unsigned, undated memo., 'Status of Air Service', with covering letter COGS to CIC, 24 Jan. 1923 (MA, A/08075).

60 J. P. Duggan, *A History of the Irish Army* (Dublin, 1991), pp. 89–91.

61 Undated map, RAF in Ireland (TNA, Air 8/49).

62 Reconnaissance report, Capt. C. F. Russell to GHQ, 22 July 1922 (UCDA, MP, P7/B/107/138).

63 M. Collins file memo., 4 Aug. 1922 (UCDA, MP, P7/B/10/29).

64 W. J. McSweeney to GHQ, 30 July 1922 (UCDA, MP, P7/10/33).

65 W. J. McSweeney to CIC, 30 July 1922 (UCDA, MP, P7/B/10/34); Complete list of landing grounds, Ireland, Capt. C. M. Pixton (in my possession).

66 AS to M. Collins, 6 Aug. 1922 (UCDA, MP, P7/B/10/26).

67 Reconnaissance report, C. F. Russell to GHQ, 10 Aug. 1922 (UCDA, MP, P7/B10/13–14).

68 E. Broy to M. Collins, 9 Aug. 1922 (UCDA, MP, P7/B/10/21).

69 Hopkinson, *Green Against Green*, pp. 162–4.

70 Reconnaissance report, C. F. Russell to GHQ, 10 Aug. 1922 (UCDA, MP, P7/B/10/13).

71 Ibid.

72 Copy message, 12 Aug. 1922 (Broy private papers, Áine Broy).

73 Copy message, 13 Aug. 1922 (Broy private papers, Áine Broy).

74 Ibid.

75 Reconnaissance report, 13 Aug. 1922 (UCDA, MP, P7/B/39/21).

76 Ibid., 14 Aug. 1922 (UCDA, MP, P7/B/39/17).

77 Ibid., 14 Aug. 1922 (UCDA, MP, P7/B/39/16).

78 Unsigned, undated memo. (UCDA, MP, P7/B/3/73).

79 M. Collins to QMG, 17 Aug. 1922 (UCDA, MP, P7/B/3/40).

80 Russell to Broy, 14 Aug. 1922 (Broy private papers, Áine Broy).

81 Broy to O/C Troops, 14 Aug. 1922 (Broy private papers, Áine Broy).

82 Collins diary, 15 Aug. 1922 (UCDA, MP, PA/62).

83 Log book, BF I, pp. 11–12 (in my custody).

84 Collins to GS, 28 July 1922 (UCDA, MP, P7/B/10/33).

85 Baldonnell return, 12–13 Nov. 1922, (MA, Army census).

86 CIC to QMG, 10 Aug. 1922 (UCDA, MP, P7/B/3/73).

87 McSweeney to Collins, 2 Aug. 1922 (UCDA, MP, P7/10/27).

88 Ibid., 7 Aug. 1922 (UCDA, MP, P7/B/10/23).

89 Ibid., 9 Aug. 1922 (UCDA, MP, P7/B/10/18).

90 McSweeney expenses 1922–3, *c.*1 Nov. 1923 (MA, AC/2/2/1); Log book, Martinsyde Scout MS I (in my custody).

91 McSweeney to CIC, 25 Aug. 1922 (UCDA, MP, P7/B/10/2).

92 Anthony P. Kearns, 'The Irish Air Corps: a history', *Scale Aircraft Modelling* 3: 10 (July 1981), p. 449; ADC to Comdt Gen. McSweeney, 16 Sept. 1922 (MA, A/06959).

93 Crossley to McSweeney, 8 Sept. 1922; Russell to McSweeney, 11 Sept. 1922 (UCDA, MP, P7/B/48/134; 119).

94 Personal notes, M. Collins, 16 Aug. 1922 (UCDA, MP, P7a/62).

95 McSweeney to GHQ, 15 Aug. 1922; Undated reconnaissance reports (UCDA, MP, P7/B/107/314; 331).

96 Collins memo., 16 Aug. 1922 (UCDA, MP, P7/10/11).

97 Collins to QMG, 17 Aug. 1922 (UCDA, MP, P7/B/3/42).

98 Collins notes, 17 Aug. 1922 (UCDA, MP, P7a/62).

99 CIC to PMG, 10 Aug. 1922 (UCDA, MP, P7/B/10/6).

100 E. Broy to PMG, 21 Aug. 1922 (UCDA, MP, P7/B/10/5).

101 Collins to QMG, 18 Aug. 1922 (UCDA, MP, P7/B/3/34).

102 CIC to GHQ, 22 Aug. 1922 (UCDA, MP, P7/B/70/37).

103 E. Dalton to COGS, 23 Aug. 1922 (UCDA, MP, P7/B/70/34).

FOUR: FROM CIVIL WAR TO ARMY MUTINY

1 Michael Hopkinson, *Green Against Green: The Irish Civil War* (Dublin, 1986), p. 172.

2 McSweeney to MFD, 22 July 1922 (UCDA, MP, P7/49/37; P7/49/40).

3 Officers' history sheets (MA, SDR550; SDR1767).

4 Aircrew service records (TNA, Air 76/115; Air 76/33).

5 ADC to McSweeney, *c.*1 Sept. 1922 (A. P. Kearns).

6 W. P. Delamere, 'Early days in the Army Air Service', *An Cosantóir* XXXII (Sept. 1972), p. 168.

7 Michael O'Malley, 'The military air service 1922–24', Appendix 4 (BA thesis, July 2002, NUIM).

8 Log books, Bristol Fighters and Martinsyde F.4s (in my custody).

9 CIC to McSweeney, 5 Sept. 1922 (UCDA, MP, P7/B/49/16).

10 CIC memo., *c.*5 Sept. 1922 (UCDA, MP, P7/B/49/15).

11 CIC to McSweeney, 11 Sept. 1922 (UCDA, MP, P7/B/49/12).

12 'McSweeney expenses', 1 Nov. 1923 (MA, AC/2/2/1); Ray Sanger, *The Martinsyde File* (Tunbridge Wells, 1999), p. 246.

13 A. P. Kearns, 'The Irish Air Corps: A History', *Scale Aircraft Modelling* 3: 10 (July. 1981), p . 449.

14 ADC to McSweeney, 16 Sept. 1922 (MA, A/06959); Log book, BF IV (In my custody). Delivery note, ADC to McSweeney, 15 Sept. 1922 (in my possession). A. P. Kearns records this aircraft as E.1959, the delivery note indicates E.1958.

15 DMA to QMG, 17 Sept. 1922 (MA, A/06959).

16 Log book, MS 1 (in my custody).

17 Teddy Fennelly, *Fitz and the Famous Flight* (Portlaoise, 1997), p. 99; AG to CIC, 18 Oct. 1922 (MA, A/07279).

18 Strength return, 21 Aug. 1923 (NAI, FIN 1/2875).

19 Statement, J. C. Fitzmaurice to W. J. Keane, 7 Dec. 1950 (MA, PC143).

20 J. C. Fitzmaurice, unpublished memoir, p. 128 (Estate of P. Selwyn-Jones).

21 Ibid.

22 File memo., 18 Oct. 1922 (MA, A/07472).

23 DMA to COS, 25 Oct. 1922 (MA, A/07435).

24 *Dáil Debates*, vol. 1, 1 Nov. 1922, 1962–63.

25 J. C. Fitzmaurice, unpublished memoir, p. 128 (Estate of P. Selwyn-Jones).

26 Log book, MS 1 (in my custody); Fitzmaurice unpublished memoir, p. 129 (Estate of P. Selwyn-Jones).

27 Log book, W. P. Delamere (Peter Delamere); J. P. Duggan, *A History of the Irish Army* (Dublin, 1991), p. 85; Hopkinson, *Green Against Green*, p. 205.

28 RAF sketch; RAF aerodrome book (in my possession).

29 Log book, W. P. Delamere.

30 Operations report, 16 Jan. 1923 (MA, CW/OPS/12/B).
31 Log book, W. P. Delamere (Peter Delamere).
32 Ibid.
33 DMA to CIC, 26 Sept. 1922 (MA, A/07041).
34 Archer to CIC, 29 Sept. 1922 (MA, A/07041).
35 CIC to Archer, 29 Sept. 1922; Archer to CIC, 3 Oct. 1922 (MA, A/07041).
36 CIC to Archer, 10 Oct. 1922; Archer to CIC, 12 Oct. 1922 (MA, A/07041).
37 QMG to CIC, 22 Aug. 1922; COS to DMA, 3 Sept. 1922 (UCDA, MP, P7/B/3/11; P7/B/49/17).
38 DMA to COS, 26 Sept. 1922 (MA, A/07189).
39 Ibid., 4 Dec. 1922; COS to DMA, 5 Dec. 1922: CIC file memo. 6 Feb. 1923 (MA, A/07189).
40 GRO No. 16, 24 Jan. 1923.
41 Log books, Avro 504K I, II, III and IV (in my custody).
42 GRO No. 9, 20 Dec. 1922.
43 Ronan Fanning, *The Irish Department of Finance 1922–58* (Dublin, 1976), pp. 114–16.
44 Duggan, *History of the Irish Army*, p. 130.
45 DF to Sec. Ministry of Defence, 20 Aug. 1923 (MA, A/09971).
46 Peter Young, 'Defence and the new Irish state, 1919–39', *Irish Sword* XIX (1993–4), p. 10.
47 Undated, unsigned memorandum, c.1939 (NAI, DT, S.11,101).
48 Strength return, June 1923 (MA, A/09971).
49 Unsigned memorandum, 15 June 1922 (MA, A/09971).
50 Minutes, Army Pay Commission, 3 May 1923 (NAI, DF, S.004/0248/24); Memo., 30 Oct. 1923 (MA, A/09971).
51 GOC AAS to GHQ, 23 Aug. 1922 (MA, A/09971).
52 Sec DF to AFO, 24 Dec. 1923 (NAI, FIN I/2975).
53 GOC, AAS, to COGS, 24 Oct. 1923 (MA, A/09971).
54 COGS to GOC, AAS, 26 Oct. 1923 (MA, A/09971).
55 Staff Duty Memo., No. 12, 29 Feb. 1924.
56 Army Estimates, 1924–5, June 1924 (NAI, DF, S.004/0005/27).
57 Duggan, *History of the Irish Army*, p. 131. Staff Duty Memo., No. 13, 6 Mar. 1924.
58 Officer's personal file (MA, SDR1182); Record of pilot intake to Air Corps (in my custody).
59 Personal comment, P. Young; Duggan, *History of the Irish Army*, pp. 130–7. See also Eunan O'Halpin, *Defending Ireland: The Irish State and its Enemies Since 1922* (Oxford, 1999), pp. 45–52.
60 List of officers, c.March 1924 (MA, A/11657).
61 Maryann Gialanella Valiulis, *Almost a Rebellion: The Irish Army Mutiny of 1924* (Cork, 1988), p. 32; Officer's history sheet (MA, SDR601).
62 Col. M. J. Costello to Army enquiry, 22 April 1924 (MA, PC586).
63 In August 1923 there were 27 line officers and 11 pilots in the Air Service.
64 Officer's history sheet (MA, SDR601).
65 Personal comment, Ms Áine Broy, 6 Feb. 2002.
66 W. J. McSweeney to AG, 18 Sept. 1922 (MA, A/06942).
67 AG to CIC, 19 Sept. 1922, Officer's personal file (MA, SDR169).
68 Certificate of military service, 29 Sept. 1926 (MA, SDR169).
69 Comdt T. Mason to COGS, 14 Apr. 1924 (MA, A/06942).
70 Personal comment Áine Broy, citing her parents' marriage certificate.

71 Undated list, resignations, dismissals and absenters [*sic*] (NAI, DT, S.3720).

72 List of additional resignations due to crisis, 19 Mar. 1924 (NAI, DT, S.3720).

73 Fitzmaurice unpublished memoir, pp. 140–1 (Estate of the late P. Selwyn-Jones).

74 Dáil Debates, vol. 6, 11 Mar. 1924, 1944.

75 W. J. McSweeney to MFD, 10 Mar. 1924; Officer's personal file (MA, SDR3718).

76 W. J. McSweeney to COS, 10 Mar. 1924, Officer's personal file (MA, SDR3718).

77 Undated 'List A', Summary of officers (MA, A/11657).

78 Patrick Mulloy, *Mutiny Without Malice* (London, 1974), p. 5.

79 J. C. Fitzmaurice, unpublished memoir, p. 121 (Estate of the late P. Selwyn-Jones).

80 Resignation, 25 Mar. 1924 (MA, SDR601).

81 Officers' history sheets (MA, SDR601; SDR3718).

82 Ibid. (MA, SDR1333; SDR1187; SDR975).

83 Signed declaration, 2 Mar. 1924 (MA, MS388).

84 J. J. Flynn to Kevin O'Higgins, 20 Apr. 1924 (MA, SDR975).

85 Undated EC note, Officer's personal file (MA, SDR975).

86 GOCF to President, 7 May 1924, Officer's personal file (MA, SDR975).

87 Officer's personal file (MA, SDR1187).

88 OC Air Service to COS, 20 June 1924, Officer's personal file (MA, SDR1187).

89 GOCF to EC, 29 May 1924 (NAI, DT, S.3720).

90 With the exception of McSweeney who was born in Manchester – his parents were from Waterford – all the ex-RAF pilots were born in Ireland.

91 J. C. Fitzmaurice, unpublished memoir, p. 143 (Estate of the late P. Selwyn-Jones).

92 Lt Col. Thomas Ryan to mutiny inquiry, 12 Apr. 1924 (UCDA, MP, P7/C/8).

93 C. F. Russell to Army inquiry, 9 May 1924 (UCDA, MP, P7/C/28).

FIVE: ORGANISATION, POLICY AND COMMAND, 1924–36

1 O'Duffy scheme, army organisation, pp. 28–9, GOCF to EC, 2 May 1924 (NAI, DT, S.3442B).

2 Ibid., explanatory notes, pp. 37–8, 1 July 1924 (NAI, DT, S.3442B).

3 Ibid., Army Organisation, pp. 2–20, 2 May 1924 (NAI, DT, S.3442B); John P. Duggan, *A History of the Irish Army* (Dublin, 1991), passim.

4 Ibid., pp. 21–53, 2 May 1924 (NAI, DT, S.3442B).

5 Ibid., explanatory notes, 1 July 1924, pp. 36–7 (NAI, DT, S.3442B).

6 GS to R. Mulcahy, 24 Jan. 1925; Mulcahy to GS, 25 Jan. 1925 (MA, SDR3718).

7 O'Duffy scheme, explanatory notes, 1 July 1924, p. 36 (NAI, DT, S.3442B).

8 Ibid., p. 23.

9 Ibid., p. 2.

10 Orders No. 3, Defence Forces (Organisation) Order, 1 Oct. 1924, pp. 34–5; Strength returns 1924 to 1927 (MA, Local Strength 8 & Local Strength 9).

11 OC AAC to GHQ, 17 Apr. 1925 (MA, DOD RM11).

12 Ibid.

13 O'Duffy scheme, explanatory notes, 1 July 1924, pp. 36–7 (NAI, DT S.3442B).

14 Ibid., p. 38.

15 Orders No. 3, Defence Forces (Organisation) Order, 1 Oct. 1924.

16 Strength returns, 29 Aug. 1924 (MA, LS9).

17 Duty officer, Sept. 1924 (MA, MS658); Ulick O'Connor, *Oliver St John Gogarty: A Poet and His Times* (London, 1981), pp. 227–35.

18 AFO to Dept. of Finance, 1 Mar. 1924 (MA, AC/2/2/2).

19 Ibid.

20 Sec Dept. of Finance to AFO, 10 Mar. 1924 (MA, AC/2/2/2).

21 Army finance minutes, 5 June 1924 (MA, A/06959).

22 Ibid.

23 AFO to Sec DF, 24 May 1924 (MA, AC/2/2/2).

24 DF to AFO, 13 June 1924 (MA, AC/2/2/2).

25 Army finance minutes, 17 July 1924; T. J. Maloney to ADC, 23 July 1924 (MA, AC/2/2/2).

26 J. F. Crowley to AFO, 30 July 1924; OC AC to AFO, 27 Sept. 1924; OC AC to AFO, 19 Dec. 1924 (MA, AC/2/2/2).

27 OC AC to COS, 31 Dec. 1924; QMG to AFO 5 Aug. 1925 (MA, AC/2/2/2).

28 OC AAC to CSO GHQ, 17 Apr. 1925 (MA, RM11).

29 Undated 'Development of the forces 1923–7' (MA, MM/1, A/0876).

30 Ibid.

31 Ibid.

32 Maloney obituary, in *An t-Oglach*, 3 Oct. 1925; Operations order, 16 Sept. 1926, P. J. Hassett Papers (Capt. Eoin Hassett).

33 COD minutes, 3 Feb. 1926; COS to OC AC, 3 Feb. 1926 (MA, MS708).

34 Anthony P. Kearns, 'The Irish Air Corps: A History', *Scale Aircraft Modelling* 3: 10 (July 1981), p. 448.

35 Log books, B17–B22 (in my custody).

36 Report of military mission to USA, 1926–7, pp. 136–7 (MA, MM/3).

37 Eunan O'Halpin, *Defending Ireland: The Irish State and its Enemies since 1922* (Oxford, 1999). p. 87; DFR 23/1929, amending orders No 3, 1 Dec. 1928; Peace Establishment (PE) 1931–2, pp. 20–3.

38 COD minutes, 4 Nov. 1929 (MA).

39 Ibid.

40 Kearns, 'Irish Air Corps', p. 449; Annex G, Report and finding, 10 Jan. 1942 (MA, ACS22–23).

41 Officers' history sheets (MA, SDR664; SDR4258).

42 Administrative order, 11 Sept. 1933, P. J. Hassett Papers.

43 Log books, Vespa I to VII (in my custody); PE, 22 Oct. 1934.

44 Duggan, *History of the Irish Army*, pp. 160–5.

45 ACS to OC AC, 25 Nov. 1931; ACS to OC AC, 28 Nov. 1931 (in my possession).

46 ACS to A/DMA, 11 Jan. 1932 (in my possession).

47 W. J. Keane to OC AC, 9 Feb. 1933 (in my possession).

48 Ibid.

49 W. J. Keane to OC AC, 20 July 1935 (in my possession).

50 COS to T. J. Maloney, 24 July 1925 (in my possession).

51 COD minutes, 3 Feb. 1926 (MA).

52 C. F. Russell to CSO GHQ, 7 Oct. 1926 (MA, 2/1113); DFR 7/1927, 18 Mar. 1927; Report and findings, 10 Jan. 1942, p. xxvii (MA, ACS22/23); Draft syllabus, 4 Nov. 1935 (in my possession).

53   Personal file (MA, SDR3693); Obituary, *Irish Independent*, 11 Mar. 1965; Personal comment, Agnes Russell, 2 June 2004.

54   CSO GHQ to OC AAC, 28 Feb. 1928 (in my possession).

55   Teddy Fennelly, *Fitz and the Famous Flight* (Portlaoise, 1997), pp. 135–51.

56   Ibid., pp. 167–82.

57   Col. W. J. Keane, 'The first class of cadets – 60 years ago', in *An Cosantóir*, 46: 3 (Mar. 1986), p. 10.

58   Officer's history sheet (MA, SDR 925).

59   Fennelly, *Fitz and the Famous Flight*, p. 279, citing no source.

60   Quigley, 'Air aspects of the emergency', *Irish Sword* xix: 75 & 76 (1993–40), p. 86, citing ACF/564 / DOD 2/49025 (MA). The investigation report of 1941–2 put the cost of the Air Corps for 1928–9 at £40,469.

61   COD minutes, 23 Mar. 1931 (MA).

62   Ibid., 23 June 1931 (MA).

63   Air Corps Routine order 243/37, 22 Oct. 1937, amending Air Corps Standing order 26 (in my possession).

64   Air Corps Routine order No. 148, Section 54, 29 June 1931 (in my possession).

65   COD minutes, 16 Nov. 1931 (MA).

66   PE 1931–2, pp. 20–3, 74; CV O/1662; CV O/287 (DFHQ, Commissioned officers record office).

67   CV O/1662, CV O/287 (DFHQ, CORO, 1 Sept. 2006); GRO 26/1932, 29 Sept. 1932.

68   Capt. P. Quinn to OC AC, 18 May 1933; Lt D. J. McKeown to OC No. 1 Sqn, 22 May 1933; 2/Lt T. J. Hanley to OC No. 1 Sqn, 22 May 1933; Lt M. J. Cumiskey to OC No. 1 Sqn, 22 May 1933; OC No. 1 Sqn to OC AC, 23 May 1933 (MA, AC/2/6/3).

69   Capt. W. P. Delamere to OC AC, 29 May 1933 (MA, AC/2/6/3).

70   OC AC to DMA, 7 July 1933 (MA, AC/2/6/3).

71   Undated tests as per DFR7/1927 (MA, AC/1/7/3); Aerial firing and bombing, July 1932 (MA, 2/30989).

72   HMSO receipt for books and manuals, 23 Sept. 1930; Saorstát Éireann, 'B' Licence No. 2, 1 Nov. 1930, P. J. Hassett Papers.

73   Log book, P. J Hassett Papers.

74   Personal comment, Pierce Cahill.

75   P. J. Hassett, unpublished memoir, *c*.1959.

76   Ibid.

77   Report and findings, p. lix, 10 Jan. 1942 (MA, ACS22/23).

78   Ibid.

79   P. J. Hassett, unpublished memoir, *c*1959.

80   Copy annual confidential report, 6 Jan. 1936.

81   Copy letter, P. J. Hassett , 13 Jan. 1936.

82   Defence order 5/1922; DFR41/1928; DFR55/1929; GRO4/1933; GRO2/1935: DFRA5, 10 Apr. 1937.

83   Files 2/30989; AC/2/6/3 (MA).

84   Section 122, Defence Forces (Temporary Provisions) Act, 1923–35.

85   Ibid..

86   Personal comment, Col. Roger McCorley.

87 Report and findings, 10 Jan. 1942, LXIV–LXV (MA, ACS22/23).
88 CV O/4431 (DFHQ, CORO).
89 Standing Orders, 3 June 1935 (in my possession).
90 AM, Intelligence notes, Nov. 1940 (TNA, Air 10/3990).
91 Appreciation by Col. C. M. Mattimoe in *An Cosantóir*, 47, No. 5 (May 1987), p. 22.
92 Col. C. F. Russell, 'The Army Air Corps', *Aviation* 1: 6 (June 1935), p. 209.
93 Ibid.
94 OC AC to COS, 23 Sept. 1935 (MA, AC/1/7/10).
95 OC AC to CSO, DOD, 4 Nov. 1935 (MA, AC/1/7/10); DFR7/1927, 18 Mar. 1927.
96 Paragraph 3, DFR40/1936, 21 May 1936.
97 T. J. Hanley to AC investigation, 17 Apr. 1941 (MA, ACS22/23).
98 Ibid.
99 COS to Major P. A. Mulcahy, 27 Aug. 1936 (MA, SDR1892); Log book Avro Cadet No. 5 (in my custody).
100 Appendix XVIII (B), Report and findings, 10 Jan. 1942 (MA, ACS22/23).

SIX: PILOT INTAKE, 1922–45

1 Maryann Gialanella Valiulis, *Almost a Rebellion: The Irish Army Mutiny of 1924* (Cork, 1988), passim.
2 Col. J. C. Fitzmaurice, unpublished memoir, p. 143 (Estate of P. Selwyn-Jones).
3 Ibid., pp. 121–3.
4 Herman Koehl, James C. Fitzmaurice and Guenther Von Huenefeld, *The Three Musketeers of the Air* (New York, 1928), p. 149.
5 Provisional Government decision, PG101, 26 Aug. 1922 (NAI, DT, S.1302).
6 Michael O'Malley, 'The military air service 1921–1924', Appendix 4 (BA thesis, NUIM, 2002).
7 GRO no. 9, 20 Dec. 1922.
8 Conference minutes, 16 Aug. 1922 (UCDA, MP, P7/B/49/48).
9 Log Book, Avro II (in my custody).
10 Record of pilot intake to Air Corps; Log Books, Avro I; II; III; IV (in my custody).
11 Ibid.
12 Log Book, Lt T. J. Nevin (Nevin family).
13 Log Book, Capt. D. J. McKeown (Mr P. Molloy, Celbridge).
14 Nevin family papers (Nevin family).
15 OC AAS report, 23 Mar. 1924 (MA, A/11270).
16 Record of pilot intake to Air Corps (in my custody); GROs 1922–4; Staff Duty memos, 1923–4.
17 Routine order, 25 Nov. 1925 (in my possession).
18 Examination results, c.June 1928 (MA, AC/1/7/3).
19 O'Duffy scheme, explanatory notes, 1 July 1924, p. 37 (NAI, S.3442B).
20 Ibid.
21 Ibid., p. 36.
22 AFO to Sec. DF, 20 Nov. 1924 (NAI, DF, S.004/383/24).

23 AFO to Sec. DF, 1 May 1925 (NAI, DF, S.004/383/24).

24 DF to AFO, 18 May 1925 (NAI, DF, S.004/383/24).

25 Draft regulation, 25 Sept. 1925 (NAI, DF, S.004/383/24).

26 Extract, *Limerick Leader*, 3 Oct. 1925 (NAI, DF, S.004/383/24).

27 Extract, *Irish Times*, 30 Sept. 1925 (NAI, DF, S.004/383/24).

28 Advertisement, *An t-Oglach* III: 20 (3 Oct. 1925), p. 17; Board report, 26 Jan. 1926 (MA, 2/1113).

29 COS to AFO, 6 Mar. 1925 (NAI, DF, S.004/383/24).

30 DF memo., 19 Apr. 1926 (NAI, DF, S.004/383/24).

31 Board report, 26 Jan. 1926 (MA, 2/1113).

32 AFO to DF, 24 Feb. 1926 (NAI, DF, S.004/383/24).

33 DF memo., 19 Apr. 1926 (NAI, DF, S.004/383/24).

34 *Defence Forces Army List and Directory, 1926 (An tOglach, 1926)*, p. 136.

35 DF to AFO, 22 Apr. 1926; AFO to DF, 22 May 1926 (NAI, DF, S.004/383/24).

36 DF to AFO, 12 Nov. 1926 (NAI, DF, S.004/383/24).

37 AFO to Sec DF, 3 Dec. 1926 (NAI, DF, S.004/383/24).

38 OC AAC to CSO GHQ, 4 June 1926 (in my possession).

39 Lt P. Quinn was born on 10 June 1899.

40 Record of pilot intake into Air Corps (in my custody); Strength return, 11 June 1926 (MA, LS8–9).

41 Defence Forces (Organisation) Order, 1 Oct. 1924, pp. 34–5; Strength returns (MA, LS8–9).

42 Undated question papers; Examination results, 1 June 1926, P. J. Hassett Papers; Record of pilot intake to Air Corps (in my custody).

43 OC AC to CSO GHQ, 7 Oct. 1926 (MA, 2/1113).

44 DFR7/1927, 18 Mar. 1927.

45 Sec. DOD to Sec. DF, 27 Oct. 1928 (MA, 2/1113).

46 Tests as per DFR7/1927, c.June 1928 (MA, AC/1/7/3); Files ACS/103; ACS/103/11/2; ACS/103/5/1; ACS/177/11; ACS/14/2; [ACS] SI 109/1(courtesy of School Commandant, 2005).

47 Lt Col. P. J. Hassett, unpublished memoir, c.1959.

48 Sec. DOD to Sec. DF, 27 Sept. 1928 (NAI, DF, S.004/383/24).

49 DFR23 of 1929, 1 Dec. 1928; PE 1931–2.

50 Col. W. J. Keane, 'The first class of cadets – 60 years ago', *An Cosantóir* No 156: 3 (Mar., 1986), p. 10.

51 Log book, Capt. M. J. O'Brien (in my possession).

52 Sec. DOD to Sec. DF, 8 May 1933; Sec. DF to Sec. DOD, 11 May 1933 (NAI, DF, S.004/0060/33).

53 Sec. DOD to Sec. DF, 13 Apr. 1933 (NAI, DF, S.004/0052/33).

54 Ibid.

55 DF memo., W. Doolin to E. O'Neill, 24 Apr. 1933 (NAI, DF, S.004/0052/33).

56 Memorandum, Army Air Corps, 16 Jan. 1934, (NAI, DF, S.004/0165/33); 'Directory of Cadet School graduates', *An Cosantoir* XXXIX (Sept. 1979), pp. 287–93.

57 'Directory of Cadet School graduates', *An Cosantóir* XXXIX: 9 (Sept. 1979), p. 288.

58 DOD to Sec. EC, 24 Aug. 1935 (MA, 2/29679).

59 DF memorandum, 31 Mar. 1933; DF to MFD, 13 May 1933 (NAI, DF, S.004/0034/33).

60 EC agenda, 23 May 1933; Sec. EC to Private Sec. MFF, 25 May 1933; Extract, Iris Oifigiúil,

26 May 1933 (NAI, DF, S.004/0034/33).

61 Sec. DOD to Sec. DF, 12 Dec. 1933 (NAI, DF, S.004/0052/33).

62 DOD to Sec., EC, 15 Nov. 1934 (MA, 2/39263).

63 Sec. DOD to Sec. DF, 11 Dec. 1933 (NAI, DF, S.004/0052/33).

64 Ibid.

65 Ibid., 31 Jan. 1935 (NAI, DF, S.004/0165/33).

66 Record of pilot intake into Air Corps (in my custody).

67 Sec. DOD to Sec. DF, 21 Mar. 1936 (NAI, DF, S.004/0165/33).

68 Ibid., 25 June 1935 (NAI, DF, S.004/0165/33; PE 1934; Record of pilot intake into Air Corps) (in my custody).

69 OC AC to COS, 23 Sept. 1935 (MA, AC/1/7/10).

70 OC AC to CSO DOD, ACF/36/24 dated 25 Sept. 1936; Draft 'Syllabus of young officers' flying training', 25 Sept. 1936 (in my possession).

71 Memo., Comdt E. Rooney to OC AC, 28 Apr. 1936 (in my possession).

72 P. A. Mulcahy to CSO, DOD, ACF/36/24 dated 25 Sept. 1936 (in my possession).

73 Memo., 7 Jan. 1937, ACS/103/11/2 (School Commandant); Record of pilot intake to Air Corps (in my custody); 'Directory of Cadet School graduates', *An Cosantóir* XXXIX: 9 (Sept. 1979), p. 288.

74 School Commandant to OC AC, 7 Feb. 1938 (MA, AC/2/6/15).

75 Minutes, 17 July 1937, Appendix I to Report and findings, 10 Jan. 1942 (MA, ACS22/23).

76 Capt. J. Devoy to AC investigation, 20 Mar. 1941 (MA, ACS22/23).

77 Lt J. Devoy to OC AC, 21 Apr. 1939, Appendix XVII (A), Report and findings 10 Jan. 1942 (MA, ACS22/23).

78 Capt. J. Devoy to AC investigation, 20 Mar. 1941 (MA, ACS22/23).

79 W. P. Delamere to OC AC, 7 July 1939, Appendix XVII (B), Report and findings, 10 Jan. 1942 (MA, ACS22/23).

80 Capt. J. Devoy to AC investigation, 20 Mar. 1941 (MA, ACS22/23).

81 P. A. Mulcahy to AC investigation, 24 Oct. 1941 (MA, ACS22/23).

82 Draft syllabus, 25 Sept. 1936 (in my possession); AP1234, *Manual of air navigation*, vol. I (London HMSO, 1935).

83 Record of pilot intake to Air Corps (in my custody); PE 1934, 22 Nov. 1934; Amendment 14 to PE 1934, 5 Apr. 1937.

84 'Record of pilot intake to Air Corps' (in my custody).

85 OC AC to COS, 10 Feb. 1937; ACF/564/1 (School Commandant).

86 Handwritten note, 12 Feb. 1937, ACF/564/1 (School Commandant).

87 P. A. Mulcahy to COS, 11 Jan. 1938 (School Commandant).

88 Ibid.

89 'Directory of Cadet School graduates', *An Cosantóir* XXXIX: 9 (Sept. 1979), pp. 288–9; Record of pilot intake to Air Corps (in my custody).

90 OC AC to COS, 19 Apr. 1938; 6 May 1937 (School Commandant).

91 Sec. DOD to Sec. DF, 19 Oct. 1938 (NAI, DF, S.004/0093/38).

92 Ibid.

93 DF memo., 25 Oct. 1938 (NAI, DF, S.004/0093/38).

94 W. Doolin to MFF, 10 Nov. 1938 (NAI, DF, S.004/0093/38).

95 Ibid.

96 Sec. DF to Sec. DOD, 23 Nov. 1938 (NAI, DF, S.004/0093/38).

97 MFF to MFD, 17 Apr. 1939 (NAI, DF, S.004/0093/38).

98 Sec. DOD to Sec. DF, 19 Apr. 1939 (NAI, DF, S.004/0093/38).

99 Ibid.

100 OC AC to CSO DOD, 10 May 1938; 16 May 1938; Draft DFR, 27 June 1938; OC AC to CSO, 30 Nov. 1938; Draft DFR, Feb. 1939; RAF advertisement, unidentified newspaper (School Commandant, ACF/564/1).

101 OC AC to COS, 15 Sept. 1938 (in my possession); OC AC to COS, 8 Nov. 1938 (School Commandant, ACF/564/1); Table 33P, PE 1939; Table 36W, WE 1940.

102 OC AC to COS, 29 Mar. 1939 (School Commandant, ACF/564/1).

103 Ibid., 28 June 1939; Convening order, 15 July 1939 (School Commandant, ACF/564/1).

104 Bernard Share, *The Flight of the Iolar: The Aer Lingus Experience 1936–1986* (Dublin, 1986), p. 34.

105 School Commandant to OC AC, 23 Jan. 1940 (School Commandant, ACF/564/1).

106 Sec. DOD to Sec. DF, 19 Feb. 1940 (NAI, DF, S.004/0093/38).

107 OIC Records to E. Comd., 4 Apr. 1940 (School Commandant, ACF/564/1).

108 Sec. DOD to Sec. DF, 10 Nov. 1939 (NAI, DF, S.004/0093/38).

109 DF to Sec. DOD, 30 Apr. 1940 (NAI, DF, S.004/0093/38).

110 Sec. DOD to Sec. DF, 9 May 1952 (NAI, DF, S.004/0093/38).

111 CFI to School Commandant, 9 June 1944, Sergeant pilots' course file (School Commandant); Table 76, 1943 establishment, 29 Mar. 1943.

112 Aer Lingus to W. J. Keane, 7 June 1951 (in my possession); Record of pilot intake to Air Corps (in my custody).

SEVEN: AVIATION POLICY AND PLANNING, 1935–40

1 P. Young, 'Defence and the new Irish state, 1919–39', *Irish Sword* XIX (1993–4), pp. 1–10, passim.

2 K. J. Meekcoms and E. B. Morgan, *The British Aircraft Specifications File* (Tonbridge, 1994), p. 86.

3 M. J. Costello to AC investigation, 18 Feb. 1941 (MA, ACS22/23).

4 Theo Farrell, 'Professionalisation and suicidal defence planning by the Irish Army, 1921–1941', *Journal of Strategic Studies* XXI: 3 (Sept. 1998), pp. 67–85, passim.

5 'Fundamental factors affecting Saorstát defence problem', May 1936 (MA, G2/0057).

6 E. O'Halpin, *Defending Ireland: The Irish State and its Enemies Since 1922* (Oxford, 1999), p. 136, citing 'Fundamental factors'.

7 'Fundamental factors', May 1936 (MA, G2/0057).

8 Ibid.

9 Farrell, 'Professionalisation and suicidal defence planning', pp. 67–85, passim.

10 COS to MFD, 22 Sept. 1936 (UCDA, MacEntee Papers, P67/191).

11 J. P. Duggan, *A History of the Irish Army* (Dublin, 1991), p .165; Annex G, Report and findings, 10 Jan. 1942 (MA, ACS22/23).

12 O'Halpin, *Defending Ireland*, p. 139. See also Theo Farrell, 'Professionalisation and suicidal defence planning'.

13 Society of British Aircraft Constructors (SBAC) Display reports, 17 July/8 Aug. 1935 (MA, AC/1/9/9).

14 QMG to OC AC, 1 Apr. 1936 (MA, AC/2/2/7).

15 OC AC to QMG, 1 Aug. 1936 (MA, AC/2/2/7).

16 T. J. Hanley to AC investigation, 23 Jan. 1941 (MA, ACS22/23).

17 A. P. Kearns, 'The Irish Air Corps: A History', *Scale Aircraft Modelling* III: 10 (July. 1981), p. 449; 459; Report and findings, 10 Jan. 1942 (MA, ACS22/23), passim.

18 Amendment 14 to PE 1934, 1 Apr. 1937.

19 Minutes, 17 July 1937, Appendix 1, Report and findings (MA, ACS22/23).

20 Ibid.

21 Ibid.

22 Ibid.

23 Ibid.

24 Estimate of defence requirements, CID, Feb. 1938; Appendix dated 21 Jan. 1938 (UCDA, MacEntee Papers, P67/192).

25 Ibid.

26 O'Halpin, *Defending Ireland*, p. 141.

27 P. A. Mulcahy to AC Investigation, 22 Jan. 1941 (MA, ACS22/23).

28 Ibid.

29 ACS to OC AC, 21 Mar. 1938, Appendix 11, Report and findings, 10 Jan. 1942 (MA, ACS22/23).

30 Ibid.

31 Ibid.

32 P. A. Mulcahy to AC investigation, 22 Jan. 1941, quoting OC AC to ACS, 21 Apr. 1938 (MA, ACS22/23).

33 P. A. Mulcahy to ACS, 21 Apr. 1938, Appendix III, Report and findings (MA, ACS22/23).

34 Air Corps – Capital Expenditure, *c.*June 1938 (UCDA, MacEntee Papers, P67/194 (34)).

35 Costello to AC Investigation, 18 Feb. 1941 (MA, ACS22/23).

36 Ibid.

37 Ibid.

38 Ibid.

39 P. A. Mulcahy to AC Investigation, 22 Oct. 1941 (MA, ACS22/23).

40 COS to MFD, 21 May 1938 (UCDA, MacEntee Papers, P67/193 (2)).

41 Ibid.

42 P. A. Mulcahy to AC Investigation, 22 Jan. 1941 (MA, ACS22/23).

43 COS to MFD, 21 May 1938 (UCDA, MacEntee Papers, P67/193 (4)).

44 Peter Young, 'Defence and the Irish State 1919–39', pp. 1–10.

45 Defence estimates, 1939–40 (NAI, DF, F102/0065/38).

46 Ibid.

47 Cost of AAC, 1939–40, Annex G, Report and findings (MA, ACS22/23).

48 Ibid., 1937–8, Annex G, Report and findings (MA, ACS22/23).

49 Estimates, 1938–9 (NAI, DF, F102/0065/38); Sec. DF to Sec. DOD, 28 June 1938 (MA, 2/54453 part 11).

50 Cost statement of AAC 1938–9, 10 Jan. 1942, Annex G, Report and findings (MA, ACS22/23).

51 OC AC to COS 12 July 1938 (MA, 2/54453 part 11); Kearns, 'Irish Air Corps', p. 459.

52 Kearns, 'Irish Air Corps', p. 459; Estimates 1938–9 (MA, AC/2/211); Cost statement of AAC, 1939–40, Annex G, Report and findings (MA, ACS22/23).

53 OC AC to COS, 14 Mar. 1938, Appendix IV, Report and findings (MA, ACS22/23).

54 ACS to OC AC, 21 Mar. 1938.

55 OC AC to ACS, 26 Mar. 1938, Appendix V, Report and findings (MA, ACS22/23).

56 Mulcahy to AC Investigation, 22 Jan. 1942 (MA, ACS22/23).

57 Meekcoms and Morgan, *British Aircraft Specifications File*, p. 206.

58 Kearns, 'Irish Air Corps', p. 449.

59 Table 32P, PE1939.

60 OC AC to COS, 24 May 193 (MA, AC/2/8/1); Mulcahy to AC Investigation, 21 Jan. 1941 (MA, ACS22/23).

61 Mulcahy to AC Investigation, 21 Jan. 1941 (MA, ACS22/23).

62 OC AC to ACS, 21 Apr. 1938, Appendix. III (H), Report and findings (MA, ACS22/23).

63 OC AC to COS, 15 Sept. 1938 (MA, 2/57617).

64 Ibid.

65 ACS to COS; COS to Sec. DOD, 29 Sept. 1938 (MA, 2/57617).

66 Table 34P, Air Corps PE, 6 Mar. 1939.

67 Sec. DOD to DF, 25 Mar. 1939 (MA, 2/57617).

68 Memorandum on Army re-organisation, 26 June 1939, quoted in Report and findings, p. viii (MA, ACS22/23).

69 Air Corps WE 1940.

70 Michael O'Malley, 'Military aviation in Ireland 1921–1945', PhD thesis, NUIM, 2007, pp. 222–49.

71 Memorandum for government, 6 Sept. 1938; File memos dated 20 Sept., 11 Oct. and 18 Oct. 1938 (NAI, DT, S.10, 823).

72 Minutes, 17 July 1937, Appendix 1, Report and findings (MA, ACS22/23).

73 ACS to OC AC, 21 Mar. 1938, Appendix 11, Report and findings (MA, ACS22/23); M. J. Costello to AC Investigation, 18 Feb. 1941 (MA, ACS22/23).

74 OC AC to ACS, 21 Apr. 1938, Appendix III (A), Report and findings (MA, ACS22/23).

75 DF memo dated 6 Feb. 1939 (NAI, DT, S.11,101).

76 Sec. DOD to Sec. DF, 17 Dec. 1938 (NAI, DF, S.007/0009/39).

77 Ibid.

78 Sec. DOD to Sec. DF, 19 Apr. 1939 (NAI, DF, S.007/0009/39).

79 Karl E. Hayes, *A History of the Royal Air Force and US Naval Air Service in Ireland 1913–23* (Irish Air Letter, 1988), p. 85.

80 Sec. OPW to Sec. DF, 4 July 1939 (NAI, DF, S.007/0009/39).

81 Ibid., 19 July 1940 (NAI, DF, S.007/0009/39).

82 OC AC to COS, 29 Aug. 1939 (MA, AC/2/9/12).

83 OPW memo, 31 Oct. 1941 (MA, 2/72465 part 111).

84 Cost of Army Air Corps 1926–7 to 1940–1, Annex G to Report and findings, 10 Jan. 1942 (MA, ACS22/23).

85 Costello to AC Investigation, 18 Feb. 1941 (MA, ACS22/23).

86 Ibid.

87 DO note, 7 May 1940 (TNA, DO 35/1078/3)

88　Duggan, *A History of the Irish Army*, p. 187.

89　OC AC to COS, ACF/631 dated 20 Sept. 1939 (in my possession). P. A. Mulcahy had been promoted to colonel on 6 March 1939.

EIGHT: SUPPORT SERVICES

1　Frank Clabby, 'The Met. Office at Baldonnel', unpublished paper, 14 Nov. 1986 (in my possession); Andy Roche, 'The Air Corps' in Lisa Shields (ed.), *The Irish Meteorological Service: The First Fifty Years 1936–1986* (Dublin, 1987), p. 83.

2　Roche, 'The Air Corps', pp. 82–4; Lt Col. M. O'Malley, 'In the beginning', in Capt. David Swan (ed.), *Irish Air Corps: Celebrating 30 Years of Helicopter Operations 1963–1993* (Defence Forces, 1993), pp. 3–4.

3　Report on meteorological services, Appendix II, 6 July, 1925 (MA, MS/418); Shields (ed.), *The Irish Meteorological Service*, p. 1.

4　Capt. J. A. McNamara to CIC, 25 July 1922 (UCDA, MP, P7/B/49/29–30).

5　Minister for Agriculture to CIC, 10 Aug. 1922 (UCDA, MP, P7/B/10/16).

6　AM to MFD, 4 May 1923 (NAI, FIN 1/2976).

7　AM to Sec. DOD, 30 June 1923 (NAI, FIN 1/2976).

8　AFO to Sec. AM, 18 July 1923 (NAI, FIN 1/2976).

9　AM to Sec. DOD, 9 Aug. 1923 (NAI, FIN 1/2976).

10　AFO to Sec. DF, 28 Aug. 1923; Sec DF to AFO, 31 Sept. 1923 (NAI, FIN/2976).

11　T. J. Maloney to MS, 2 June 1924; MS to T. J. Maloney, 6 June 1924 (MA, MS/418).

12　Memorandum, T. J. Maloney, May 1925 (MA, MS/418).

13　Memorandum, Rev. W. O'Riordan, 3 Oct. 1935, quoting Report of Inter-Departmental Committee on Meteorology, 7 May 1925 (MA, 2/27175).

14　E. Cannon to OC AC, 18 Aug. 1925 (in my possession).

15　File memo, 11 Sept. 1925 (in my possession).

16　Standing order, Sept. 1924 (MA, MS/685).

17　Duty Officer's Report, 30 June 1925 (MA, MS/685).

18　A/OC AC to DOD, 21 July 1928; OC AC to Rev. W. O'Riordan, 9 Mar. 1936 (in my possession).

19　Rev. W. M. O'Riordan to A/OC AC, 30 June 1928; A/OC AC to CSO DOD, 2 July 1928 (ACF/338, in my possession).

20　Sec. DOD to Sec. DF, 25 Nov. 1929; Sec. DF to Sec. DOD, 27 Nov. 1929 (NAI, DF, S.008/0076/29).

21　Memorandum, 2 Dec. 1933 (NAI, DF, S.008/0076/29).

22　Sec. DOD to Sec. DF, 18 Aug. 1936; Sec. DF to Sec. DOD, 21 Aug. 1936 (NAI, DF, S.008/0076/29).

23　J. P. Twohig to OC AAC, 20 Mar. 1930 (in my possession).

24　OC AC to Signals' Corps, DOD, 31 July 1930 (in my possession).

25　DMA to DOD, 25 Jan. 1932 (in my possession).

26　DS to DMA, 16 Nov. 1932; DMA to Valentia Observatory, 24 Feb. 1932; M. T. Spence to DMA, 26 Feb. 1932; DMA to ACS, 7 Mar. 1932 (in my possession).

27　O'Riordan to Adjt AC, 23 July 1935; O'Riordan to OC AC, 1 Mar. 1934; O'Riordan to OC AC, 16 Mar. 1934 (in my possession).

28 O'Riordan to Adjt, 23 July 1935; OC AC to OC AC Depot, 23 July 1935; O'Riordan to Adjt, 25 July 1935; Adjt to OC AC, 27 July 1935 (in my possession).

29 Memorandum, Rev. W. O'Riordan, 3 Oct. 1935 (MA, 2/27175).

30 Ibid.

31 Ibid.

32 P. A. Mulcahy to COS, 4 Oct. 1935 (MA, 2/27175).

33 File memo, P. A. Mulcahy, 16 Oct. 1935 (in my possession).

34 COS to Comdt G. Carroll, 31 Oct. 1935 (in my possession).

35 O'Riordan to OC AC, 8 Mar. 1936 (in my possession).

36 Ibid.

37 OC AC to Rev. W. O'Riordan, 9 Mar. 1936) (in my possession).

38 Ibid.

39 Dermot O'Connor, 'A brief history of the meteorological service', in Shields (ed.), *The Irish Meteorological Service*, p. 1.

40 Shields (ed.), *The Irish Meteorological Service*, passim.

41 COE to OC AC, 21 Jan. 1942 (MA, AC/2/4/6).

42 File memo, 24 May 1943 (MA, AC/2/4/6).

43 1943 WE; Baldonnel – Wind analysis, 1 July 1944–30 Sept. 1948 (in my possession); File memo, 1 Oct. 1943 (in my possession); Marty Keane, 'The weather observation network', in Shields (ed.), *The Irish Meteorological Service*, p. 26.

44 R. W. O'Sullivan to OC AC, 12 Feb. 1944 (in my possession).

45 Ibid.

46 A. H. Nagle to OC AC, 28 Oct. 1944 (in my possession).

47 Ibid., 15 Jan. 1945 (in my possession).

48 OC AC to ACS, 24 July 1957 (DOD, 2/93247).

49 T. J. Gray to Sec DOD, 29 Aug. 1957, quoting Sec. I & C to Sec. DOD, 27 July 1945 (DOD, 2/93247).

50 Ibid.

51 Duties of Duty Officer, Sept. 1924 (MA, MS/685).

52 Duties of Air Officer, Standing orders, 1 Feb. 1927, P. J. Hassett Papers.

53 Standing Orders, Baldonnell Aerodrome, 1 Jan. 1928, P. J. Hassett Papers.

54 *Baldonnel: Dublin's Airport 1919–1939* (Irish Air Letter, 1989), passim.

55 Air Corps Standing orders, 1935 (in my possession).

56 Corps Routine Order No. 243, 22 Oct. 1937 (in my possession).

57 Sec. DOD to Sec. DEA, 9 Nov. 1935 (MA, AC/1/1/27).

58 HC to Sec DEA, 5 Dec. 1935 (MA, AC/1/1/27).

59 Lt E. F. Stapleton and Lt M. J. Cumiskey to A/OC AC, 30 Jan. 1936, AC/ 1/1/27, MA.

60 Standing Orders, 1935, amended 22 Oct. 1937 (in my possession).

61 Control Officers, Foynes, 5 Jan. 1944; Control Officers, Aug. 1943: ATC Officer to OC AC, 4 Jan. 1946; Record of Control Officers, 5 Jan. 1944 (in my possession).

62 Peter Tormey and Kevin Byrne, *Irish Air Corps: A View from the Tower* (Defence Forces, 1988) p. 25.

63 DFR 7/1927, 18 Mar. 1927.

64 PE, 1931–2.

65 OC AC to Director of Training, 27 Aug. 1932 (MA, 2/30989).

66 Tactical exercise, 25 July 1932 (MA, 230989); Air patrols, 21 Aug. 1933, P. J. Hassett Papers.

67 Seán Ó hAllmhúráin (ed.), *Aviation Communications Service 1936–1986* (Department of Communications, 1986), pp. 10–12.

68 Civil Aviation Notice No. 3 of 1936, 15 May 1936.

69 Committee of Investigation to P. J. Murphy, 30 Jan. 1941 (MA, ACS22/23).

70 P. J. Murphy to AC investigation, 30 Jan. 1941 (MA, ACS22/23); Annex III to Operations Order 1/1940, 28 May 1940 (MA, EDP 1/1); Fighter Squadron, 16 Dec. 1940 (in my possession).

71 P. J. Murphy to AC Investigation, 30 Jan. 1941 (MA, ACS22/23).

72 Ibid; 1940 WE.

73 P. J. Murphy to AC Investigation, 30 Jan. 1941 (MA, ACS22/23).

74 T. J. Hanley to AC Investigation, 23 Jan. 1941 (MA, ACS22/23).

75 P. J. Murphy to AC Investigation, 30 Jan. 1941 (MA, ACS22/23).

76 Carmella Corbett, 'History of the service 1936–1986', in Ó hAllmhúráin (ed.), *Aviation Communications Service*, pp. 6–17; Radio stations, Shannon, 14 July 1939 (MA, EDP/30).

77 J. Devoy to OC AC, 21 Apr. 1939, Report and finding (MA, ACS22/23).

78 Ibid.

79 T. J. Hanley to AC Investigation, 23 Jan. 1941 (MA, ACS22/23).

80 Report to COS, 22 Dec. 1939 (MA, AC/2/8/4).

81 Report and findings, 10 Jan. 1942, p. liv (MA, ACS22/23).

82 T. J. Hanley to OC AC, 6 Jan. 1940 (MA, AC/2/8/4).

83 Ibid.

84 A/OC AC to DS, 17 Jan. 1940 (MA, AC/2/8/4).

85 P&T to DOD, 4 Mar. 1940 (MA, AC/2/8/4).

86 OC AC to DS, 26 Mar. 1940 (MA, AC/2/8/4).

87 P. J. Murphy to OC AC, 19 Apr. 1940 (MA, AC/2/8/4).

88 I & C to OC AC, 8 Oct. 1940, (MA, AC/2/4/29).

89 OC AC to DS, 12 Oct. 1940 (MA, AC/2/4/29).

90 I & C to OC AC, 6 Nov. 1940 (MA, AC/2/4/29).

91 OC AC to I & C, 13 Nov. 1940 (MA, AC/2/4/29).

92 OC AC Coy, SC, to OC AC, 12 Sept. 1941 (MA, AC/2/4/29).

93 A. C. Woods to OC CP Sqn, 20 Nov. 1940; OC CP Sqn to AC Signals' Officer, 20 Nov. 1940 (MA, AC/2/4/29).

94 T. J. Hanley to OC AC, 9 Jan. 1941 (MA, AC/2/4/29).

95 DF log, 21 Nov. 1940; A. C. Woods to OC CP Sqn, 29 Nov. 1940 (MA, AC/2/4/29).

96 AC Signals' Officer to OC AC, 23 Nov. 1940 (MA, AC/2/4/29).

97 OC AC to OC FS, 28 Nov. 1940 (MA, AC/2/4/29).

98 OC AC to OC CP Sqn 28 Nov. 1940 (MA, AC/2/4/29).

99 A. C. Woods to OC CP Sqn 29 Nov. 1940 (MA, AC/2/4/29).

100 Capt. T. J. Hanley to OC AC, 9 Jan. 1941 (MA, AC/2/4/29).

101 T. J. Hanley to AC Investigation, 23 Jan. 1941 (MA, ACS22/23).

102 Report and findings, 10 Jan. 1942, pp. lxix–lxx (MA, ACS22/23).

103 I & C, 'Civil Aviation Notice, No. 3 of 1936, 15 May 1936.

104 Radio stations for Shannon Airport, 14 July 1939 (MA, EDP/30).

105 P. J. Murphy to AC Investigation, 30 Jan. 1941 (MA, ACS22/23).

106 Appendix to report, 18 Mar. 1944 (TNA, PREM 3/133/3).

107 Memo, 21 Feb. 1945 (TNA, DO 114/117).

108 Maffey to Machtig, 6 Mar. 1945 (TNA, DO 35/2117).

NINE: THE AIR CORPS' EMERGENCY

1 OC AC to COS, ACF/631 dated 20 Sept. 1939 (in my possession); Record of pilot intake into Air Corps (in my custody); T. J. Hanley to AC investigation, 17 Apr. 1941 (MA, ACS22/23).

2 Officer's history sheet O/287 (COMO, DFHQ); T. J. Hanley to AC investigation, 12 Nov. 1941 (MA, ACS22/23); OC AC to OC E. Comd., ACF/495/1 dated 2 Jan. 1946 (in my possession).

3 Appointments officers, 5 May 1939; OC AC to COS, 20 Sept., 1940; Organisation charts, Fighter Sqn, 16 Dec. 1940; R & MB Sqn, 14 Jan. 1941; CP Sqn, Dec. 1940; Corps HQ, Workshops and Depot, 18 Dec. 1940; AC Schools, Dec. 1940 (in my possession).

4 OC AC to COS, ACF/631 dated 20 Sept. 1939 (in my possession).

5 Services rendered by Air Corps, 1 July 1937 to 31 Dec. 1944 (MA, EDP23/3); Record of control officers at Foynes, 1937 to 2 Jan. 1946, ACF/503/2; 1943 officers' appointments, undated (in my possession).

6 OC AC to COS, 29 Aug. 1939 (MA, AC 2/9/12).

7 P. A. Mulcahy to AC Investigation, 23 Oct. 1941 (MA, ACS22/23).

8 W. J. Keane to OC S. Comd, 12 Apr. 1940, Appendix XXII, Report and findings, 10 Jan. 1942 (MA, ACS22/23); Air Defence, Operations Order No 1/1940, 25 May 1940 (MA, EDP 1/1). Key to air and marine intelligence special map, G2 Branch, GHQ (in my possession).

9 File notes, J. E. Stephenson, 3 May 1940 (TNA, DO 35/1078/3).

10 Sec. DOD to Sec. DF, 30 Sept. 1939 (NAI, DF S.105/0048/38).

11 Michael Kennedy, *Guarding Neutral Ireland: The Coast Watching Service and Military Intelligence, 1939–1945* (Dublin, 2008), pp. 27–8.

12 Amendment 14 to 1934 PE, 1 Apr. 1937.

13 OC AC to ACS, 26 Mar. 1938, Appendix v, Report and findings (MA, ACS22/23).

14 Maj. P. A. Mulcahy to ACS, 21 Apr. 1938, Report and findings Appendix III (A) (MA, ACS22/23). The figures in brackets are estimated.

15 1940 WE.

16 Keane report, 12 Apr. 1940 (MA, ACS22/23).

17 Nine Ansons were distributed between R & MB Sqn, CP Sqn and AC Schools.

18 Strength return, 14 Jan. 1941 (in my possession).

19 Keane report, 12 Apr. 1940 (MA, ACS22/23).

20 Administrative diary, 30 Aug.; 1 Sept. 1939 (in my possession).

21 *Shannon Airport, 50 years of engineering, 1937–1987* (Aer Rianta, 1987), passim.

22 Administrative diary, 1 Sept. 1939 (in my possession).

23 Keane report, 12 Apr. 1940 (MA, ACS22/23); Gun History Sheet, B.18064 (in my possession).

24 Sec. DOD to Sec. DF, 22 Sept. 1939; Memo, 29 Sept. 1939 (NAI, DF, S. 007/0024/39).

25 Sec. DF to Sec. OPW, 30 Sept. 1939 (NAI, DF, S.007/0024/39).

26 Sec. OPW to Sec. DF, 25 Oct. 1939 (NAI, DF, S.007/0024/39).

27 DF file memo, 3 Nov. 1939 (NAI, DF, S.007/0024/39).

28 Sec. DF to Sec. OPW, 31 Oct. 1939; Sec. DF to Sec. OPW, 4 Nov. 1939 (NAI, DF, S. 007/0024/39).

29 OPW, Annual estimates, 1940–4, 3 Nov. 1939 (NAI, S.007/0024/39).

30 Extract, list of OPW contracts, 16 Jan. 1940 (NAI, DF, S.007/0024/39).

31 Airport Construction Committee (ACC) to Sec. I & C, 26 Oct. 1939 (NAI, DF, S.007/0024/39).

32 Sec. DOD to Sec. DF, 26 Feb. 1940 (NAI, DF, S.007/0024/39).

33 DF to OPW, 11 Mar. 1940 (NAI, DF, S.007/0024/39).

34 Sec. OPW to Sec. DOD, 11 Apr. 1940 (NAI, DF, S.007/0024/39).

35 Sec. DF to Sec. OPW, 11 Apr. 1940 (NAI, DF, S.007/0024/39).

36 Sec. OPW to Sec. DF, 4 Mar. 1941; Sec. DF to Sec. OPW, 25 Mar. 1941 (NAI, DF, S.007/0024/39).

37 Sec. DF to Sec. OPW, 4 July 1941 (NAI, DF, S.007/0024/39).

38 Accommodation Rineanna, undated (MA, File AC/2/9/12).

39 OPW memo, 31 Oct. 1941 (MA, 2/72456 part III).

40 Keane report, 12 Apr. 1940 (MA, ACS22/23).

41 K. J. Meekcoms and E. B. Morgan (eds), *The British Aircraft Specifications File: British Military And Commercial Aircraft Specifications 1920–1949* (Tonbridge, 1994), p. 213.

42 J. J. Halley, *Squadrons of the Royal Air Force* (Tonbridge, 1985), passim; Ray Sturtivant, *The History of Britain's Military Training Aircraft* (Yeovil, 1987), pp. 77–87; Michael Armitage, *The Royal Air Force: An Illustrated History* (London, 1993), p. 75.

43 Meekcoms and Morgan, *Specifications File*, p. 206.

44 Report and findings, pp. vii, ix (MA, ACS22/23).

45 Administrative diary, 31 Aug.; 4 Sept. 1939 (in my possession).

46 Ibid., 4 Sept. 1939.

47 DS to Capt. P. J. Murphy, 7 Sept. 1939 (MA, AC/2/8/4).

48 Ibid.; Capt. W. J. Keane to OC AC, 8 Sept. 1939; OC AC to ACS, 11 Sept. 1939 (MA, AC/2/8/4).

49 Signals' Technical Instruction No 24, 18 June 1941 (in my possession).

50 Keane report, 12 Apr. 1940 (MA, ACS22/23).

51 Sec. DOD to Sec. DF, 25 May 1940 (NAI, DF, S.008/0029/39).

52 Keane report, 12 Apr. 1940 (MA, ACS22/23); Sec. DOD to Sec. DF, 27 May 1940; Sec. DOD to Sec. DF, 16 Sept. 1941 (NAI, DF, S.008/0029/39).

53 OC AC to COS, 23 Nov. 1939 (MA, AC/2/2/35).

54 P. A. Mulcahy to AC investigation, 21 Jan. 1941 (MA, ACS22/23).

55 Ibid.

56 Keane report, 12 Apr. 1940 (MA, ACS22/23).

57 Sec. DEA to DO, 6 May 1940 (TNA, DO35/1078/3).

58 Costello to investigation, 18 Feb. 1941 (MA, SCS22/230).

59 OC S. Comd. to COS, 3 May 1940 (MA, PC586).

60 Ibid.

61 COS to OC S. Comd, 10 May 1940 (MA, PC586).

62 P. A. Mulcahy to AC investigation, 19 Nov. 1941 (MA, ACS22/23).

63 Capt. T. J. Hanley to AC investigation, 23 Jan. 1941 (MA, ACS22/23).

64 Report, 2 June 1940 (MA, AC/2/2/41).

65 Minutes of meeting, 10 May 1941; Secret Order, Col. P. A. Mulcahy, 22 May 1941; Movement Order 1/1941, W. J. Keane, 23 May 1941 (MA, AC/2/8/3).

66 OC AC to CSO DOD, 20 Sept. 1939 (in my possession).

67 Ibid., ACF/631 dated 20 Sept. 1939 (in my possession).

68 1940 WE, 13 June 1940.

69 Organisational, FS, 12 Dec. 1940 (in my possession).

70 Meekcoms and Morgan, *Specifications File*, pp. 211, 255; Armitage, *Royal Air Force*, pp. 78–9.

71 Halley, *Squadrons of the RAF*, passim.

72 Meekcoms and Morgan, *Specifications File*, p. 201.

73 Armitage, *Royal Air Force*, p. 83.

74 Halley, *Squadrons of the RAF*, p. 185.

75 P. A. Mulcahy to AC Investigation, 23 Oct. 1941 (MA, ACS22/23).

76 Ibid.

77 Report and findings, p. xx (MA, ACS22/23).

78 Meekcoms and Morgan, *Specifications File*, p. 269.

79 Sub-head 'O', Defence estimates 1939–40 (NAI, DF, F.102/0065/38).

80 Meekcoms and Morgan, *Specifications File*, p. 235.

81 Officer's history sheet, O/287, (COMO, DF HQ); P. A. Mulcahy to AC investigation, 21 Jan. 1941 (MA, ACS22/23).

82 Annex III to Operation Order No. 1/1940, 28 May 1940 (MA, EDP 1/1); Mulcahy to AC investigation, 21 Jan. 1941 (MA, ACS22/23).

83 'Outline of Emergency defence plan No. 1, May 1940 (MA, EDP 1/1).

84 Colm Mangan, 'Plans and operations', *Irish Sword* XIX: 75 & 76 (1993–4), pp. 48–9, citing 1 May 1940 (MA, EDP 1/1).

85 Operations Order No. 2/1940, 1st AA Brigade, 25 May 1940 (MA, EDP 1).

86 'Air Defence, Annex No. VI, Operations Order No. 1', 25 May 1940 (MA, EDP 1/1).

87 Ibid.

88 Armitage, *Royal Air Force*, pp. 95–112.

89 OC AC to COS, 16 July 1940, EDP/21; Appendix 1 (MA, General report on the Defence Forces, 1940–1).

90 Addendum No. 1 to Operations Order No. 1, 8 Aug. 1940 (MA, EDP 1/1).

91 Archer summary report, March 1944 (MA, SCS/14); Firing Record, Browning No. B19064 (in my possession).

92 A/OC AC to COS, 11 Sept. 1940 (MA, EDP/21).

93 Michael Kennedy, *Guarding Neutral Ireland* (Dublin, 2008) p. 162; 180.

94 Ibid., p. 166.

95 'Air Defence, Annex No. 2 to Operations Order 4/1941', 8 July 1941 (MA, EDP/21).

96 Air Commodore T. N. Carr to AM, 14 Oct. 1940 (TNA, Air 2/5130).

97 Annex No. III, 28 May 1940, Operations Order No. 1/1940, 24 May 1940 (MA, EDP 1/1).

98 Ibid.

99 Ibid.

100 'Landing fields', OC AC to Commands, 30 May 1940 (MA, EDP/4).

101 G1, GHQ to commands, 28 May 1940 (MA, EDP/4).

102 List of officers, 23 Dec. 1940 (MA, EDP/1/2); P. A. Mulcahy to AC Investigation, 22 Oct. 1941 (MA, ACS22/23).

103 'Supplementary to Annex III' to Operations Order 1/1940, *c.*15 June 1940 (MA, EDP/4 (1940)).

104 P. A. Mulcahy to AC Investigation, 21 Jan. 1941.

105 Ibid.

106 Robert Fisk, *In Time of War: Ireland, Ulster and the Price of Neutrality 1939–45* (London, 1983), pp. 234–40.

107 Estimate of the situation No. 2, 30 Oct. 1940 (MA, EDP/19).

108 P. A. Mulcahy to AC investigation, 21 Nov. 1941 (MA, ACS22/23).

109 Report and findings, 10 Jan. 1941, passim (MA, ACS22/23).

110 Ibid., p. lxix, 10 Jan. 1941 (ACS22/23).

TEN: SERVICES RENDERED

1 Facilities obtained from the Government of Éire during the war, 21 Feb. 1945 (TNA, DO 114–7).

2 Archer summary, Mar. 1944; Childers report, 17 Oct. 1947 (MA, SCS/14; SCS/1).

3 Sec DOD to Sec DF, 10 Dec. 1938 (NAI, DF, S.105/0048/38).

4 E. O'Halpin (ed.), *MI5 and Ireland, 1939–1945: The Official History* (Dublin, 2003), p. 22.

5 Note on meetings with Mr Dulanty and Mr Walshe on 11 and 12 Oct. 1938 (TNA, CAB 104/23).

6 Archer summary, Mar. 1944, (MA, SCS/14); 'Notes on the origin and development of contacts with British Army 1940–1945', Lt Col. R. A. Childers, 17 Oct. 1947 (MA, SCS/1). Hereafter Childers report.

7 Ibid.; See Robert Fisk, *In Time of War: Ireland, Ulster and the Price of Neutrality 1939–45* (London, 1983), pp. 233–6.

8 Cecil Liddell to CID, 20 May 1940 (TNA, CAB 104/184); O'Halpin, *MI5 and Ireland*, p. 53.

9 Cecil Liddell to CID, 20 May 1940 (TNA, CAB 104/184); Nigel West (ed.), *The Guy Liddell Diaries, vol. I: 1939–1942* (Abingdon, 2005), p. 79.

10 Archer summary, Mar. 1944 (MA, SCS/14); Childers report, 17 Oct. 1947 (MA, SCS/1).

11 'Report of 27 June 1940; 'Report No. 1', 11 June 1940 (TNA, Air 2/5129).

12 Childers report, 17 Oct. 1947 (MA, SCS/1).

13 Ibid.

14 Archer summary, Mar. 1944 (MA, SCS/14).

15 Childers report, 17 Oct. 1947 (MA, SCS/1).

16 Archer summary, Mar. 1944 (MA, SCS/14); Files 2/78084 and S.194 (MA).

17 Childers report, 17 Oct. 1947 (MA, SCS/1).

18 Maffey to DO, 29 May 1940 (TNA, Air 2/5129).

19 DO to Maffey, 28 May 1940 (TNA, Air 2/5129).

20 Undated wireless net, *c.*June 1940 (TNA, Air 2/4601); AM to AOC, Fighter Command, 11 July 1940 (TNA, Air 2/5185).

21 Memorandum, 27 June 1940 (TNA, CAB 104/184).

22 AM to Station AA 11 July 1940; Lywood to AM, 13 July 1940 (TNA, Air 2/5129).

23 E. O'Halpin, 'Aspects of intelligence', *Irish Sword* XIX: 75 & 76 (1993–4), p. 64; O'Halpin (ed.), *MI5 and Ireland*, p. 21, n. 3 citing DO130/4 and DO130/14 (TNA).

24 Lywood Report No. 1, 11 June 1940 (TNA, Air 2/5130).

25 Ibid.

26 Ibid.

27 Lywood Report No. 2, 18 June 1940 (TNA, Air 2/5130).

28 Fuel stocks, 29 May 1940 (MA, AC/2/8/3).

29 Lywood to AM, 11 June 1940 (TNA, Air 2/5130).

30 Annex V, Operations Order No. 1, 29 May 1940 (MA, EDP 1/1).

31 OC AC to Commands, 30 May 1940 (MA, EDP/4).

32 Operations Order 3/1940, 17 Dec. 1940 (MA, EDP 1/2).

33 Ibid.

34 Ibid.

35 Lywood to AM, 17 June 1940 (TNA, Air 2/5130).

36 AM to Lywood, 6 July 1940 (TNA, Air 2/5130).

37 T. N. Carr to AM, 14 Oct. 1940 (TNA, Air 2/5130).

38 RAF memo to CAS, 16 Oct. 1940 (TNA, Air 2/5130).

39 DCAS to OC RAF Aldergrove, 21 June 1940 (TNA, CAB 104/184).

40 Fighter Command RAF to Deputy CAS, 19 July 1940 (TNA, Air 2/5185).

41 Ibid.

42 Preparation of air forces, Éire or Northern Ireland, 4 Aug. 1940 (TNA, Air 2/5130).

43 AM agenda, 29 Dec. 1940 (TNA, Air 20/2073).

44 Air Defence of Ireland, Oct./ Dec. 1940 (TNA, Air 2/5172); HQ RAF NI to HQ Fighter Command, 28 Sept. 1940 (TNA, Air 16/530).

45 RAF NI to AM, 30 July 1940 (TNA, Air 2/5130).

46 AM to Station AA, 11 July 1940 (TNA, Air 2/5129).

47 Minute 36, 8 Feb. 1941; Minute 37, 12 Feb. 1941; Minute 39, 15 Feb. 1941, AM file S.5503 (TNA, Air 2/5172).

48 AM to Lywood, 14 Mar. 1941 (TNA, Air 2/5172).

49 Supply of aircraft to Éire, 26 Jan. 1941 (TNA, Air 8/361).

50 VCAS memorandum, 24 Jan. 1941 (TNA, Air 8/361).

51 OC AC to squadrons, 26 Mar. 1941 (MA, AC/2/9/14).

52 OC AC to COE E. Comd., 12 July 1941 (MA, AC/2/9/14).

53 COS to OC AC, 21 Aug. 1941 (MA, AC/2/9/14).

54 OC AC to COS, 16 July 1941 (MA, AC/2/9/14).

55 COS to OC AC, 21 Aug. 1941; Memo, 25 Aug. 1941 (MA, AC/2/9/14).

56 OC AC to COS, 23 Aug. 1941 (MA, AC/2/9/14).

57 Ibid., 29 Aug. 1941 (MA, AC/2/9/14).

58 DOD to DF, 4 Sept. 1941 (MA, AC/2/9/14).

59 OC AC to COS, 3 Dec. 1941 (MA, AC/2/9/14).

60 Sec. DOD to Sec. DF, 4 Sept. 1941 (MA, AC/2/9/14).

61 Major J. Gleeson to OC AC, 9 Oct. 1941; OC AC to COS, 13 Oct. 1941; OC AC to COS, 17 Oct. 1941 (MA, AC/2/9/14).

62 Report on Army exercises 1942, Sept. 1942 (MA).

63 A. P. Kearns, 'The Irish Air Corps: A History', *Scale Aircraft Modelling* III: 10 (July. 1981), p. 459; Sec. DOD to Sec. DF, 6 Jan. 1943 (NAI, DF, S/008/0029/39).

64 Kearns, 'Air Corps', p. 459.

65 O'Halpin, 'Aspects of intelligence', p. 64.

66 AM to Station AA, 11 June 1940; Lywood to AM, 13 June 1940; AM minute, 12 June 1940 (TNA, Air 2/5129).

67 O'Halpin, 'Aspects of intelligence', p. 64.

68 Childers report, 17 Oct. 1947 (MA, SCS/1).

69 'Facilities obtained from the government of Éire during the war', 21 Feb. 1945 (TNA, DO 114/117).

70 Capt. T. F. Doherty report, 3 Sept. 1939; G2 Journal, 3 Sept. 1939 (MA, G2/X/1224); HQ S. Comdt. to G2 GHQ, 21 Sept. 1939 (MA, SI/319).

71 Lt P. Swan to OC AC, 16 May 1940 (MA, ACF/S/36).

72 Memorandum for the Government, DOD 3/2314, May 1949 (in my possession).

73 File, Reports of force landed or crashed aircraft (foreign) (MA, no reference).

74 Crashes and forced landings, 1939–45 (MA, PC143).

75 Comdt D. Mackey to CSO G2, 2 Oct. 1940 (MA, G2/X/0513); Kearns, 'Irish Air Corps', p. 459.

76 Lt J. Teague to OC AC Depot, 2 Feb. 1941; File note, Comdt P. Quinn, 4 Feb. 1941; Ted Hoctor to R.W. O'Sullivan, 3 Feb. 1941; File note, Col. P. A. Mulcahy, 6 Feb. 1941 (in my possession).

77 Undated note; Receipt, 11 Feb. 1941; P. A. Mulcahy to COS 12 Feb. 1941 (in my possession).

78 File memos, 5 Mar.; 6 Mar.; 7 Mar.; 8 Mar. 1941 (in my possession).

79 OC AC to COS, 2 Apr. 1941 (in my possession).

80 A. P. Kearns, 'Irish Air Corps', p. 446.

81 Comdt P. Quinn to OC AC, 18 Apr. 1941 (MA, AC/2/11/20).

82 Ibid.; Estimates 1940–1 (MA, AC/2/2/34).

83 Undated list, Aircraft salvaged and returned, W. J. Keane (MA, PC143).

84 Memorandum for the Government, DOD 3/2314, May 1949 (in my possession).

85 File G2/X/1407 (MA). The version of this file presented in MA had been photocopied selectively to exclude any reference to British authorities, civil or military.

86 Eric Brown, *Wings on My Sleeve* (London, 2007) pp. 101–2; photographs in my possession.

87 Memorandum for the Government, DOD 3/2314, May 1949 (in my possession).

88 OC AC to A/COS, 26 June 1944 (MA, AC/2/10/9).

89 OC AC to COS, 2 Nov. 1942; G2, W. Comd to OC AC, 13 May 1943 (in my possession); C. F. Shores, 'Lockheed Hudson MKs I to VI', in *Aircraft Profile* 253 (Apr. 1973), p. 174.

90 Fisk, *In Time of War*, p. 176.

91 Intelligence file, summary of chronological list of forced landings or crashes of belligerent aircraft from the outbreak of war to 30 June 1945 (MA, no reference).

92 Fisk, *In Time of War*, pp. 327–30.

93 ACS to W. P. Delamere, 28 Jan. 1942 (MA, G2/X/0961).

94 Capt. M. Cumiskey to CSO G2, 5 Sept. 1941 (MA, G2/X/0827); Fisk, *In Time of War*, pp. 327–30.

95 Press cutting, *Irish Press*, 22 Aug. 1941 (MA, G2/X/0827).

96 Press cutting, *Daily Mirror*, 22 Aug. 1941 (MA, G2/X/0827).

97 W. P. Delamere to COS, 30 Dec. 1942 (MA, EDP 24/2/1).

98 Tables 30 w and 30a w, 1943 Establishment.

99 Summary of chronological list of forced landings or crashes of belligerent aircraft from the outbreak of war to 30 June 1945 (MA, no reference).

ELEVEN: THE AIR CORPS' INVESTIGATION OF 1941–2

1 Report and findings, 10 Jan. 1942, pp. i–ii (MA, ACS22/23).

2 List of files, Appendix No. XLII, Report and findings, 10 Jan. 1942 (MA, ACS22/23).

3 P. A. Mulcahy to AC investigation, 21 Nov. 1941 (MA, ACS22/23).

4 T. J. Hanley to AC investigation, 23 Jan.; 17 Apr.; 12 Nov. 1941 (MA, ACS22/23).

5 P. A. Mulcahy to AC investigation, 19 Nov. 1941 (MA, ACS22/23).

6 T. J. Hanley to AC investigation, 23 Jan. 1941 (MA, ACS22/23).

7 P. A. Mulcahy to AC investigation, 19 Nov. 1941 (MA, ACS22/23).

8 T. J. Hanley to AC investigation, 23 Jan. 1941 (MA, ACS22/23).

9 P. A. Mulcahy to AC investigation, 21 Nov. 1941 (MA, ACS22/23).

10 Ibid.

11 Report and findings, 10 Jan. 1942, p. lxix (MA, ACS22/23).

12 Ibid., p. vi.

13 Ibid.

14 Ibid.

15 Ibid., p. iv.

16 A. P. Kearns, 'The Irish Air Corps: A History', *Scale Aircraft Modelling* III: 10 (July. 1981), p. 459.

17 Report and findings, 10 Jan. 1942, p. vi (MA, ACS22/23).

18 J. J. Halley, *Squadrons of the Royal Air Force* (Tonbridge, 1985) shows that most squadrons had no more than a single aircraft type except when being re-equipped. Eighteen or 24 aircraft would be a normal complement.

19 Report and findings, 10 Jan. 1942, p. viii (MA, ACS22/23).

20 Ibid., pp. v–vi.

21 Ibid., p. v.

22 Ibid., p. ix.

23 Ibid., pp. ix–x.

24 Ibid., p. x.

25 Ibid., p. xi, quoting OC AC to COS, 28 Sept. 1937. While this submission was cited by Mulcahy a number of times, he was unable to produce a copy for the committee.

26 Report and findings, 10 Jan. 1942, pp. xi–xii, quoting DOD S/157 (MA, ACS22/23).

27 Ibid., pp. xi–xii.

28 Ibid.

29 Ibid., p. xiv.

30 Ibid.

31 Ibid., p. xv.

32 Ibid., pp. x–xvii.

33 Ibid., p. xx.

34 Ibid., pp. xviii; xxiii.

35 Ibid., p. xviii.

36 Ibid., Annex G.

37 Ibid., p. xxiv.

38 Ibid., pp. xxiii–xxiv.

39 Ibid., p. xxv.

40 Ibid., p. xxvi.

41 Ibid.

42 Ibid., p. xxx.

43 'Explanatory notes', O'Duffy scheme, 1 July 1924 (NAI, DT, S.3442B).

44 Report and findings, 10 Jan. 1942, p. xxx (MA, ACS22/23).

45 Ibid., p. xxvii.

46 Delamere and Curran to OC AC, 17 Feb. 1939 (MA, AC/2/6/16).

47 Appendix XLII, Report and findings, 10 Jan. 1942 (MA, ACS22/23).

48 Report and findings, 10 Jan. 1942, pp. xxvii–xxix (MA, ACS22/23).

49 Ibid.

50 Ibid., p. xxxiv.

51 Ibid., p. xxxiii.

52 Ibid., p. xxxv.

53 Ibid., p. xxvi.

54 Ibid., p. xliii.

55 T. J. Hanley to AC investigation, 17 Apr. 1941 (MA, ACS22/23).

56 Report and findings, 10 Jan. 1942, p. xliii (MA, ACS22/23).

57 Ibid., p. xliv.

58 Ibid.

59 Ibid., p. lxi.

60 Ibid., pp. lxv–xlix.

61 Ibid., pp. xlix–xl.

62 P. J. Hassett to Committee of Investigation, 8 Mar. 1941, P. J. Hassett Papers.

63 Report and findings, 10 Jan. 1942, p. lii (MA, ACS22/23).

64 Ibid.

65 Ibid., p. liii.

66 Ibid., p. liv.

67 Section 23, 1 Jan. 1929, AC Standing orders, 1929–35 (in my possession).

68 Sixty years later, on 16 Oct. 2001, in accordance with the COS's administrative instruction of that date, the Signals' Corps unit at Baldonnell became an integral part of the Air Corps, and came under the command of GOC Air Corps for the first time.

69 Report and findings, 10 Jan. 1942, p. lxi (MA, ACS22/23).

70 Ibid., pp. lxiii–lxvi.

71 Ibid., p. lxi.

72 Ibid., p. lxx.

73 Ibid.

TWELVE: RE-EQUIPPING, REORGANISATION AND DEMOBILISATION

1   FS Sqn Organisation, 12 Dec. 1940 (in my possession).

2   AC HQ Aircraft State, 16 Aug. 1940 (in my possession).

3   General report on Defence Forces, 1942–3 (MA).

4   AC HQ Aircraft State, 5 Nov. 1943 (in my possession).

5   Aircraft State, 3 Jan. 1945 (MA, EDP/24). As unserviceable aircraft were retÁined on charge until formally boarded the number of aircraft in actual service at any one time is not easily determined.

6   Memorandum, visit to England, 21 May 1940 (MA, AC/2/2/41).

7   COS Committee Minutes, 24 Jan. 1941, 3 Feb. 1941; 6 Feb. 1942 (TNA, CAB 121/335).

8   COS Committee, 29 Jan. 1941 (TNA, CAB 121/335).

9   CAS, Minute, 8 Jan. 1943 (TNA, Air 20/2442).

10   Aircraft flying times, Lt Col. J. Teague, 1923–70 (in my possession).

11   Aircraft State, 3 Jan. 1945 (MA, EDP/24).

12   General reports on the Defence Forces, 1 Apr. 1941 to 31 Mar. 1942 (MA).

13   Conference Minutes, 10 May 1941 (MA, AC/2/8/3).

14   A. J. Thornton's Log book, Sept. to Nov. 1941 (in my possession).

15   Interviews, Tommy McKeown, Denis Cousins, 31 Mar. 2005; Alphonsus J. Thornton, 1 Apr. 2005 (Marion Finucane Show, RTÉ). Sound track courtesy of Ronan Kelly.

16   Rifles inspected by Comdt M. Kelly, 24 Apr. 1944 (MA, AC/2/9/19).

17   1940 War Establishment (MA); Ordnance, Air Corps, 4 Mar. 1939; 'Location of aerial and land machine guns', 20 May 1941 (MA, AC/2/9/19).

18   COS Report, 6 Aug. 1942 (TNA, Microfilm CAB 66/27–342).

19   Delamere to COS, 30 Dec. 1942 (MA, EDP 24/2/1).

20   Curriculum vitae, O/644; O/2826, courtesy of Officers' Records Section, DFHQ; Record of pilot intake to Air Corps (AC Museum).

21   1940 War Establishment; 1943 Establishment; Strength returns 1940, 1941, 1942 and 1943 (MA, SR.1).

22   Tables 29W–33W, Air Corps 1943 Establishment; Instruction No 1/1943, Reorganisation Air Corps, OC AC, 3 Apr. 1943 (copy in my possession).

23   Sergeant Pilot Course file (School Commandant, 2005).

24   Lt L. O'R[iain], 'A pilot looks down; an Air Corps officers impressions of the 1942 exercises' in *An Cosantóir* 111: 3 (Mar. 1943), pp. 163–8, passim.

25   GOC 1st Div. to COS 19 Sept. 1942, Army exercises 1942 file (MA, no reference).

26   Report on Army exercises 1942 (MA); O'Riain, 'A pilot looks down', pp. 163–8.

27   Training Inspection, 1 May 1944 (MA, AC/2/9/19).

28   DF file 'Damage to army aircraft resulting from accidents during exercises – cost of repairs' 1938–45 (NAI, S008/0029/39).

29   Confidential memorandum, 29 Nov. 1943; Aircraft flying time 1943–8 (in my possession). The style and content of both documents identifies Lt Jim Teague as the compiler of the statistical information. In later years Lt Col. Jim Teague was scathing in his comments on P. A. Mulcahy and the adverse effect he had on pilot morale and flight safety.

30   D. K. Johnston's log book and photographs (Annette Peard).

31   Report and findings, 10 Jan. 1942, p. lviii (MA, ACS22/23).

32 AOC RAF NI to AM, 14 Oct. 1940 (TNA, Air 2/5130).

33 RAF memo to CAS, 16 Oct. 1940 (TNA, Air 2/5130).

34 A. P. Kearns, 'The Irish Air Corps: A History', *Scale Aircraft Modelling* III: 10 (July. 1981), p. 459.

35 Operational Instruction No 1/1943, Reorganisation, Air Corps, 3 Apr. 1943; Table 31W, 1943 Establishment (MA); No. 1 Squadron Movement Order 1/1943, 11 Apr. 1943; Air Corps-appointments officers, 18 Mar. 1943 (in my possession); Aircraft, 3 Jan. 1945 (MA, EDP/24). Report and findings, 10 Jan. 1942, p. xxiii (MA, ACS22/23).

36 Aidan A. Quigley, *Green is My Sky* (Dublin, 1983), p. 152.

37 Kearns, 'Irish Air Corps', p. 459; Lt Col. J. Teague, 'Irish Air Corps aircraft registrations, 1921–1974', in my possession; A. P. Kearns, 'The Air Corps 1939–1945', *An Cosantóir* no 49: 9 (Sept. 1989), p. 19.

38 M. J. Noone, 'Air Corps operations 1939–1945' (MA thesis, NUIM, 2000), pp. 31–42, citing W. J. Keane to OC AC, 24 July 1943 (MA, ACF/750/17).

39 Noone, 'Air Corps operations', p. 42, citing interview with Capt. A. A. Quigley, 6 Dec. 1999.

40 R. W. O'Sullivan to OC AC, 12 Apr. 1944 (MA, AC/2/9/19).

41 Kearns, 'The Air Corps 1939–1945', p. 19.

42 Lt Col. J. Teague, 'Aircraft flying time 1943–8' (in my possession).

43 General reports on the Defence Forces, 1 Apr. 1944 to 31 Mar. 1945 (MA).

44 Lt Col. M. O'Malley, *Gormanston Camp 1917–1986* (Defence Forces, 1986), pp. 17–19; FS order 1/45, 17 Apr. 1945 (copy in my possession).

45 Strength returns, 1940–1945 (MA, SR.1).

46 List of apprentice classes (in my possession).

47 AC HQ to E. Comd, ACF/707, 26 May 1944 (copy in my possession).

48 List of the Defence Forces personnel dismissed for desertion (MA).

49 Record of pilot intake to Air Corps (in my custody).

Bibliography

—

PRIMARY SOURCES

PUBLIC RECORDS

THE NATIONAL ARCHIVES
Air Ministry: Air 2, Air 5, Air 8, Air 9, Air 10, Air 16, Air 20, Air 76
Meteorology: BJ 5
Cabinet Office: CAB 16, CAB 21, CAB Microfilms 23, 66, CAB 104, CAB 115, CAB 120, CAB 121, CAB 123
Colonial Office: CO 904
Dominions Office: DO 35, DO 114
Foreign Office: FO 371
Prime Minister's Office: PREM 1, PREM 3, PREM 131
War Office: WO 32, WO 208

MILITARY ARCHIVES
Air Corps
Army census, 12–13 Nov. 1922
Army crisis, 1924
Army Estimates 1922–45
Army mission to USA 1926–7
Council of Defence minutes
Department of Defence – 2 Bar
Early Department of Defence – A/series
Emergency Defence Plans
General reports on the Defence Forces
Liaison Papers
Local Strength Returns
Memorandum on the Defence Forces, August 1944
Minister's Secretary
Officers' Personal files
Personal Collection 586: Gen. M. J. Costello
Personal Collection 143: Col. W. J. Keane
Proceedings of Committee of Investigation into the effectiveness, organisation, equipment, training and administration of the Air Corps. Report and findings of the Committee and Annexes thereto

NATIONAL ARCHIVES
Dáil Éireann Accounts
Department of Finance Supply
Department of the Taoiseach
Early Department of Finance
Office of Public Works

COLLECTIONS OF PRIVATE PAPERS

NATIONAL LIBRARY OF IRELAND
Art O'Brien Papers

UNIVERSITY COLLEGE DUBLIN ARCHIVES
Eamon de Valera Papers
Seán McEntee Papers
Richard Mulcahy Papers

PAPERS IN MY CUSTODY
Aircraft log books: Bristol Fighters, Avro 504K, Martinsyde Scouts, Martinsyde Type A, MK
    II, Vickers Vespa
Manuscript, 'Record of pilot intake to Air Corps'

PAPERS IN MY POSSESSION
Aerodromes book RFC/RAF Ireland
Air Corps files; ACF/36/8, ACF/144/1, ACF/338, ACF/465, ACF/503/2, ACF/S/67
Air Corps Standing Orders 1927, 1929, 1935
Army Book 129, Complete list of landing grounds – Ireland, Capt. C. M. Pixton, RAF.
Establishments Air Corps: Peace 1924, 1931–2, 1934, 1937, 1939, 1946. War 1940, 1943; Peace 1946
Flying Time & Aircraft Accidents statistics 1923–63 (Lt Col. J. Teague)
Key to air and marine special map
Maps/drawings: Baldonnell Aerodrome 1927, Fermoy Aerodrome 1922, Tralee Landing
    Ground 1922, Rathduff 1942
Memorandum for the Government, DOD, 3/2314, May 1949
Miscellaneous documents from Air Corps files: ACF/36/34, ACF/150, ACF/336/5, ACF/338/5,
    ACF/564/5, ACF/631
Nominal rolls – Air Corps officers: 12 July 1928, 3 June 1935, 1 Apr. 1937, 5 May 1939, 16 Aug. 1941,
    18 Apr. 1943, 2 Dec. 1943, 14 Dec. 1944, 4 Jan. 1945
Orders: Fighter Squadron, Operational Instruction No. 1/1943, Movement Order 1/1943,
    Movement Order, 17 Apr. 1945
Unit organisational charts & nominal rolls, Dec. 1940/Jan. 1941; Air Corps HQ & Depot, Air
    Corps School, Fighter Squadron, Reconnaissance & Medium Bomber Squadron, Coastal
    Patrol Squadron

OTHER PAPERS IN PRIVATE KEEPING

Col. W. P. Delamere – Log book (Peter Delamere, Killiney, Co. Dublin)
Lt Col. P. J. Hassett – Log book and papers (Capt. Eoin Hassett, Skerries, Co. Dublin)
Capt. D. J. McKeown – Log book (Padraic Molloy, Celbridge, Co. Kildare)
Lt T. J. Nevin Log book and Papers (G. M. Nevin, Loughrea, Co. Galway)
On loan from School Commandant, Air Corps Flying School: ACF/36/23, ACF/564/1,
    ACS/14/2, ACS/103, ACS/103/5/1, ACS/103/11/2, ACF/109/1, ACS/177/11

PRINTED PRIMARY SOURCES

Ronan Fanning et al. (eds), *Documents on Irish Foreign Policy: vol. I 1919–22* (Dublin, 1998)
—— (eds), *Documents on Irish Foreign Policy: vol. II 1922–6* (Dublin, 2000)
——(eds), *Documents on Irish Foreign Policy: vol. III 1926–32* (Dublin, 2000)
Catriona Crowe, et al. (eds), *Documents on Irish Foreign Policy: vol. V 1937–9* (Dublin, 2006)

NEWSPAPERS AND CONTEMPORARY PERIODICALS

Aeroplane
An Cosantóir
An t-Oglagh
Aviation (1935–7)
Evening Mail
Flight
Freeman's Journal
Irish Times

DEFENCE FORCES PUBLICATIONS

Defence Forces Handbook (1982)
Irish Air Corps: Celebrating Thirty Years of Helicopter Operations 1963–1993
Irish Defence Forces Handbook 1968
Irish Defence Forces Handbook 1974
The Army To-day (1945)
The Irish Air Corps 1922–1997
The Irish Defence Forces – A Handbook (1988)

WORK OF REFERENCE

Lalor, Brian (ed.), *The Encyclopaedia of Ireland* (Dublin, 2003)

OFFICIAL PUBLICATIONS, IRELAND

A Handbook on the Identification of Aircraft (Department of Defence, Mar. 1941).
Constitution of Ireland, 1 July 1937
Dáil Éireann Parliamentary Debates

Defence Forces (Temporary Provisions) Act, 1923–45
Defence Forces Act, 1937
Ministers and Secretaries Act, 1924

OFFICIAL PUBLICATIONS, UK

Air publication 129, Royal Air Force Flying Training Manual, Part I, Flying Instruction, (Air
  Ministry, 1923)
Air publication 129, RAF Flying Training Manual, Part I, Flying Instruction (Air Ministry, 1931)
Air publication 1234, Manual of Air Navigation, vol. I (Air Ministry, 1936)
Air Publication 1525A, Pilot's Notes for Anson I (Air Ministry, 1943)

MILITARY REGULATIONS AND ORDERS

Defence Forces Regulations
Defence Orders, October 1922 to November 1924
General Routine Orders 1922–3
Orders No. 3, Defence Forces (Organisation) Order, 1924
Staff Duty Memos 1923–4

SECONDARY SOURCES

BOOKS AND ARTICLES

Brown, Eric, *Wings on My Sleeve* (London, 2007)
Byrne, Liam, *History of Aviation in Ireland* (Dublin, 1980).
Doherty, Gabriel and Keogh, Dermot, (eds), *Michael Collins and the Making of the Irish State*
  (Dublin, 1998).
Duggan, J. P., *A History of the Irish Army* (Dublin, 1991).
Dwyer, T. Ryle, *Guests of the State* (Dingle, 1994).
Fanning, Ronan, *Independent Ireland* (Dublin, 1983).
——*The Irish Department of Finance 1922–58* (Dublin, 1978).
——'Neutral Ireland?' *An Cosantóir* 49: 9 (Sept. 1989).
Farrar-Hockley, Sir Anthony, *The Army in the Air: The History of the Army Air Corps* (Stroud,
  1994).
Farrell, Theo, 'The Model Army: military imitation and the enfeeblement of the Army in
  post-revolutionary Ireland', *Irish Studies in International Affairs* 8 (1997), pp. 111–27.
——'Professionalisation and suicidal defence planning by the Irish Army, 1921–1941', *Journal of
  Strategic Studies* 21: 3 (Sept. 1998), pp. 67–85.
Fisk, Robert, *In Time of War: Ireland, Ulster and the Price of Neutrality 1939–45* (London, 1983).
Girvin, Brian, *The Emergency: Neutral Ireland 1939–1945* (London, 2006).
Hayes, Karl E., *A History of the Royal Air Force and US Naval Air Service in Ireland
  1913–23* (Irish Air Letter, 1988).
Heron, Capt. O. A., 'Ireland and aviation', *An t-Oglach*, 11: 3 (Oct. 1929), pp. 6–12.

Hopkinson, Michael, *Green Against Green: The Irish Civil War* (Dublin, 1988).

—— (ed.), *The Last Days of Dublin Castle: The Mark Sturgis Diaries* (Dublin, 1999).

Joubert de la Ferté, Sir P. *The Third Service: The Story Behind the Royal Air Force* (London, 1955).

Kearns, A. P., 'The Irish Air Corps: A History', *Scale aircraft modelling* III: 10 (July 1981), pp. 440–61.

——'The Irish Air Corps 1939–1945', *An Cosantóir* XLIX: 9 (Sept. 1989), pp 13–19.

Kennedy, Michael, *Guarding Neutral Ireland: The Coast Watching Service and Military Intelligence* (Dublin, 2008).

Mangan, Colm, 'Plans and operations', *Irish Sword* XIX: 75 & 76 (1993–4), pp. 47–56.

McCarron, Donal, *Wings Over Ireland: The Story of the Irish Air Corps* (Leicester, 1996).

McCarthy, Patrick J., 'The RAF and Ireland, 1920–22', *Irish Sword* XVII (1989), pp. 174–88.

McCartney, Donal, 'From Parnell to Pearse (1891–1921)', in Moody, T. W. and Martin, F. X. (eds), *The Course of Irish history* (Cork, 1984), pp. 294–312.

O'Carroll, Donal, 'The emergency army', *Irish Sword* XIX: 75 & 76 (1993–4), pp. 19–46.

O'Farrell; Padraic, *Who's Who in the Irish War of Independence and Civil War* (Dublin, 1997).

O'Halpin, Eunan, 'Aspects of intelligence', *Irish Sword* XIX: 75 & 76 (1993/94), pp 57–65.

——*Defending Ireland: The Irish State and its Enemies since 1922* (Oxford, 1999).

——(ed.), *MI5 and Ireland, 1939–1945* (Dublin, 2003).

O'Malley, Lt Col. Michael C., 'Baldonnell Aerodrome 1917–1957', *Dublin Historical Record* LVI: 2 (Autumn 2002), pp. 170–81.

——'The Officers' Mess and other works of W. H. Howard Cooke at Baldonnell Aerodrome', unpublished undergraduate essay (NUIM, 2001).

——'The Military Air Service 1921–24', unpublished BA thesis (NUIM, 2002).

——'Military aviation in Ireland 1921–1945', PhD thesis (NUIM, 2007).

Oram, Hugh, *Dublin Airport: The History* (Aer Rianta, 1990).

O'Riain, Lt Liam, 'A pilot looks down; an Air Corps officer's impressions of the 1942 exercises', *An Cosantoir* III: 3 (Mar. 1943), pp. 163–8.

O'Rourke, Madeleine, *Air Spectaculars: Air Displays in Ireland* (Dublin, 1989).

O'Sullivan, R. W., *An Irishman's Aviation Sketchbook* (Dublin, 1988).

Pakenham, Frank, *Peace by Ordeal: The Negotiation of the Anglo-Irish Treaty, 1921* (London, 1972).

Parsons, Denis, 'Mobilisation and expansion 1939–40', *Irish Sword* XIX: 75 & 76 (1993–4), pp. 11–18.

Quigley, Aidan, 'Air aspects of the emergency', *Irish Sword*, XIX: 75 & 76 (1993–4), pp. 86–90.

Ryan, Meda, *The Day Michael Collins Was Shot* (Swords, 1989).

Share, Bernard, *The Flight of the Iolar: The Aer Lingus Experience 1936–1986* (Dublin, 1986).

Swan, Capt. Patrick, 'The Air Corps', *An Cosantóir* VII: 4 (Apr. 1947), pp. 199–202.

Valiulis, Maryann Gialanella, *Almost a Rebellion: The Irish Army Mutiny of 1924* (Cork, 1988).

West, Nigel, *MI5 British Security Service Operations 1909–1945* (London, 1981).

—— (ed.), *The Guy Liddell Diaries, vol. I: 1939–1942* (Abingdon, 2005).

Young, Peter, 'The way we were', *An Cosantóir* 49: 9 (Sept. 1989), pp. 33–8.
——'Defence and the new Irish state 1919–23', *Irish Sword* XIX: 75 & 76 (1993–4), pp. 1–10.

BIOGRAPHIES

Fennelly, Teddy, *Fitz and the Famous Flight* (Portlaoise, 1997).
Lewis, Cecil, *Sagittarius Rising* (London, 2003).
Pinkman, John A., *In the Legion of the Vanguard*, ed. Maguire, Francis E. (Cork, 1998).
Ring, Jim, *Erskine Childers* (London, 1996).

SPECIAL SUBJECTS

hAllmhurain, Seán (ed.), *Aviation Communications Service, 1936–1986* (Dublin, 1986).
Bowyer, Chaz, *Bristol F2B Fighter: King of Two Seaters* (Shepperton, 1985).
Byrne, Kevin and Tormey, Peter, *Irish Air Corps: A View from the Tower* (Defence Forces Printing Press, 1991).
Cambridge University, *Aircraft Navigation* (Cambridge, 1943).
Collins, C. B., 'Inter-aerodrome navigation', *Flight*, 7 Dec. 1933.
Dunne, Tom et al., *Shannon Airport: 50 Years of Engineering 1937–1987* (Aer Rianta, 1987).
Erecting and Aligning Avro Biplanes. 3rd. edn A. V. Roe & Co. Ltd (Manchester, 1918).
Halley, J. J., *Squadrons of the Royal Air Force* (Tonbridge, 1985).
Hughes, A. J., *History of Air Navigation* (Woking, 1946).
Irish Air Letter, *Baldonnel: Dublin's Civil Airport 1919 to 1939* (Dublin, 1989).
——*Aviation on the Shannon* (Dublin, 1985).
——*The Flying Fields of Cork* (Dublin, 1988).
Jackson, A. J., *Avro Aircraft since 1909* (London, 1990).
Jordan, David, 'War in the air: the fighter pilot', in Liddle, Peter, Bourne, John and Whitehead, Ian (eds), *The Great World War 1914–45: vol. I: Lightning Strikes Twice* (London, 2000).
Meekcoms, K. J. and Morgan, E. B. (eds), *The British Aircraft Specifications File: British Military and Commercial Aircraft Specifications 1920–1949* (Tonbridge, 1994).
Sanger, Ray, *The Martinsyde File* (Tunbridge Wells, 1999).
Shields, Lisa (ed.), *The Irish Meteorological Service: The First Fifty Years* (Dublin, 1987).
Sinclair, Duncan, 'Airport communications', *Flight*, 7 Dec. 1933.

Index

340